WICKED INTELLIGENCE

Wicked Intelligence

Visual Art and the Science of
Experiment in Restoration London

MATTHEW C. HUNTER

THE UNIVERSITY OF CHICAGO PRESS * *Chicago & London*

Matthew C. Hunter is assistant professor in the Department of Art History and Communication Studies at McGill University. He is coeditor of *Beyond Mimesis and Convention: Representation in Art and Science* and *The Clever Object*, and an editor of *Grey Room*.

The University of Chicago Press, Chicago 60637
The University of Chicago Press, Ltd., London

Printed in the United States of America

22 21 20 19 18 17 16 15 14 13 1 2 3 4 5

ISBN-13: 978-0-226-01729-7 (cloth)
ISBN-13: 978-0-226-01732-7 (e-book)

Library of Congress Cataloging-in-Publication Data

Hunter, Matthew C.
Wicked intelligence : visual art and the science of experiment in Restoration London / Matthew C. Hunter.
pages : illustrations ; cm
Includes bibliographical references and index.
ISBN 978-0-226-01729-7 (hardcover : alkaline paper) — ISBN 978-0-226-01732-7 (e-book)
1. Scientific illustration—England—London—History—17th century. 2. Art and science—England—London—History—17th century. 3. Art, British—England—London—17th century. 4. Architectural drawing—England—London—17th century. 5. Hooke, Robert, 1635–1703—Criticism and interpretation. 6. Lely, Peter, 1618–1680—Criticism and interpretation. 7. Royal Society (Great Britain)—History—17th century. I. Title.
Q222.H86 2013
509.421'09032—dc23

2012044463

♾ This paper meets the requirements of ANSI/NISO Z39.48-1992 (Permanence of Paper).

CONTENTS

ILLUSTRATIONS

COLOR PLATES (FOLLOWING PAGE 102)

BLACK-AND-WHITE FIGURES

ACKNOWLEDGMENTS

I AM DELIGHTED TO THANK the numerous individuals, institutions, and groups from whom I have learned during the decade plus when researching, writing, and revising *Wicked Intelligence*. My thanks go first of all to Joel Snyder, Kim Rorschach, and Rachael Z. DeLue, who made this book a possibility, and to Christine Stevenson, who acted as my generous mentor in London. I am pleased to acknowledge the support of the Kress Foundation, the Social Science Research Council, the Whiting Foundation, the Courtauld Institute of Art's Research Forum, California Institute of Technology, the Huntington Library, and McGill University for enabling the book's research and writing phases. As the project developed, I learned much from conversations with John Brewer, Adrian Johns, Byron Hamann, Allison Morehead, Craig Hanson, Josh Ellenbogen, Michael Hunter, Peter Parshall, Rob Iliffe, Ludmilla Jordanova, Ken Arnold, Roman Frigg, Jim Elkins, Francesco Lucchini, Caroline Arscott, Katie Scott, David Solkin, Christine Stevenson, Rick Brettell, Nicole Ryder, Charles Ford, Alicia Weisberg-Roberts, Alex Marr, Nick Dew, Susana Soares, Daniela Bleichmar, Sachiko Kusukawa, and Simon Dickie. I thank them all, heartily.

Writing this book developed in part through "The Clever Object Research Project," a seminar cosponsored by the Research Forum of the Courtauld Institute of Art and California Institute of Technology. I thank Pat Rubin for believing in that project, and seminar co-organizer Francesco Lucchini along with all its participants, who generated such a stimulating environment for thinking: Caroline Arscott, Byron Hamann, Ian Kiaer, Christiane Rekade, Katie Scott, Simon Starling, and Rachel Wells. An earlier version of this book's second chapter developed in that project has since appeared as "Picture, Object, Puzzle, Prompter: Devilish Cleverness in Restoration London," *Art History* 36, no. 3 (June 2013): 546–67; I thank the Association of Art Historians for granting permission to reproduce it here. I am grateful to the following scholars for generously offering valuable feedback on portions of the present text: Katherine Acheson, Paroma Chatterjee, Moti Feingold, Jennifer Greenhill, Chris Hunter, Dana Katz, Tarja Knuutila, Gideon Manning, Lia Markey, Alex Marr, Kris Neville, Dawn Odell, Claudia Swan, and Mary Terrall, along with three anonymous referees. I'd also like to thank the "Theorizing Early Modern Studies" seminar at the University of Minnesota (especially Michael Gaudio and J. B. Shank), where I was able to present a draft of the book's introduction in April 2010. Jason LaFountain made incisive comments on several chapters, while John Brewer gave ongoing crucial support to the whole enterprise. Byron Hamann has been a fabulous interlocutor from start to finish; I'm sure this book would have been much better had I taken on more of his challenging suggestions. I thank Marty Ward for organizing a book-writing workshop in the spring of 2009, and my fellow alums from Chicago's Art History Department who made that occasion so productive. Working with Susan Bielstein has been an absolute, supercharged delight. I struggle to find the words with which to thank Rachael Z. DeLue for the years of brilliant criticism and incredible generosity she has offered. I hope she can see her hand in the book's good bits, and I look forward to learning from her critique of the other two hundred-odd pages. Thanks to Anthony Burton at the University of Chicago Press for his expert handling of various preproduction matters, and to David Vann, who introduced me to Marilynne Robinson's *Housekeeping* at a crucial moment.

Much of the research for *Wicked Intelligence* was undertaken at archives and collections in the United Kingdom. I would like to thank the curators and librarians at the Royal Society Library (especially Rupert Baker, Felicity Henderson, Jo Hopkins, and Keith Moore), British Library (particularly Giles Mandelbrot, formerly thereof), British Museum, Wellcome Library (especially William Schupbach, Ross Macfarlane, and Phoebe Harkness), Dulwich College Library, Dulwich Picture Gallery, Bodleian Library, Ashmolean Museum, and Trinity College Library. In Southern California, I thank the Huntington Library and the Norton Simon Museum, along with David Wilson at the Museum of Jurassic Technology. Thanks to Susan Davis, Lindsay Cleary, and Sini Elvington for their incredible support over three luxurious years at California Institute of Technology.

The following individuals have helped to make the project of writing this book such a total hoot: John Urang, Fredrik Albritton Jonsson, Jeremy Biles, Tom Cummins, Kerry Boeye, Richard Clay, Nick Grindle, Rebecca Givan, Catherine Lock, Gillian Buchan, Jeffrey Saletnik, Dawna Schuld, Nick Popper, Mark Allen, John Harwood, Zeynep Celik Alexander, Chris Lakey, Sarah Hamill, Jon Sachs, and all those I corralled into the Colorado Bar on perfectly useful afternoons, including

Sean Roberts, Marius Stan, Peter Stallybrass, John Krige, and Marita Huebner. I thank Charles Gagnon, Ara Osterweil, Nick Dew, and Andrew Piper, along with Cecily Hilsdale, Angela Vanhaelen, Amelia Jones, Christine Ross, Jeff Moser, and all my colleagues in McGill's Department of Art History and Communication Studies, who have made Montreal feel like home. Neither my father, Graham C. Hunter II (1947–2011), nor my grandmother, Edith F. Hunter (1919–2012), lived to see this book finished; I'd like to think that at least one of them would have wanted to read it. Thanks to Susan W. Hunter and Patrick W. Hunter, along with Danae and Nicholas Kouretas, for staying strong during some difficult times. To anyone who knows, this book is obviously for Daphne Kouretas, the model of the reader whose heart I want to win (and maybe also seduce).

NOTE ON CONVENTIONS AND DATES

MANY OF THE QUOTATIONS IN this book are taken from my own transcriptions of early modern manuscripts. Whenever possible, I have tried to preserve period spelling and grammar, confining my interventions to minor expansions (such as "ye" to *the*) or conventional alterations (such as "v" to *u*). In the rare cases in which I have introduced punctuation into a quotation to facilitate comprehension, I have signaled it with square brackets. I use *City* as shorthand for the City of London or the area of the English capital within the ancient Roman walls, governed by the Lord Mayor and Corporation. Unless otherwise specified, *London* refers to the greater metropolitan region of the capital, including the City. All dates are given as if the seventeenth-century English calendar year had begun on January 1.

INTRODUCTION

"Very Able, Very Sordid, Cynical, Wrong Headed and Whimsical"

ONCE UPON A TIME in later seventeenth-century London, a merchant paced the floor of his shop in utter agony.[1] A dealer in dyes and pigments, this sufferer lived on Snowhill Street, just beyond the western edge of the ancient Roman walls marking the City of London's official boundaries (fig. 0.1). In the early modern property game, his was an advantageous location. With the emerging middle-class neighborhood of Holborn to the north and the fashionable "entertainment district" of Covent Garden to the west, the merchant's premises faced in the directions toward which the metastasizing metropolis had been expanding exponentially since the Restoration of Stuart monarch Charles II in 1660. This color merchant must also have commanded some administrative authority, as he is described as a "Deputy." But despite his favorable urban and social positions, our colorman was made acutely vulnerable, wracked by a terrible toothache. And seeing the opportunity created by piercing pain, "a Waggish Painter" offered an unusual remedy.

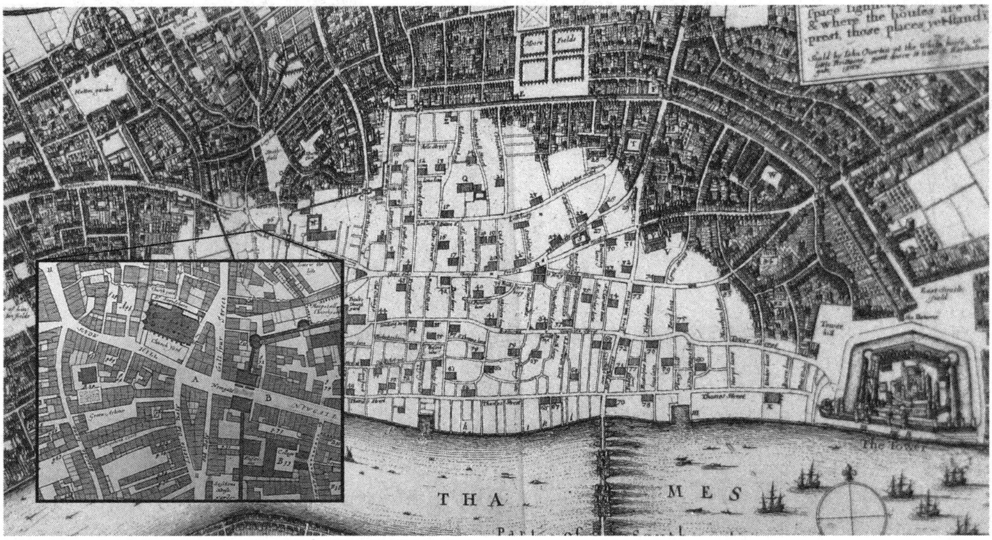

FIGURE 0.1 * Detail of the location of Snowhill Street (from John Ogilby and William Morgan, *A Large and Accurate Map of the City of London* [1676]) set into a detail in Wenceslaus Hollar, *A Map or Grovndplot of the Citty of London . . . by which is exactly demonstrated the present condition thereof since the last sad accident of fire* (London: John Overton, 1666); Huntington Library, RBN 1068. Reproduced by permission of The Huntington Library, San Marino, California.

Instructing the merchant to start a fire and await his return, the painter darts off to prepare the cure. This is what happens next:

> Bringing with him a little fair water a hammer and some nailes the painter perswades the deputy to goe with him and not to think much of his Directions though they might seem extravagant. In fine he perswades him to let downe his breeches and leane his breech against the fire[.] He in the mean time tacking up his shirt to the mantletree with his nailes. Then bidding him fill his mouth with the water & keep it in till such a time as it began to scald his mouth he advised him to have patience and hold by his breech soe till he found the effect. In the mean time he [the painter] pretends to step downe for somewhat and slipt away.

His nailed, half-naked body made the very kettle for boiling his liquid "cure," the deputy's plight only worsens as a visitor arrives for official consultation. Discovered burned and bare by his maid, the suffering official becomes desperate. He stamps his feet at the maid in mute, incomprehensible communication, "mummd with his mouth full of water." From here, the narrative descends into the stuff of farce. The terrified maid runs to the mistress of the house, convinced the master has gone insane. Seeing her stripped, singed, and babbling husband tacked before a roaring fire, the mistress flies into "a great passion fearing he was mad[;] indeed kept soe great a coyle as made him spit out his water." In the end, the maid cries, administrative business is forgotten, and the wife gets berated for causing the deputy to "spit up his water before it boyled and soe could not be cured of the tooth ake." No doubt from some safe distance down Snowhill Street, the hidden painter savors the confusion he had brought into the public duties and private sphere of the merchant on whom his own art depended.

What are we to make of this strange tale? Recent approaches to early modern London life offer several attractive possibilities. The historically minded reader might be tempted to identify this tormented colorman-deputy as Charles Beale. Husband of portrait painter Mary Beale, Beale was both a dealer in pigments and a deputy clerk of the patent office.[2] Assigning identities this way could place the events of our otherwise undated story sometime around 1660–64, when the Beales lived on Hind Court, less than a mile from Snowhill Street. It would also position our protagonists within the greater ambit of Peter Lely (1618–1680), Restoration London's leading portrait painter—an artist with whom Mary Beale studied and a painter to whom Charles Beale sold pigments.[3] These suggestions are made all the more intriguing since they corroborate what we know unequivocally about this odd narrative; that is, the story was recorded by English experimental philosopher, architect, and general polymath Robert Hooke (1635–1703).[4] Former apprentice to Lely and accomplished picture-maker himself, Hooke had heard this tale from an apothecary named Mr. Whitchurch.[5] Probably also a pigment dealer, this Whitchurch was clearly a peddler of the kinds of curious information Hooke gathered compulsively during the near-daily rounds he made through the City, often with his friend and collaborator Christopher Wren (1632–1723), as architects charged with surveying and reconstructing London's built environment after the devastating fire of 1666.[6]

Read less for its elusive facts than for its mythic evocations, Robert Hooke's tale suggests much about the urban geography, polymathic agents, and innovative visual practices at the core of this book. We see the learned philosopher circulating through a bustling metropolis whose very walls he was employed (and handsomely rewarded) to reconfigure. He is not cloistered in a provincial university college but trafficking strange knowledge with enterprising purveyors of far-flung commodities. We can also witness that experimentalist transcribing for posterity the deeds of an adept in an ingenious art to which his own keen visualizing skills were indebted:

painting. For early modern England, all this borders on radical novelty. Whereas Leon Battista Alberti and other learned "Renaissance men" had fused *perspectiva* optics with picture-making as they promoted a return to the antique practice of gentlemanly painting in 1430s Florence, drawing was scarcely domesticated into the leisured pastimes of the English gentry before the seventeenth century. Only in 1670 did Hooke's friend Joseph Moxon publish the first English treatise on pictorial perspective.[7]

Among the traditions of mathematicized natural philosophy in which Hooke and Wren moved, England's artistic eccentricities had clear epistemic consequences.[8] If Elizabethan mathematician Thomas Harriot famously used a telescope to draw the moon months before Galileo did so, the Englishman's tortured drawings display not only his ignorance of the principles of Florentine *disegno* but his incomprehension of the magnified lunar craters depicted and understood so deftly by Galileo.[9] Even on its own terms, the English artistic tradition of Inigo Jones, Nicholas Hilliard, and their contemporaries has cut a modest profile against the seventeenth century's unfolding "golden ages"—Velázquez and Zubarán in Spain, Rembrandt and Vermeer in the neighboring Dutch Republic. At best, art historians have argued, the Stuart court of Charles I glorified its own image by recruiting the foreign talents of Peter Paul Rubens, Anthony van Dyck, and Peter Lely.[10] At worst, the civil wars that waged across the British Isles between 1642 and 1651 compounded legacies of Tudor iconoclasm, scattering and rupturing England's fledgling traditions of artistic production and patronage.[11] Most telling is the contrast with the rising rival across the English Channel. At nearly the same moment in the late 1640s when artists in Louis XIV's France founded an institution as crucial to modern art as would become the Académie royale de peinture et de sculpture, England's victorious Parliament was ordering the execution of Charles I and auctioning his magnificent artistic collection in what recent scholars have dubbed "the sale of the century."[12]

Hooke's story might aptly indicate how and by whom those tumultuous artistic tides were turned in Restoration London. By the last decades of the seventeenth century, so social historians argue, emerging middle classes made wealthy through incipient industrialization and expanding commerce with the colonies and the Continent were commanding a market for visual art unprecedented in England.[13] Anticipating and encouraging the eighteenth century's "birth of a commercial society," art auctions exploded in their frequency and traffic.[14] Sold in new social spaces such as the coffeehouse, paintings and prints could now be found packing the walls of urban dwellings up and down London's tightly packed social ladder. Drawing academies proliferated as the hold of conservative artistic authorities such as the Painter-Stainers' Company waned following a general relaxation of guild restrictions devised to lure that massive force of skilled labor needed to rebuild the City after the Great Fire of 1666.[15]

Growing in patrons and practitioners, the arts of painting, printmaking, architecture, and sculpture also began to enjoy key support from noble bodies, themselves new to England. Most important among these is the Royal Society of London, the institution founded in 1660 under protection from Charles II, which soon became later seventeenth-century Europe's leading scientific academy. Claiming among its members Isaac Newton, Robert Boyle, and John Locke—with Wren as its President (1680–82), Hooke as its longtime "Curator of Experiments"— not only did the Fellowship of the prestigious Royal Society admire and collect art as it penned treatises advancing visual art's theory and practice, but Fellows also actively deployed artistic techniques in the production and circulation of their innovative philosophical work.[16] Historians of science have put the point this way: because so few practitioners in seventeenth-century Europe possessed the expensive instruments and specialized skills required to perform the scientific experiments devised by Hooke and his colleagues, Restoration experimentalists needed visualizations acutely (fig. 0.2). Appropriating techniques of detail-laden depiction perfected by early modern painters (as we see in Hooke's own ink study of microscopically magnified urine crystals), Royal Society Fellows privileged prints and drawings that could generate a mental "image of an experimental scene as obviates the necessity for either direct witness or replication."[17]

The spectacle on Snowhill Street, we might thus say, dramatizes a fateful nexus. Experimentalist-narrator, color merchant, and painter embody crucially generative fields of force that were bringing the Scientific Revolution, protoimperial commercial culture, and England's burgeoning discovery of the visual arts into profound symbiosis on the streets of later seventeenth-century London. It is only fitting that such a story was faithfully recorded by that doyen of "the art of describing," Robert Hooke.[18]

Yet such a reading surely chafes against the thrust of the story itself. The cosmology imagined by Hooke's yarn is less a mutually reinforcing trinity than a pantheon of tricksters. What commands the experimentalist's attention in this tale told by the merchant from whom Hooke bought medicaments for his niece (and lover) has nothing to do with the painter's ability to render natural specimens or to depict wondrous spectacles in ways potentially useful for natural philosophy.[19] Instead, the narrative turns on that painter's audacious, merciless exploitation of vulnerabilities besetting a colleague to whom the artist's own professional livelihood was intricately networked. So, what if we were to see the crossroads of collective interest triangulated between Restoration experimental philosophy, commerce, and visual art as styled

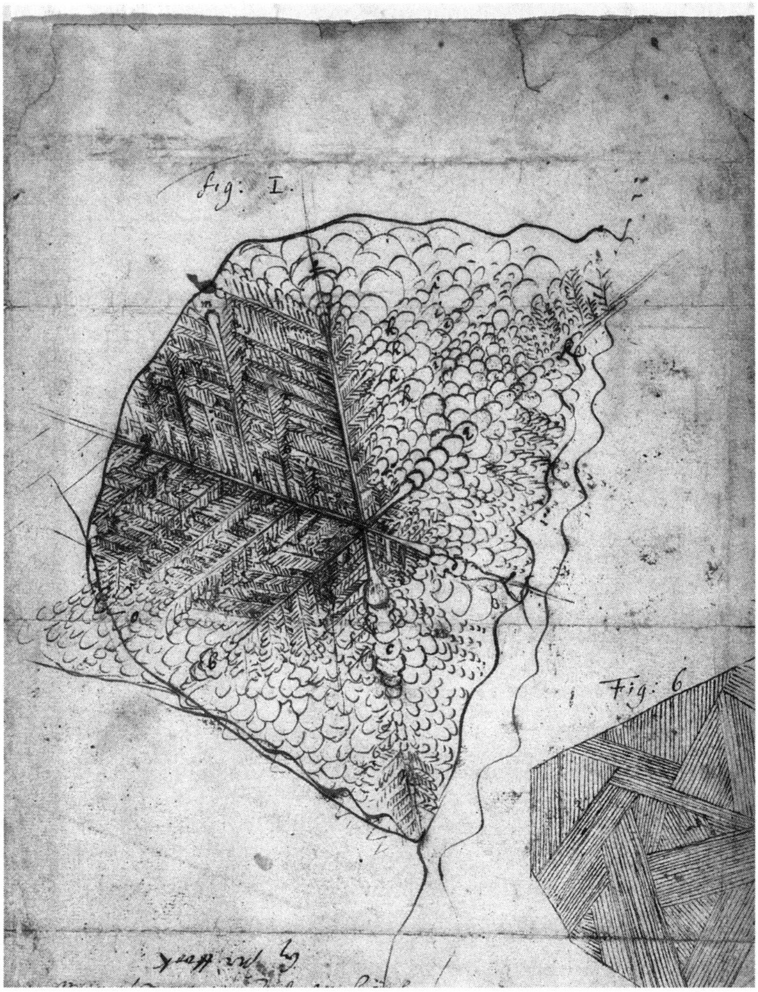

FIGURE 0.2 * Robert Hooke's ink drawing of frozen urine crystals (circa December 1662); Royal Society Classified Papers, vol. 20, fol. 6. © The Royal Society. Reproduced by permission.

in a commensurate chiaroscuro? That is, if Restoration experimental philosophy and emergent communities of visual art really did need one another—just as painter and colorman, Hooke and Whitchurch relied on each other's business—might we see those relations as equally implicated in the matrix of cunning, cruelty, and cold-blooded skullduggery that transpired on Snowhill Street? And even as we can acknowledge early modern England's unsettling predilection for "an almost unquestioned pleasure at the sight of deformity or misery—an automatic and apparently unreflective urge to laugh at weakness simply because it is weak," just why is it that

Robert Hooke placed this story of artistic torture at the head (or conclusion) of his famous narrative of his own life?[20]

Wicked Intelligence takes up that gambit. Diving deep into the visual archive of experimental-philosophical practice, this book offers an unprecedented exploration of the stunning cognitive techniques and stylized strategies through which London's experimentalists pursued knowledge. Organized around the nested actions of drawing, collecting, and building circa 1650 through circa 1720, it demonstrates how the project of synthesizing the sprawling domains of visualization produced by experimental-philosophical collaborators came to require a peculiar mode of intelligence—a ruthless cleverness theorized by and keenly embodied in the images, artifacts, and baroque architectural monuments designed by Wren and Hooke themselves. Bringing to analysis this largely forgotten world of visualization and the deft cunning required to manage it, I offer an interpretive framework with which to rethink the parameters of visual art and experimental philosophy on the cusp of England's commercial efflorescence. But the story I tell is no inexorable march from the founding of the Royal Society of London in 1660 to the opening of the Royal Academy of Arts in 1768. Instead, what I show is that just as the craft and craftiness materialized in experimental visual practice both promoted *and* liquidated the artistic traditions on which it drew, so the vexed, internally divided project of Hooke, Wren, and their colleagues would be blackened—defaced—in the eighteenth century.[21]

This talk of tricks, treachery, and disfiguration may sound a discordant note to readers versed in the recent historiography of seventeenth-century science. After all, the early Royal Society of London has served for three decades as *the* test case demonstrating the force of aristocratic gentility in the construction of scientific knowledge. Restoration experimentalists, we read, fashioned a revolutionary social epistemology that located knowledge in consensual agreements formed from publicly observed, experimentally generated matters of fact. In place of logical deductions about nature drawn by a privileged, isolated scholar, the early Royal Society made good on the promises of Francis Bacon (1561–1626). Embracing inductive inferences based on collected empirical data, the institution privileged "horizontal" collaborations between philosophers whose membership in the scientific community now required fluency in codes of gentlemanly trustworthiness, moderation, and sociability. Polite and reasonable, these philosophers aimed to reconcile the practice of science with the abstemious morality and fervent Protestantism of the "Christian virtuoso" exemplified by chemist Robert Boyle (1627–1691).[22]

Even the strongest advocates of these positions have had to acknowledge figures like Robert Hooke as exceptions. Son of a minor cleric from the Isle of Wight, a paid employee of the Royal Society, and the brother of a suicide, Hooke's social status was, so historian of science Steven Shapin has influentially argued, liminal, fluid, problematic.[23] Unlike class-conscious Fellows of the Royal Society such as John Evelyn (1620–1706), Hooke appears in the archive as equally keen to mingle with elites at the court of Charles II as to milk information from "the dark shops of Mechanicks."[24] Although recent scholarship has begun to reconsider the question of Hooke's social status, the force of his proximity to shadowy mechanism remains.[25] His body mythically deformed by excessive use of a lathe, Hooke has been cast as an inheritor of Renaissance magic in whom an active conception of matter's "occult qualities" survived alongside the inert physics of post-Cartesian mechanical philosophy; he has even been imagined as a natural magician *tout court*.[26] But while recent biographers have trawled through Hooke's infamous conflicts with Christiaan Huygens, Isaac Newton, and a parade of other scientific luminaries, our broader picture of intellectual life in the early Royal Society has only waxed commensurately darker. Just as Michael Hunter's meticulous research has demonstrated how far the institution stood in practice from its lofty, theoretical ambitions, so the acrimonious, litigious experimental underworlds detailed by Adrian Johns, Rob Iliffe, Ofer Gal, and others have come to recall the "flagrant disregard for communal norms, the outrageous, egregious behavior so often displayed by scholars . . . to the delighted disgust of their colleagues" found in the broader, trans-European Republic of Letters.[27] If Restoration England can be described as "a world of change and uncertainty, of sensational plots and conspiracies, of endless personal intrigue and maneuvering, of widespread corruption and almost universal cynicism," then the Royal Society envisaged in recent historiography appears not much better.[28]

Wicked Intelligence argues that these are matters of fundamental concern to historians of early modern art and architecture as well as to students of the history and philosophy of science. Not only does attention to the hurly-burly of experimental visual practice enable us to see the masterpieces of baroque architecture designed by Wren and Hooke in fundamentally new ways, but it discloses crucial dimensions of what those experimentalists understood the enterprise of architecture to be. Tellingly coeval with John Milton's *Paradise Lost* (1667), the Royal Society all too keenly recognized the frailties of the fallen human body and its ever-wayward mind for the exacting, exhausting work of natural philosophy. Innovative instruments, far-flung pharmaceuticals, experimental artifacts, and the built environment itself could all, as we will see, diagnose and partially redress these congenital failings. Only by dividing its sprawling intellectual project between the specialized competencies and diversified resources of a variously skilled, geographically distributed collective, however, could philosophy possibly advance beyond its seemingly perpetual infancy. And key questions remained: by what means would the dazzlingly diverse array of drawings, diagrams, models, specimens, samples, and other visualizations "crowd-sourced" from the experimental collective possibly be synthesized into reliable knowledge? More elementally, in an era that increasingly understood self-interest as a given of the human condition, how could the recognized need for trusting, philosophical collaboration be reconciled with the tempting opportunities for personal gain and individual advancement endlessly afforded by that networked community?[29] If the coming consumer culture of the early eighteenth century would find one solution in Bernard Mandeville's notorious dictum of "private vices, public benefits," I build on the work of J. A. Bennett, Hentie Louw, and others to show how architecture came to provide Restoration experimentalists with both a crucial, theoretical model and a top-down, "vertical" method for alternatively addressing this dilemma in practice.[30] Baroque monuments such as St. Paul's Cathedral need to be seen, I suggest, as products of, models for, and agents in that fundamental, contested dynamic.

Generative and influential as they were, the ways of making and knowing crafted at this powerful nexus of experimental philosophy, visual practice, and commercial efflorescence also garnered many enemies. When made the victim of plagiarism by Edmond Halley in 1686, Astronomer Royal John Flamsteed quickly identified the true source of such devious behavior. Halley, Flamsteed claimed, "is got into Mr. Hooke's acquaintance, has been his long intimate, and from him he has learnt these and some other disingenuous tricks."[31] If the accelerating, posthumous defacement of Hooke's reputation by the followers of his greatest rival, Newton, is now well known, we need to remember that even Christopher Wren stood accused of fraud and graft before his summary dismissal as Surveyor of the King's Works in 1718. Recognizing that troubled legacy can help to clarify the contribution of my book not only to histories of science, art, and architecture but also to visual studies—to thinking about images that are not art. What I mean is that relations between Restoration science and architecture have frequently been staged as a conflict, a "tyranny of the intellect" in the famous phrase of John Summerson.[32] When they have been considered at all, visual strategies produced in experimental contexts have, at best, served historians of British architecture as tokens of the curiously expansive Restoration mind. At worst, they have figured as "empirical" evidence of its perniciously intellectualizing tendencies.[33]

The situation has been far less charitable in histories of British art, where the Restoration experimentalists have sat either at a convenient *terminus ad quem* or as a messy, elliptical digression only from which narration of art proper can begin.[34] Pungently expressive of experimentalism's uneasy place in that artistic canon is Horace Walpole's assessment of the "celebrated astronomer and miser Robert

Hooke." Summarizing a long line of bad blood in his seminal *Anecdotes of Painting* (1763), Walpole wrote of Hooke:

> He gave a plan for rebuilding London after the fire, but though it was not accepted, he got a large sum of money, as one of the commissioners, from the persons who claimed the several distributions of ground, and this money he locked up in an iron chest for thirty years. . . . He was very able, very sordid, cynical, wrong headed and whimsical. Proof enough of the last, was his maintaining that Ovid's Metamorphoses was an allegoric account of earthquakes.[35]

A pecunious speculator and a builder of schemes, Walpole's Hooke is—like Hooke's own rascal dentist-painter—a knave, and a fool to boot. He is literally a footnote in the history of British art.

Must we right these historiographical wrongs? Although my fundamental claim is that historians of early modern art and architecture (especially as practiced on the British Isles) critically need to engage with the worlds of experimental visualization analyzed in this book, I am proposing that we do so neither by ennobling all the weird, provisional images made, remade, destroyed, and thought with in philosophical contexts as "art," nor by replacing art with general talk of "visual culture." Art, as I will argue, was a subject of fascination, anxiety, even tactical resistance among Restoration experimentalists, and reckoning with those conflicted attachments is deeply instructive. As much as I mean to bring to light the productive, concrete contacts between London's experimentalist and emergent artistic communities, I also aim to show where and how those borrowings were strategically constrained, their artistic allurements (at least theoretically) counteracted. In a certain sense, then, I think that Walpole was right; and I call this sense "wicked intelligence." To explain what that intelligence would entail, we need now to turn from the painter's story to a cryptographer's tale. Following it around a scenic "plateau," we can then descend back into the means and mechanics of the book itself.

"To Apprehend the Severall Waies whereby They May Be Expressed"

In mid-July 1683, the armies of Ottoman sultan Mehmed IV under command of Grand Vizier Kara Mustafa besieged Vienna, imperial capital of Habsburg emperor Leopold I. Broken by a coalition of Christian princes after nearly two months, the siege galvanized an aggressive campaign of territorial reclamation in eastern Europe at the hands of a "holy league" financed by Pope Innocent XI.[36] For those in western Europe menaced by the far greater threat of Louis XIV's France, the siege of Vienna demonstrated the pressing need for a means of rapid, long-distance communication—what Robert Hooke called in a public lecture from the spring of 1684 a way "to convey Intelligence, almost in a Moment, to twice, thrice, or more Times that Distance [of forty miles], with as great a Certainty, as by Writing."[37] The term *intelligence*, as Hooke uses it here, signals a valuable and increasingly desired commodity in seventeenth-century England: news.[38] But the agonistic situation prompting Hooke's lecture, combined with the ingenious visual solution he devised to deliver that new knowledge, can help to illuminate the contours of the peculiar intelligence at the heart of this book.

As a solution to Vienna's entrapment, Hooke imagined a semaphore network stretching between Europe's capitals: a system of signal towers strung like jewels on a necklace between Paris, London, and Amsterdam. Rising high above the coal-choked haze of the seventeenth-century city, these towers were to be manned by teams of two or three attendants per station, equipped with reliable timekeeping instruments and dedicated, precision-crafted telescopes (fig. 0.3 *a*, *b*). Relaying information back and forth, day and night, these communicators would be kept constantly busy in translating messages into and out of encrypted characters.

To craft these codes, Hooke could draw on the prodigious catalogue of cryptography set out by an early mentor at Oxford, John Wilkins (1614–1672).[39] Writing on the cusp of the English Civil Wars and informed by his own experiences at the exiled court of the Elector Palatine displaced to The Hague by the Thirty Years' War, Wilkins demonstrated how to deploy a stunning array of materials—from invisible ink to parables, gifts, landscape paintings, and even feasts—as vehicles for encryption. With Vienna's siege of 1683 in mind, consider an example from Wilkins's *Mercury; or, The Secret and Swift Messenger* (1641): "Wee prosper still in our affaires and shall (without having any further helpe) endure the siege" (fig. 0.4).[40] Meticulously inspecting the typographical characters of this apparently cheery report (as Wilkins's cryptographers would know to do by previous compact), we can observe that certain letters have either been printed in slightly larger, incrementally more ornate typefaces or spaced at nonstandard intervals. Like the bulbous curve from which the *W* of "Wee" unfurls, its paired, upright *E*'s differ from the modest, lowercase character printed as *e* directly below them. This robust, rampant *E* repeats again in "pros*p* Er," where it is grouped slightly apart with an *r* and framed by a *p* that bears an additional decorative swag absent from the letterform with which the word begins. Having identified these minor visual excesses within the message, the interpreter can patiently scrutinize the code to extract the latent content from manifest appearance (fig. 0.5). Stitching those modestly enhanced characters together, the dismal, hidden report becomes perceptible: "Wee perish with hunger helpe us."

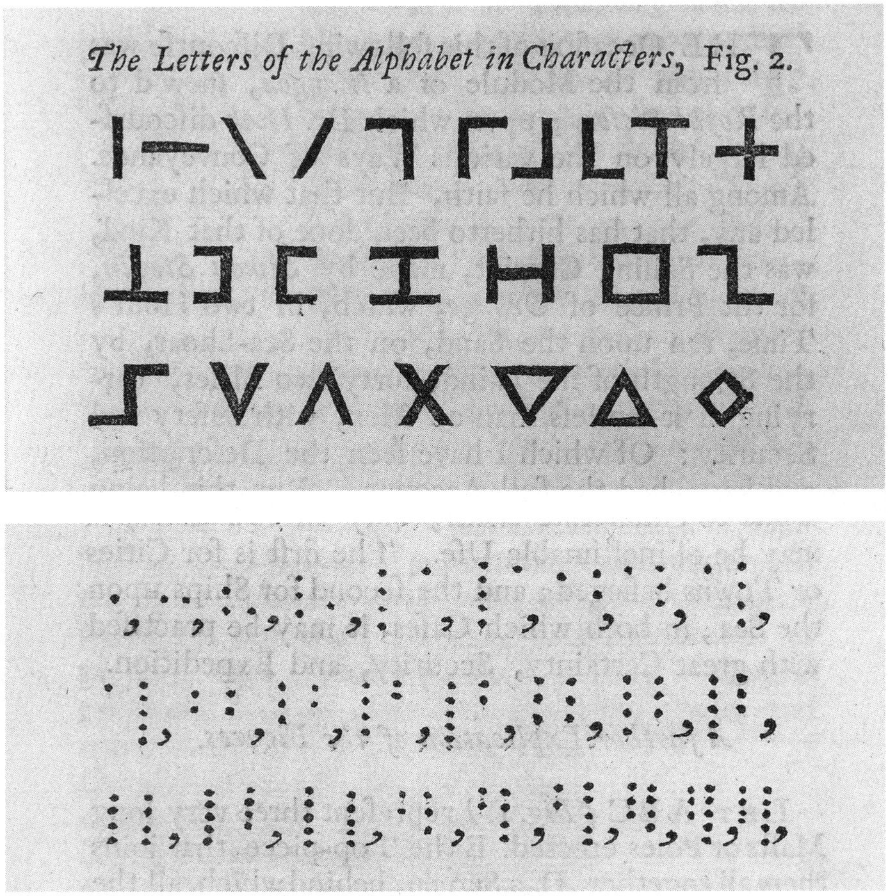

FIGURE 0.3*a*, *b* * Robert Hooke's designs for semaphore codes as printed in "Dr. Hooke's Discourse to the Royal Society, May 21, 1684. Shewing a Way How to communicate one's Mind at great Distances"; in *Philosophical Experiments and Observations*, ed. William Derham (London: W. and J. Innys, 1726); see 149–50. © The Royal Society. Reproduced by permission.

In his influential theory of "multiple intelligences," psychologist Howard Gardner stipulates that an intelligence requires "a set of skills of problem solving—enabling the individual *to resolve genuine problems or difficulties* that he or she encounters and, when appropriate, to create an effective product—and must also entail the potential for *finding or creating problems*—thereby laying the groundwork of the acquisition of new knowledge."[41] A keen problem-solving tool used among philosophers navigating the religious strife, brutal warfare, and political upheavals of mid-seventeenth-century Europe, cryptographic intelligence also became an important mechanism for

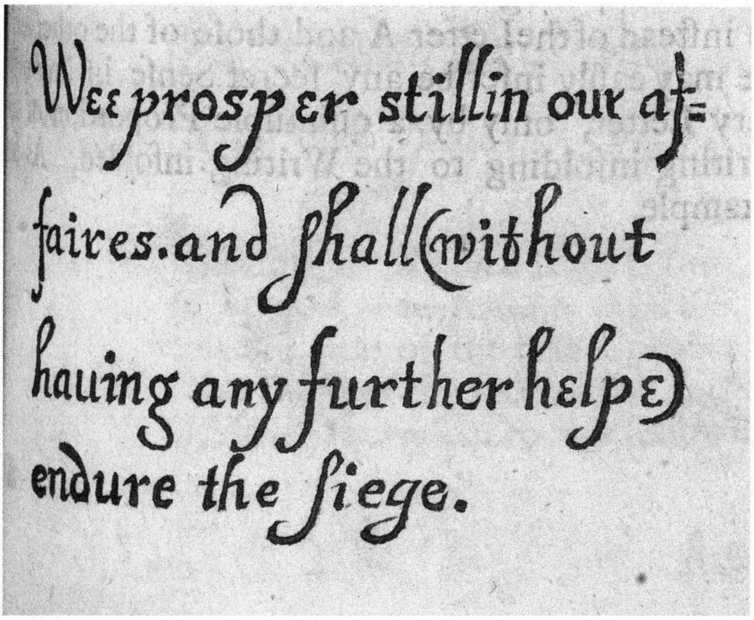
Wee prosper still in our af=
faires. and shall (without
hauing any further helpe)
endure the siege.

FIGURE 0.4 * Typographical code from John Wilkins's *Mercury; or, The Secret and Swift Messenger*, in *The Mathematical and Philosophical Works of John Wilkins* (London: J. Nicholson, 1708), 37. © The Royal Society. Reproduced by permission.

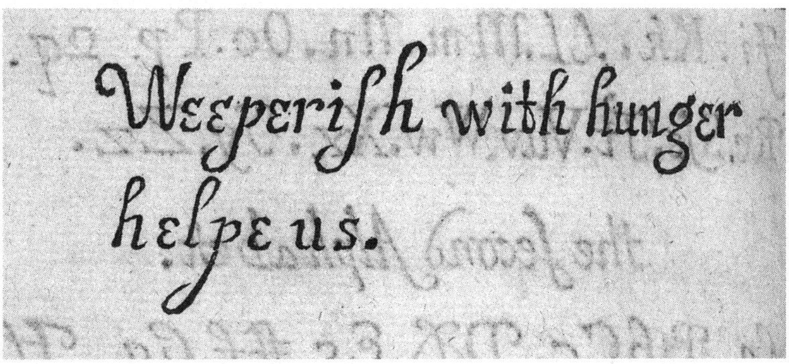
Wee perish with hunger
helpe us.

FIGURE 0.5 * Encoded message from Wilkins's *Mercury; or, The Secret and Swift Messenger*, in *The Mathematical and Philosophical Works of John Wilkins* (London: J. Nicholson, 1708), 38. © The Royal Society. Reproduced by permission.

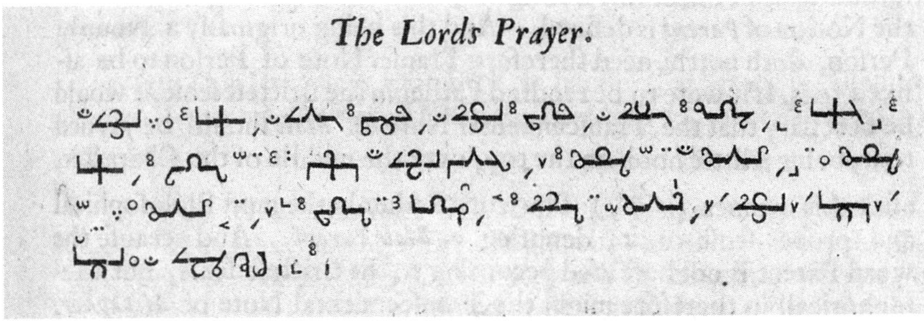

FIGURE 0.6 * The Lord's Prayer rendered in philosophical script as printed in John Wilkins's *An Essay towards a Real Character and a Philosophical Language* (London: S. Gellibrand and J. Martyn, 1668), p. 395; Huntington Library, RB 601596. Reproduced by permission of The Huntington Library, San Marino, California.

knowledge creation.[42] Wilkins's own *An Essay towards a Real Character and a Philosophical Language* (1668) is not only the best known among the period's numerous attempts at linguistic reform but an exemplary instance where code was made agent for rectifying thought[43] (fig. 06). Wilkins designed an elegant script whose "real characters" would clearly, unambiguously denote things and concepts catalogued in a massive taxonomy assembled collaboratively by leading Royal Society Fellows.[44] Fluent in Wilkins's philosophical language, Hooke rated that tongue favorably to the Mandarin Chinese he learned from books and coffeehouse conversation. In a fine piece of preaching to the choir, he also extolled to Gottfried Leibniz the manifold advantages of such a language not only "for expressing and remembering of things and notions but to direct, regulate, assist and . . . compel the mind to find out and comprehend what soever is knowable."[45] Plainness, economy, clarity: Michel Foucault has observed that seventeenth-century knowledge is remarkable less for what is new in it than for "what is *missing*."[46] Cleaved and cleared by the knife of philosophical language, the experimental mind (like those dissected brains drawn by Wren for Thomas Willis's seminal *Cerebri Anatomae* of 1664 [fig. 0.7]) would be made as elegant, spare, and ordered as Wilkins's characters themselves.

From pictorial plenitude to visual attenuation, from mimesis to code—juxtapositions like these figure prominently in recent approaches to narrating scientific visuality. Historian of science Peter Galison has interpreted twentieth-century particle physics as a pitched struggle between "image and logic," between traditions of research that aim to capture "natural processes in all their fullness and complexity" and those that forego mimetic aspiration in favor of amassing voluminous numerical data on target-phenomena.[47] Art historian David Freedberg pursues a commensurate

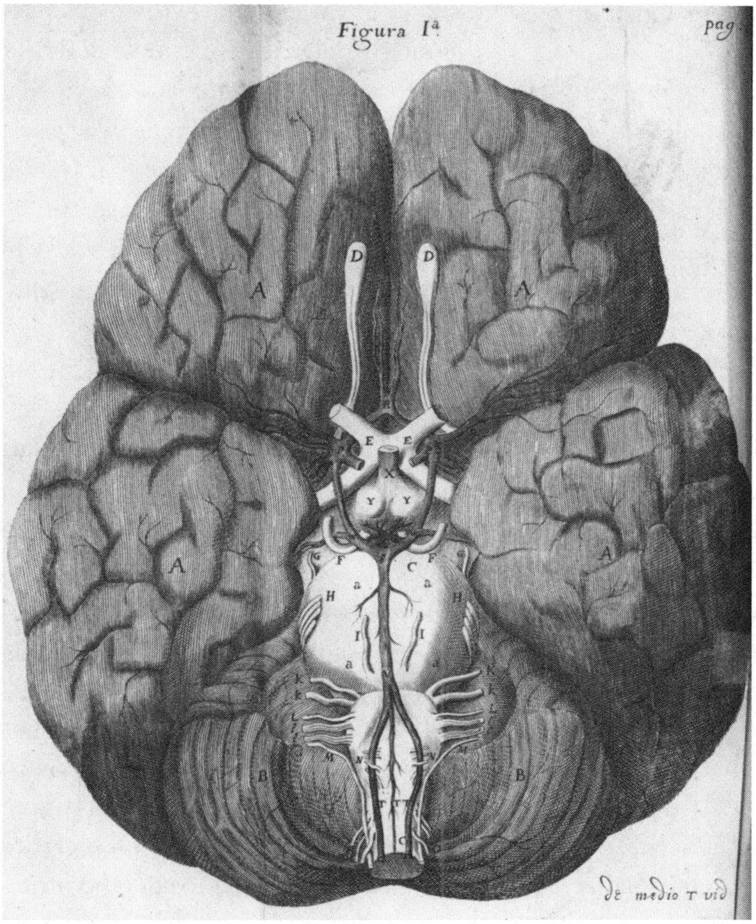

FIGURE 0.7 * Engraving after Christopher Wren's drawing of the "Circle of Willis"; from Thomas Willis's *Cerebri Anatome* (Amsterdam: G. Schagen, 1664), fig. 1a. © The Royal Society. Reproduced by permission.

strategy in his reading of seventeenth-century Italian natural history as a decline of the picture and a rise of the diagram.[48] By the mid-seventeenth century, pictorial schemata derived from artists' studios could no longer "yield the principles of order; these could only be achieved by penetrating beneath the surface, by counting, and by reducing the fullness of pictorial description to their essential geometrical abstractions."[49] For Freedberg, picture and diagram not only map onto, respectively, the resemblances of the Renaissance *episteme* and the representational signs of classical knowledge as envisioned by Foucault, but they constitute a "clear, serious, and instructive" polarity.[50]

My approach is more pragmatic. What I see evolving in the experimental entrepôt of Restoration London is less a semiotically polarized epistemological field than intelligence fashioned and refashioned through manifold hybrid fusions of visualization techniques—and often deployed for far less enlightened purposes than those claimed by their advocates.[51] Consider the case of Elizabeth Travers, a woman "aged 23 or 24 fair of complexion, brown hair'd, of an healthy constitution, small of stature, of honest repute, but of poor and mean parentage," who awoke in Plymouth one morning in 1669 to find her breasts swollen to pathological (and, ultimately, fatal) proportion[52] (figs. 0.8, 0.9). After inspecting the patient, local physician William Durston wrote a detailed case study, a *text* he had countersigned to affirm the veracity of his report. With that text, Durston also transmitted to London a company of linen *tapes* cut to the length and circumference of each monstrous breast.[53] Products of Foucault's measurement-hungry Classical *mathesis*, we might say, Durston's tapes sacrifice all lingering pictorial density and repleteness for a severe visual economy worthy of Wilkins's philosophical characters. Yet in addition to his text and tapes, Durston also enlisted the talents of a local draftsman versed in the rudiments of portrait painting (fig. 0.10). In that *depiction*, we see Travers facing demurely outward, her bedclothes primly cinched at their apertures and pulled apart to reveal her tragic malady.

Representational diversity was crucial to the experimental philosopher. When addressing the philosopher-cum-cryptographer in 1641, John Wilkins had advanced the model of the merchant. "As it will concerne a man that deals in traficke, to understand the severall kinds of money, and that it may be framed of other materialls besides silver and gold," he proposed, "so likewise do's it behove them, who professe the knowledge of nature or reason, rightly to apprehend the severall waies whereby they may be expressed."[54] Text, tapes, depiction: William Durston used all three modes of reference to communicate intelligence (that is, news) about his case to the Royal Society. As with the fragmented, fictionalized strategies of modeling now studied by philosophers of science, moreover, Durston deployed an array of representational techniques, each of which was capable of targeting specific aspects of his case in ways advantageous to his visual, epistemological, and professional interests.[55] Given the ubiquity of such strange, hybrid cases in the experimental archive, I have found James Elkins's notion of a "domain of images" immensely useful for framing my project. Instead of conceptualizing scientific representation after the manner of "illustrations" infused with artistic mimesis or as bifurcated into oppositions between iconic and conventionalized symbols, I deploy Elkins's basic model of a continuous spectrum of representational possibility triangulated between the always-impure fields of writing, picture, and notation.[56] The domain

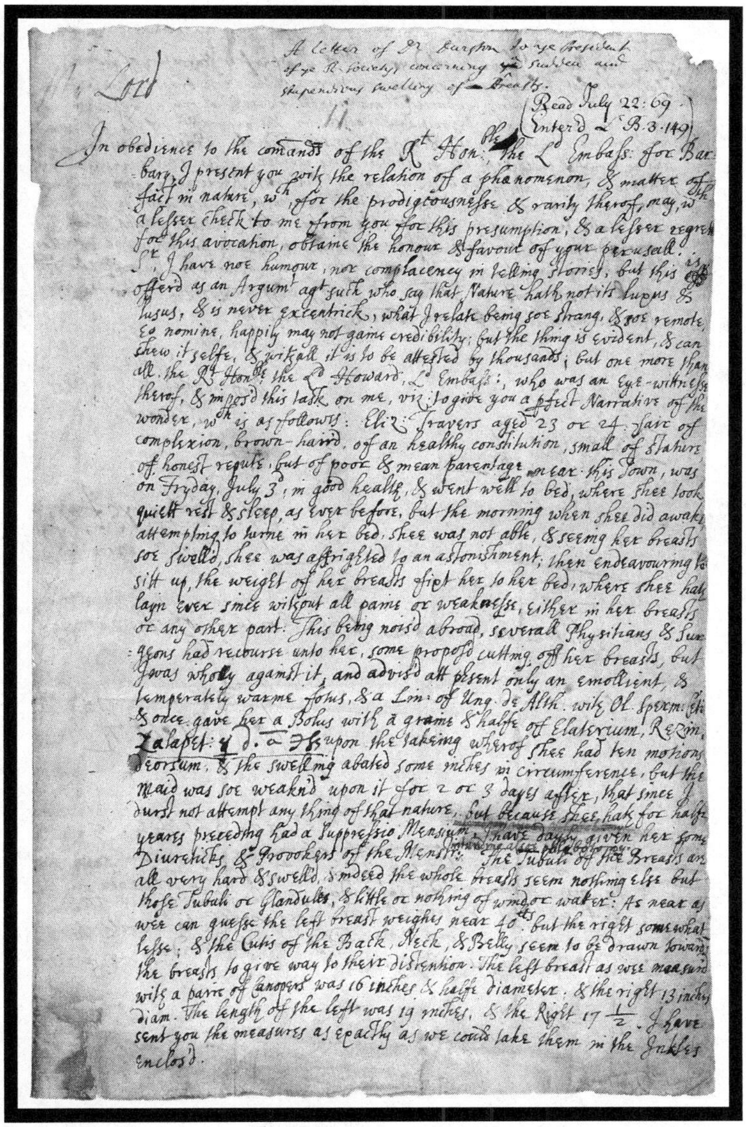
A letter of Dr Durston to ye President of ye R. Society concerning ye sudden and stupendious swelling of Breasts.

11

Read July 22: 69
Enter'd L. B. 3. 149

My Lord

In obedience to the comands of the Rt Honble the Ld Embass: for Bar-
bary, I present you with the relation of a phænomenon, & matter of
fact in nature, wch, for the prodigeousnesse & rarity therof, may, wth
a lesser check to me from you for this presumption, & a lesser regret
for this avocation, obtaine the honour & favour of your perusall.
Sr, I have noe humour, nor complacency in telling stories, but this is
offer'd as an Argumt agt such who say that Nature hath not its lupus &
lusus, & is never excentrick, what I relate being soe strang, & soe remote,
eo nomine, happily may not gaine credibility; but the thing is evident, & can
shew it selfe, & withall it is to be attested by thousands; but one more than
all, the Rt Honble the Ld Howard, Ld Embass:, who was an Eye-witnesse
therof, & impos'd this task on me, viz: to give you a pfect Narrative of the
wonder, wch is as follows: Eliz: Travers aged 23 or 24, fair of
complexion, brown-hair'd, of an healthy constitution, small of stature
of honest repute, but of poor & mean parentage near this Town, was
on Fryday, July 3, in good health, & went well to bed, where shee took
quiett rest & sleep, as ever before, but the morning when shee did awake
attempting to turne in her bed, shee was not able, & seeing her breasts
soe swell'd, shee was affrighted to an astonishment; then endeavouring to
sitt up, the weight of her breasts fixt her to her bed, where shee hath
layn ever since without all paine or weaknesse, either in her breasts
or any other part. This being nois'd abroad, severall Physitians & Sur-
geons had recourse unto her, some propos'd cutting off her breasts, but
I was wholly against it, and advis'd att psent only an emollient, &
temperately warme fotus, & a Lin: of Ung. de Alth. with Ol. Sperm: Cti
& once gave her a Bolus with a graine & halfe of Elaterium, Rezin.
Zalapet: [illegible] upon the takeing wherof shee had ten motions
deorsum, & the swelling abated some inches in circumference, but the
Maid was soe weakn'd upon it for 2 or 3 dayes after, that since I
durst not attempt any thing of that nature, but because shee hath for halfe
yeares preceeding had a Suppressio Mensium, I have daily given her some
Diureticks & Provokers of the Menstr: [illegible] The Tubuli of the Breasts are
all very hard & swell'd, & indeed the whole breasts seem nothing else but
those Tubuli or Glandules, & little or nothing of wind or water: As near as
wee can guesse the left breast weighes near 40lbs but the right somewhat
lesse; & the Cutis of the Back, Neck, & Belly seem to be drawn towards
the breasts to give way to their distention. The left breast as wee measur'd
with a paire of Canopers was 16 inches & halfe diameter, & the right 13 inches
diam. The length of the left was 19 inches, & the Right 17 1/2. I have
sent you the measures as exactly as we could take them in the Inkles
enclos'd.

FIGURE 0.8 * Letter detailing the case of Elizabeth Travers sent by William Durston to Lord Brouncker, July 22, 1669; Royal Society Early Letters, vol. D1, fol. 11.

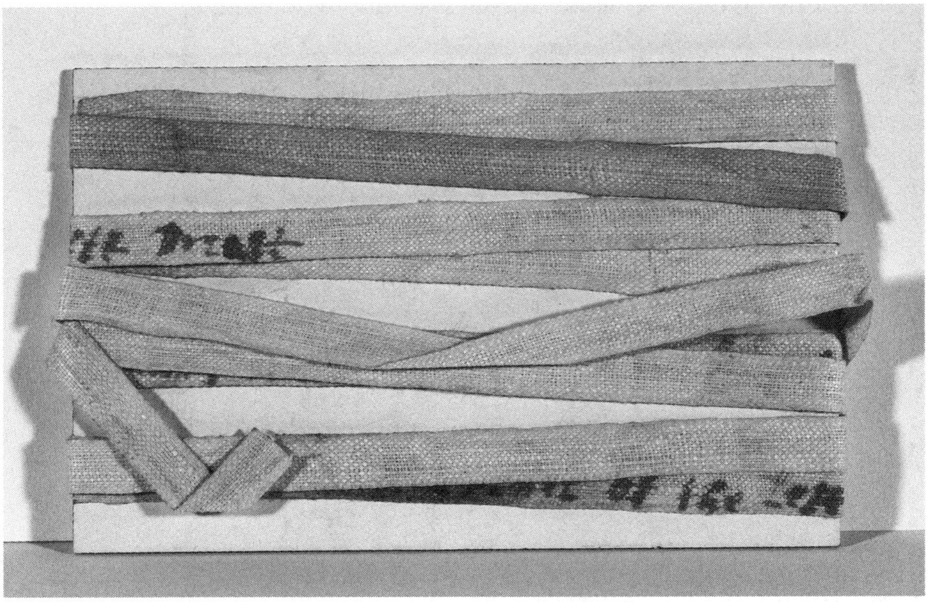

FIGURE 0.9 * Linen tapes measuring the breasts of Elizabeth Travers sent by Dr. William Durston to Henry Oldenburg, November 2, 1669; Royal Society Early Letters, vol. D1, fol. 14. © The Royal Society. Reproduced by permission.

of experimental visibility analyzed in this book thus stretches from Durston's written texts, cut tapes, and pen drawing to samples and diagrams, prints, models, oil paintings, instruments, museum artifacts, architectural spaces, and back again.

Among the networks of Royal Society informants moving through Turkey, Egypt, South Asia, the American colonies, and numerous points in between, we will see that few correspondents made images themselves or actually knew how to draw. In exceptional cases, investigators such as Antoni van Leeuwenhoek employed highly skilled, artistically trained draftsmen to produce their visualizations. (See figure 3.1 and figure 4.7*a*, *b*.) Most, like Durston, improvised with the indifferent talents at their disposal. Rarely did they bother to articulate—let alone theorize—the visual bricolage they mailed off to London. Yet reckoning with the material form of these variegated representational techniques was of fundamental practical and conceptual importance to the elite, polymathic figures who unpacked those transmissions in the capital and who sit at the center of my story. First of all, apprehending the physical properties, production history, and practical applications of a given material was standard procedure for the broadly Baconian brand of mechanically inclined experimental philosophy favored by Hooke, Wren, and their close associates. Not only did inside-and-out knowledge of materials ensure that these "Archietonical"

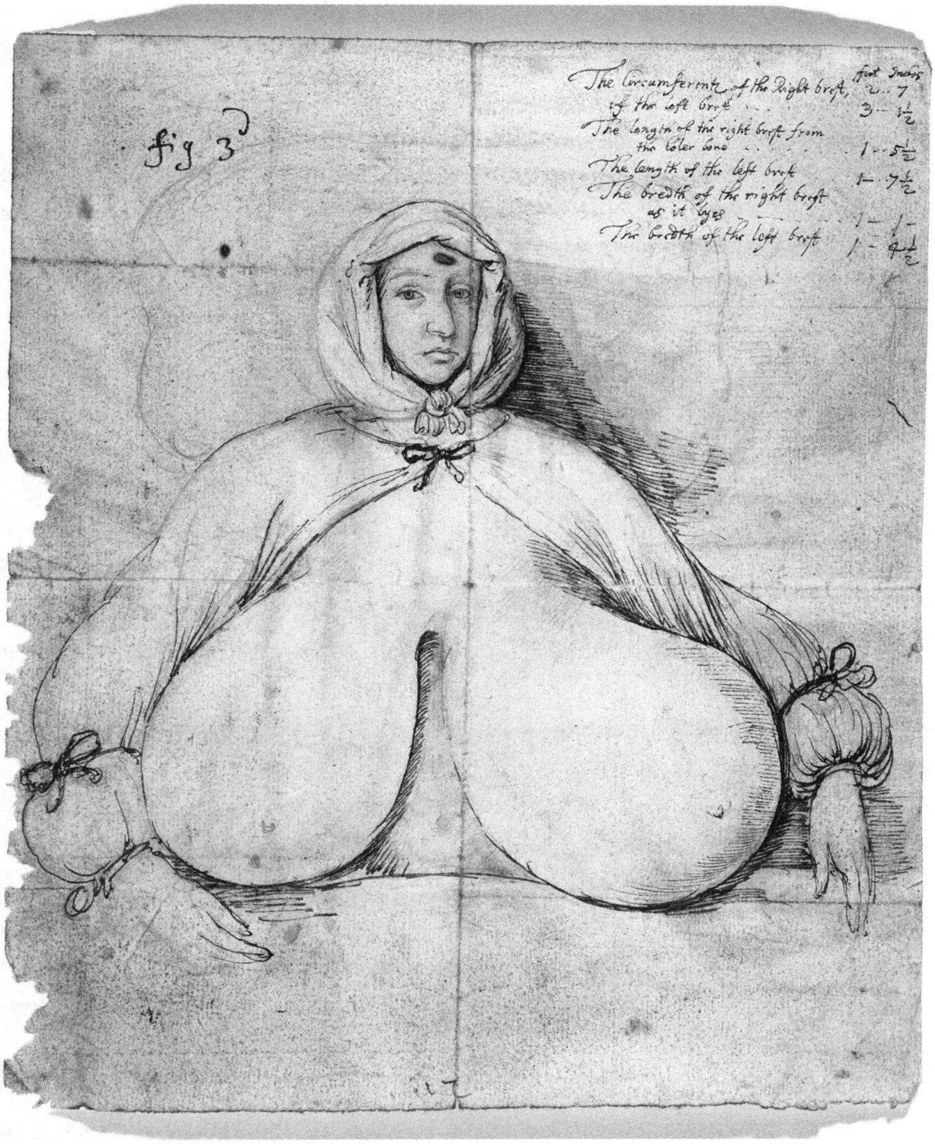

FIGURE 0.10 * Anonymous portrait of Elizabeth Travers sent by William Durston to Lord Brouncker, July 22, 1669; Royal Society Early Letters, vol. D1, fol. 11. © The Royal Society. Reproduced by permission.

philosophers (as we will see them called) could direct others in the performance of a representational task, but it meant that those materials could potentially be harnessed as conceptual resources for understanding, even guiding, philosophical problems. Knowing how paper is made, how ink is manufactured, and by what means ink sinks into paper would, as we will see, prove keenly instructive for Robert Hooke as he drew, cut up, pasted, folded, and *dreamed with* his paper model of a telescopic micrometer while thinking about the interiors of animal bodies in the later 1660s.

Several important studies have also recently turned to languages of intelligence when grappling with such thinking with materials. Michael Cole has found dazzling articulation of the "goldsmith's intelligence" in Benvenuto Cellini's masterworks, which advance sophisticated arguments about their maker's power and invention in a miraculous, marginalized medium.[57] Among the artisans and craftsmen of the Northern Renaissance, Pamela H. Smith traces a similarly assertive "artisanal epistemology," as it located authority not in ancient texts but in the encounter between maker's body and nature's materials.[58] Inspired by a painter who played as fast and loose with the matter of art as Giambattista Tiepolo, Svetlana Alpers and Michael Baxandall have imagined a "pictorial intelligence" so virtuosic and luscious that it amounts to an ongoing act of world-making.[59] And below the surface of his medium's dry, mechanized sight, contemporary photographer Jeff Wall hears the murmurings of an atavistic consciousness, a "liquid intelligence."[60] Made with and through things, the experimental intelligence advanced in this book shares the rustling, rupturing engagement with materials emphasized by these scintillating studies.

Yet, already embracing knowledge forged by controversial fusion of "*Mechanical Hand*, and . . . *Philosophical* . . . *Mind*" in an age that measured social status by its distance from manual labor, Restoration experimentalism practically invited sinister associations.[61] As a way of returning to the codes with which I began, consider Hooke's idiosyncratic reading of a figure who will surface at key junctures in my narrative: Elizabethan philosopher John Dee (1527–circa 1609).[62] To later seventeenth-century audiences, Dee had been cast as a tragic, Faustian figure whose failed attempts to learn the secrets of nature through communication with angels told an important moral lesson. According to conservative Anglican cleric Meric Casaubon, who published the "angel diaries" in 1659, Dee anticipated the dangerous partisanship of the English Civil War by valuing private "Instinct and Inspiration, as they call it" above the stable, steadying traditions of secular and ecclesiastical authority.[63] Robert Hooke took a very different line. At first glance, he acknowledged, Dee's diaries seem like "a Rapsody of incoherent and unintelligible Whimsies of Prayers and Praises, Invocations and Apparitions of Spirits, strange Characters, uncouth and unintelligible Names, Words, and Sentences"[64] (fig. 11). Yet the odd

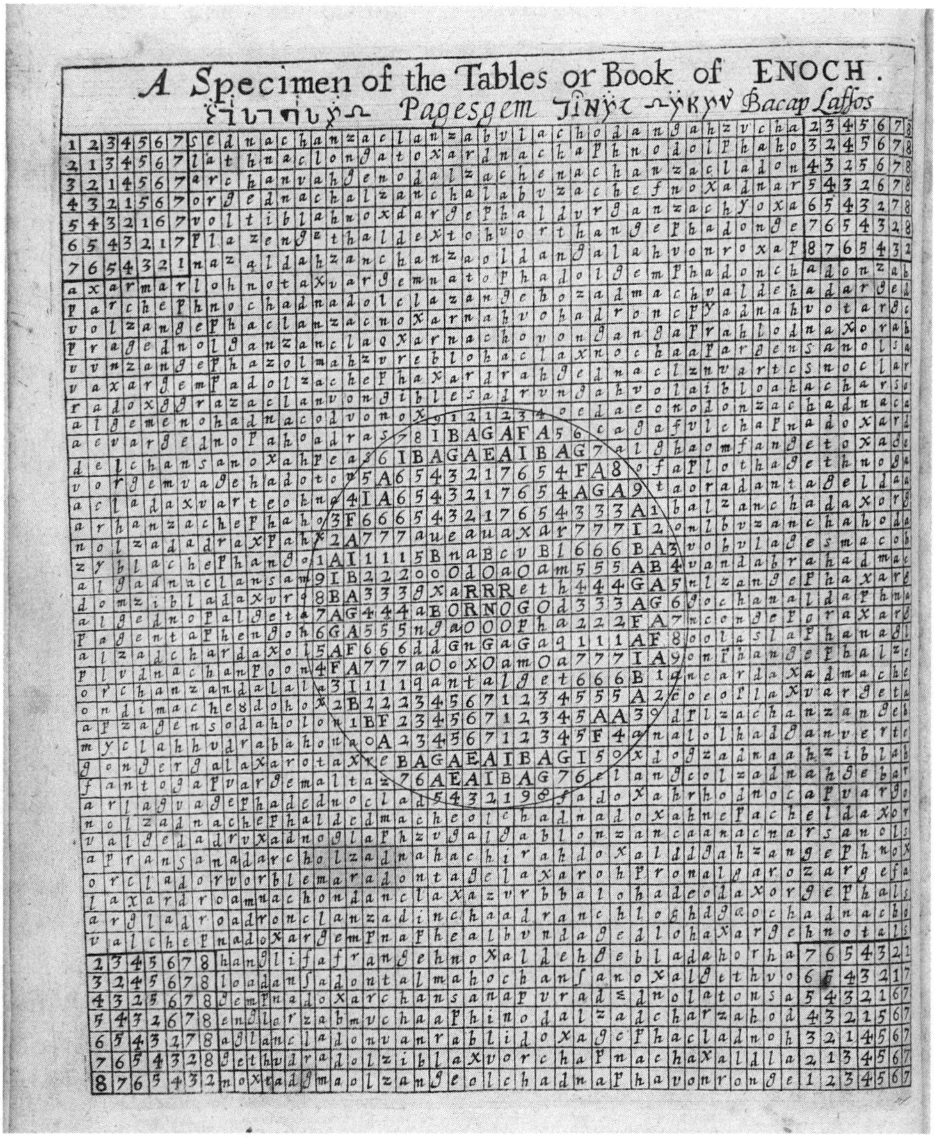

FIGURE 0.11 ✻ "A Specimen of the Tables or Book of Enoch" as printed in Meric Casaubon's *A True & Faithful Relation of what Passed for Many Yeers between Dr. John Dee and Some Spirits* (London: D. Maxwell, 1659), n.p.; Huntington Library, RB 601446. Reproduced by permission of The Huntington Library, San Marino, California.

alphanumeric tables and scripts scattered throughout Dee's work signaled the need for a different kind of interpretation.[65] Simply put, the angel diaries were "Instances of Cryptography."[66] Hooke's Dee knew exactly what he was doing. Traipsing around the courts of late sixteenth-century central Europe under orders from Elizabeth I, Dee had shrewdly recognized that "if he should fall under suspicion as to the true designs of his travels . . . the Inquisition that should be made, or Prosecution, if discovered, would be more gentle for a Pretended Enthusiast, than for a real Spy."[67] Like the cryptography he practiced, so the philosopher's intelligence also had to carefully conceal its true, cunning intents, biding its time to strike.

In their classic study of ancient Greek literature, Marcel Detienne and Jean-Pierre Vernant suggest that *metis* names a species of intelligence situated in and directed toward practical efficacy: "It may involve multiple skills useful in life, the mastery of the artisan in his craft, magic tricks, the use of philtres and herbs, the cunning stratagems of war, frauds, deceits, resourcefulness of every kind."[68] Handy and underhanded, *metis* could be a dirty cheat or the savvy act by which a weaker agent bests a superior rival by seizing a momentary advantage. Like "the glittering of a weapon, the dappled hide of a fawn or the shining back of a snake mottled with darker patches," as Detienne and Vernant put it, the intelligence of *metis* is constantly in flux, changing in response to its situation.[69] I see the Dee that Hooke imagined—and, indeed, the wicked intelligence Hooke and his colleagues used Dee to craft—as working along similar lines. Let me say something about what I mean with that expression. "Wicked intelligence" is my descriptive phrase for an interrelated set of patterns recurrent through the visual archive.[70] First, the phrase aims to capture the readiness with which the ingenious ways of working and thinking with visual materials practiced in Restoration London's collaborative experimentalist circles also frequently alienated and enraged contemporaries. When he discovered that the calculating machine he had just donated to the Royal Society's Repository had been reverse engineered by the museum's curator, Gottfried Leibniz declared the tactics of that curator (Robert Hooke) to be "unworthy of his nation, and unworthy of the Royal Society."[71] Historians of science have cited clashes of personalities, institutional limitations, or conflicting views of scientific community as explanations for these torrents of dispute. "Wicked intelligence," in the first instance, means to signal just how deeply these acrimonious dynamics and nefarious tactics permeated the visual enterprises generated by Hooke and his milieu.

Second, I use "wicked intelligence" to highlight the ways in which this group of London experimentalists effectively courted a shady reputation, not only through fascination with heterodox intellectuals like Dee who informed their visual work or with the occulted powers they attributed to artists and artworks, but through their seemingly outrageous claims as makers. Hooke's assertion of the standard magician's claim that he possessed dozens of techniques for human flight springs directly to mind.[72] Third, I see the phrase as a useful means for recalling the fragility of the internally divided visual project engineered by Hooke, Wren, and their associates, along with the tarnish their reputations quickly gained in the eighteenth century. Fourth, with "wicked intelligence" I mean to push on the tension in recent historiography that has alternately plotted Hooke as an incestuous, low-status mechanic or ennobled him as a "gentleman of science" who serves as an important font of eighteenth-century British art. It's not the answers I want to question, but the questions. We have benefited from asking who they were; it is now time for asking *how* Hooke, Wren, and their associates practiced, organized, and conceptualized a visual enterprise that stretched from the microscope to the macrocosm of what was soon to be Europe's largest city; *where* those visual practices intersected and departed from artistic (and other) traditions; and *why* they made people so angry.

Ultimately, though, *Wicked Intelligence* seeks to make the London experimentalists' explosively clever, freewheeling styles of working and thinking with visual materials accessible to a range of contemporary readers—even to practicing artists. I move toward that aim guided by the pragmatic model of Hooke and Wren themselves, using historical arguments and analytic techniques as needed while always seeking to keep the narrative voice as open, generous, and, frankly, engaging as I can. This fifth sense is thus unapologetically contemporary. If the practices analyzed are not always evil or sinister as the term implied in "Olde England," I mean to foreground deeds and doings of that excessive, flabbergasting awesomeness signaled by *wicked* in modern New England.

I tell this story based, first and foremost, on the visual archive of Restoration London's experimental community that I was able to examine in situ for four years. Despite the Royal Society's widely acknowledged importance to modern science, our understanding of how images were actually produced, processed, and performed within the institution is only beginning. To bring that crucial image ecology and its stakes into focus, my book draws on an array of interdisciplinary theoretical approaches and questions. But because I want to keep the visual archive in the foreground, Robert Hooke figures prominently in my narrative, and it is important to explain why. Coming of age in London and then Oxford, Hooke won influential friends through whom he parlayed his exceptional fluency in drawing, instrument-making, and other skilled manipulations of matter into a paid position within the fledgling Royal Society. Since he maintained an active role within that bureaucratic institution for the remainder of his life, we possess an unusually comprehensive and diverse record of his activities. Living in the Elizabethan college where the Royal Society held its weekly meetings and serving from late 1662 as Curator of Experiments, Hooke was literally central to the institution in its early decades. In addition to his surviving drawings, artifacts, and buildings, we also have his published works, extensive manuscripts, correspondence, reading notes, catalogues of his book library, and—nearly uniquely among seventeenth-century experimental scientists—his famous diaries. While, especially in the later chapters of this book, I interpret those sources against an increasingly countervailing range of period voices, Hooke's works and deeds frequently supply material evidence by which I connect the activities of a mason laboring on the fabric of St Paul's Cathedral to the pictorial concerns of Peter Lely and even to the visual proclivities of the Stuart monarchs themselves.

This raises a sequence of methodological questions with which I open the first chapter, "'I Resolved to Throw Aside All Manner of Hypotheses . . . and to Attend Wholly to What the Appearances Themselves Would Teach Me.'" I begin with *Micrographia* of 1665, which is not only Robert Hooke's most famous work but a totemic project for the early Royal Society. In engraved and etched plates that positively bristle with detail, Hooke depicted microscopic molds flowering into putrid bloom, crystals protruding like warts from mineral skins, and—for the first time in history—cells made visible to the eyes of a general viewership. Published when its author was not yet thirty years old, *Micrographia* manifests the pronounced influence of lessons Hooke had learned in the painting studio of Peter Lely and from collaborative work with Oxford's scientific circles. Yet the confident pictorial strategies and robust epistemological value that Hooke theorized for *Micrographia*'s stunning images also depart massively from the fragmentary, tortured visual forms of his later draftsmanship, particularly his astronomical drawings of the early 1680s. That late graphic work has been much less well known because it was actively suppressed by Hooke's posthumous editor—this despite the fact that Hooke assigned it even stronger cognitive force than *Micrographia*'s plates. How do we account for these strange shifts and discrepancies between graphic form and cognitive function? Are they susceptible to the art-historical narratives of the "history of the eye" in which Hooke has effectively been implicated for the past three decades? And what can we possibly make of the mess that is the creation, style, and publication history of these late drawings? By working between these paired groupings of drawings from the 1660s and 1680s, I show how Hooke's weird pictorial project effectively explodes a sequence of carefully crafted observational protocols and models of the experimental-philosophical self, as well as the narratives by which recent interdisciplinary interpretation would bind them.

The second chapter, "Knives Out: Thinking On, With, Through, and Against Paper in the Mid-1660s," takes the traffic between graphic making and experimental knowing right down into the fabric of the page. The chapter focuses on a single artifact: the paper model of astronomer Richard Towneley's telescopic micrometer that Robert Hooke fashioned in the fall of 1667. Apparently fragile and wounded, Hooke's paper model needs to be seen, I argue, as positively kaleidoscopic in its generativity. I show how it began life as a picture, then matured as an object at a nexus of technological competition, artistic skill, and frankly wild speculation among leading French and English experimentalists before giving birth to varieties of conceptual shape-shifting that targeted and liquidated nothing less than artistic objecthood itself. Having set out its microhistorical backstory and conflicted relations with art, I then elaborate the philosophical force of the procedures by which Hooke drafted, cut apart, and pasted his paper micrometer while fantasizing about machines, the machinelike bodies of animals he was then dissecting in landmark anatomical experiments, and the genesis of those animal bodies he was modeling through studies of papermaking. The chapter concludes by showing how Hooke attempted to theorize the cognitive agency of artifacts like his paper micrometer by redeploying his own pivotal ideas on celestial mechanics and attraction at a distance. Distinctly different from both art objects and scientific images as recent scholarship conceives them, Hooke's ever-evolving paper artifact becomes an instructive site for grasping the objecthood favorable to the experimental intelligence.

Pressed by the rupturing ways of working and destabilized object ontologies progressively exposited in the first two chapters, "Pictorial Intelligence: Peter Lely, Experimental Culture, and the Parameters of Painting" then forces the issue, revisiting the relationship between Hooke and his former mentor Lely. If Restoration experimentalists were so intent on transforming and defacing the pictorial heritage in which they were abundantly learned and skilled, should we not follow their interpretive cue? In other words, why shouldn't we take the course proposed by theorists such as James Elkins and simply let the fetishized relationship between painting and science go?[73] This chapter aims to show that just as Peter Lely and his period commentators can disclose fundamental concerns shared among Robert Hooke's circles, so the experimental archive provides valuable resources for rethinking Lely, the preeminent visual artist of Restoration London. I begin by using Lely's historical "subject pictures" to surreptitiously illuminate the contours of the Restoration-era philosophical beholder who stood over and against experimental objecthood. The second section turns the interpretive direction around, using the Royal Society's experimental research on the visible bodily effects produced by exotic intoxicants to reexamine a defining stylistic feature of Lely's Restoration portraiture: the "sleepy-eyed look." I conclude synthetically by considering Lely's pictures as made things whose complex, collaborative production presages problems and strategies central to the experimental community, especially from the 1670s through the early 1680s. Ann Bermingham describes the structure of one of her books as "two funnels joined at their narrowest points"; mine might be thought of as a fan held by a slim handle. In this chapter, the story splays outward from drawing and Hooke toward acts of collecting negotiated among a broader, hotter experimental collectivity.[74]

By necessity, the philosophy of experiment was to be a collaborative project. As Royal Society Fellows repeatedly emphasized, images were frequently much more useful for conveying information to fellow experimentalists than words alone. But how was such a scientific collectivity to be organized? And by exactly what means were images to serve the advancement of experimental knowledge? Drawing on a telling selection of materials submitted to the Royal Society, "Cascade, Copper, Collection: Constellations of Images in 1670s Experimental Philosophy" elucidates just how broad a range of visual strategies was required for doing experimental philosophy in the later Restoration by examining the epistemological claims and other interests invested in them. Centered on the *Philosophical Collections* (the scientific periodical Robert Hooke edited between 1679 and 1682), this fourth chapter reads the evidence of how these diverse visual materials were made, used, and understood to be susceptible to incorporation as scientific knowledge. Tracing microhistories of the journal's contributors and placing them against Hooke's contemporaneous writings on the ideal organization of the experimental community, I show how the management of diverse agents and information was, in the later 1670s, both a pressing problem for Royal Society Fellows and increasingly discussed through languages of architecture. But as Hooke's contributors keenly recognized, his vision for a vertically integrated "philosophical army" was often directly opposed to the individual interests they aimed to pursue with and through images.

From networked correspondents conjoined on paper, chapter 5 then turns to the institution's physical spaces of collecting and the ways they were imagined. "'The Wonderful Elaboratory of the Animal Body': The Royal Society's Repository at Work" focuses on the Royal Society's early museum at Gresham College, one of the few spaces of experimental-philosophical practice open to a (paying) London public. Building from the framework of chapter 4, I show how central, London-based Fellows of the Royal Society employed strategies of patronage and manipulation of scientific networks to obtain objects from contacts near to hand and at distant outposts. Yet by examining the ways in which these valuable artifacts were physically disassembled, reconfigured, and recoded with meaning (often many times over) in the meetings of the organization, the chapter demonstrates how the working collection came to serve as a powerful model for the faculties of cognition—particularly for Robert Hooke, Keeper of the museum itself. Hooke's writings on cognition from the early 1680s articulate a telling epistemological geography. Distributed networks of informants became the senses of the experimental body which would deliver ontologically fragile, unreliable objects to the centralized laboratory of the mind. There the countervailing agency of what Hooke would call (through renewed engagement with John Dee) "Archietonical Power" remakes them in relatively stable, rational order and feeds intelligence back out to the periphery. By shifting between the Royal Society's constant bricolage of museum artifacts and Hooke's conception of reason, the chapter sheds raking light on the darker textures of experimental intelligence.

My sixth chapter, "The Architecture of Science and the Science of Architecture," suggests how thinking about the archietonical powers necessary to all cognition would evolve in and as London's built environment itself. I begin by positioning the plans of Wren, Hooke, Evelyn, and others for rebuilding the City after the Great Fire of 1666 within a broader conversation that traced causal relations between urban, architectural space and the capital's peculiar intelligence of "wit." Yet this discourse was crucially contested by enterprising property developers like Dr. Nicholas Barbon, who looked upon London's post-Fire built environment as a speculative market that spoke not to the mind but to consumer desire. These conflicts between heart and mind, between emergent consumer culture and civic governance fell directly on Surveyor General Christopher Wren, who was charged with enforcing property laws and building codes. Using the strange, highly pragmatic interpretation of the Egyptian pyramids he penned in the mid-1670s, I show how Wren came to understand architecture as a massively socially distributed, vertically integrated project commanded by a polymathic intelligence—an enterprise that could uniquely counteract the social problems generated by excessive consumption. If the innovative strategies for fabrication and the collaboratively built form of St. Paul's Cathedral enable us to see that model put into practice, I suggest that they also reveal instructive clefts between Wren and Hooke, whose conception of philosophical architecture remained in dynamic evolution as the walls of the cathedral rose. My conclusion gestures to when and how the project of Wren, Hooke, and their experimentalist colleagues was effectively undermined and defaced in the early decades of the eighteenth century.

Painter, cryptographer, collector, architect: these identities converge at the nexus of protoimperial commercial ebullience, emergent artistic culture, and revolutionary science that met in the rebuilt streets of later seventeenth-century London. My story aims to bring those interlocking worlds into renewed visibility by tracking the trail of the wicked intelligence permeating, refracting, and crafted through them all.

CHAPTER ONE

"I Resolved to Throw Aside All Manner of Hypotheses . . . and to Attend Wholly to What the Appearances Themselves Would Teach Me"

LATE INTO THE SUMMER nights of 1683, observers from Europe to China tracked the appearance of a newly sighted comet. From a turret on Bishopsgate Street in the heart of the City of London, experimental philosopher Robert Hooke first sighted the object on July 16. Using an arsenal of telescopes he had largely designed and built himself, Hooke made his observations at London's Gresham College, the Elizabethan compound that served as his home and as meeting place for the Royal Society of London.[1] And while he published amply in his lifetime, the most direct evidence of Hooke's engagements with this and other comets of the early 1680s comes from a collection of drawings now held in the Royal Society's library.[2]

Taking a page almost at random, the severe interpretive challenges presented to the beholder by these drawings become readily, unpleasantly clear (fig. 1.1). Subdivided into two distinct registers by a horizontal fold of the folio-sized sheet, Hooke's notes from July 16–24, 1683, mix markings promiscuously. From the upper left-hand corner in both registers, we can read off his terse narration of the time and

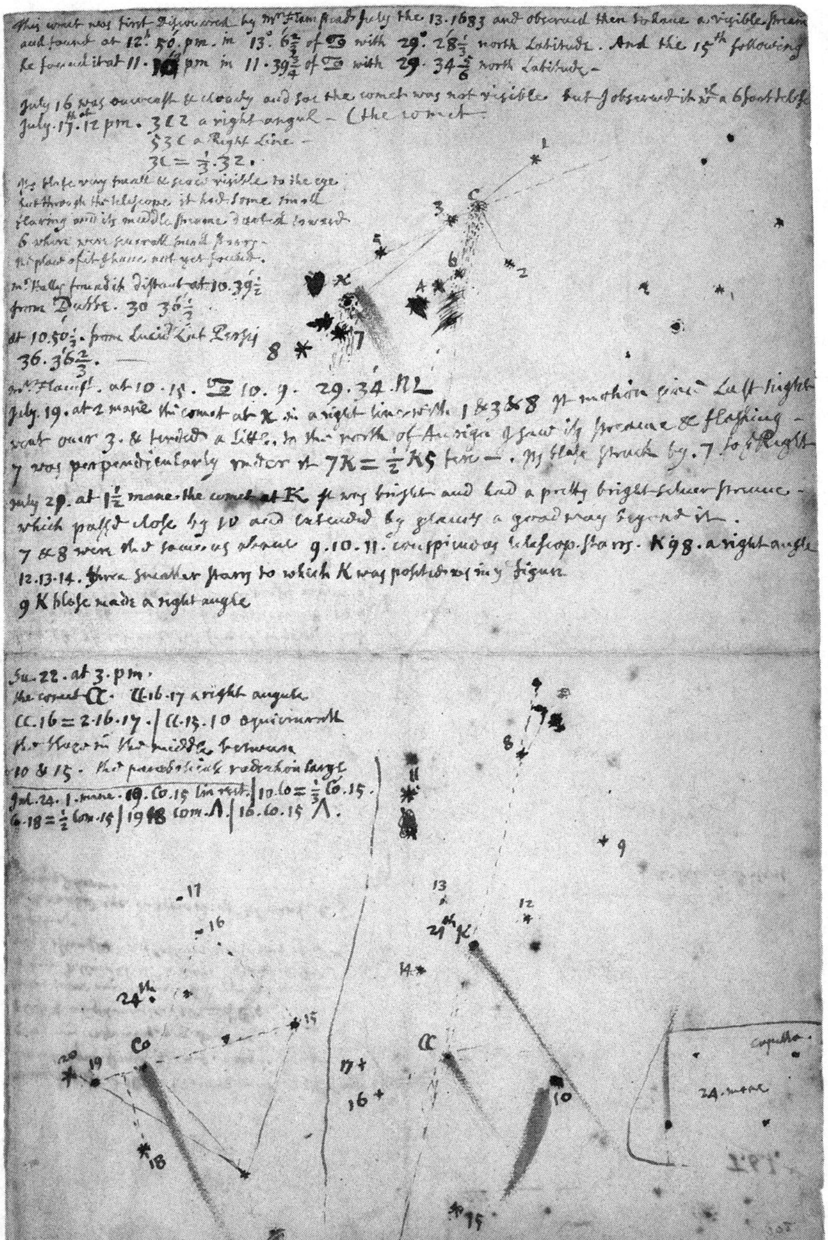

FIGURE 1.1 * Robert Hooke's pen-and-ink drawings and notations on comets, July 16–24, 1683; Royal Society Classified Papers, vol. 24, fol. 308. © The Royal Society. Reproduced by permission.

celestial place of each night's meteoric visit. Encoded in planetary symbols, alphanumeric scripts, and the chicken scratch of his own handwriting, these coordinates plot Hooke's observations against those made simultaneously by his philosophical contemporaries, such as Edmond Halley and John Flamsteed, Astronomer Royal at King Charles II's then newly constructed observatory at Greenwich in southeast London.[3] As throughout this collection of difficult drawings, Hooke's texts also serve to visually frame the depictions of comets to which they repeatedly refer. Consider the image in the page's upper register, where the comet floats upward like a jellyfish trailing shimmering, stippled streamers as it moves between and beyond the smudgy, enumerated crosses Hooke uses to designate stars. Applying these swirling dots of ink to the page in stipples ("pricks," Hooke calls them in his accompanying notes) was clearly a messy job.[4] If corrections, residual blots, and accidental markings litter these pages, Hooke began to make a virtue of his medium's unruly proclivities as the observations progressed. Gradually, he came to denote the comet by inking its stippled head and dragging the viscous substance across the page. When drawing celestial sights on July 24, Hooke indicated the meteor's tail at the page's lower left with sequential smudges of still-tacky ink ground into the coarse weave of the paper.

Unknown to and actively concealed from modern interpretation, these drawings from the 1680s appear worlds away from the elegant renderings of comets apparently produced under Hooke's auspices less than two decades earlier. More problematically, they betray few of the salient properties interpreters have particularly come to value in the crisp depictions of natural phenomena that Hooke oversaw as the Royal Society's de facto engine of graphic production from the early 1660s through the mid-1680s[5] (fig. 1.2). As seen in drafts from the winter of 1664–65 that have been attributed to Hooke, comets were conventionally depicted in mid-seventeenth-century philosophical circles as gem-like collections of starry forms arrayed beneath spare, dated legends.[6] Allowing that we reverse the optical values of these inky stars and the dried bone of their environing page, these markings might suggest how a comet looked in the less densely light-polluted heavens above later seventeenth-century Europe. Precisely outlined, the comet's Catherine-wheel head in the drawing from December 12, 1664, casts its stippled spray diagonally back toward the Cor Hydra, the sole astronomical location named here. Like contemporaneous renderings of that meteoric object sent to the Royal Society from Paris by Christiaan Huygens (fig. 1.3) or from Johannes Hevelius in Danzig, these images are precise, neat, and economical in their notation.[7] Where Hooke's images from the 1680s would be crammed with multiple nights of observations (or multiple observations from the same night), these drawings of 1664–65 present one astronomical sighting per spare, luminous page.

How do we account for the profound differences obtaining between these two

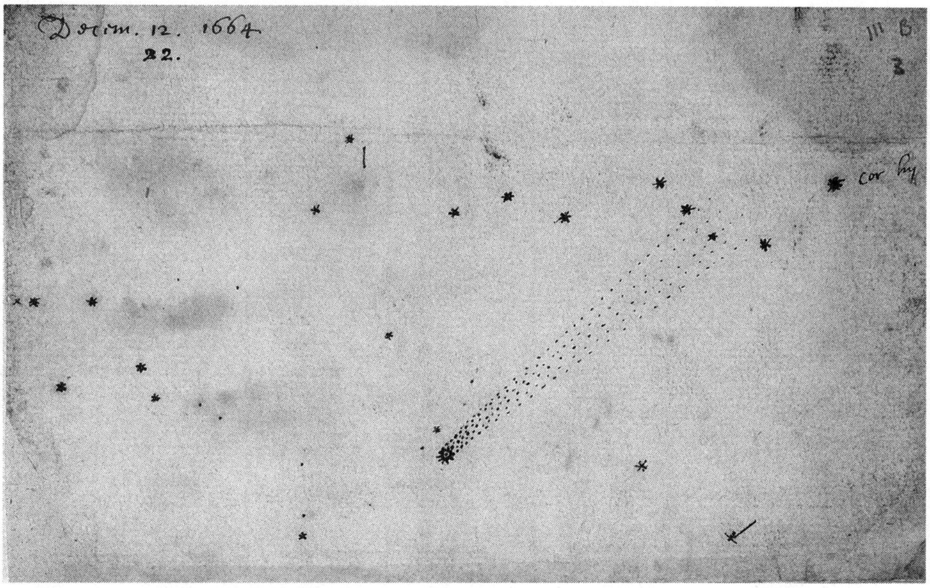

FIGURE 1.2 * Comet drawing from December 12, 1664, previously attributed to Robert Hooke; MS 131 ("Waller Scrapbook"), fol. 111B 3. © The Royal Society. Reproduced by permission.

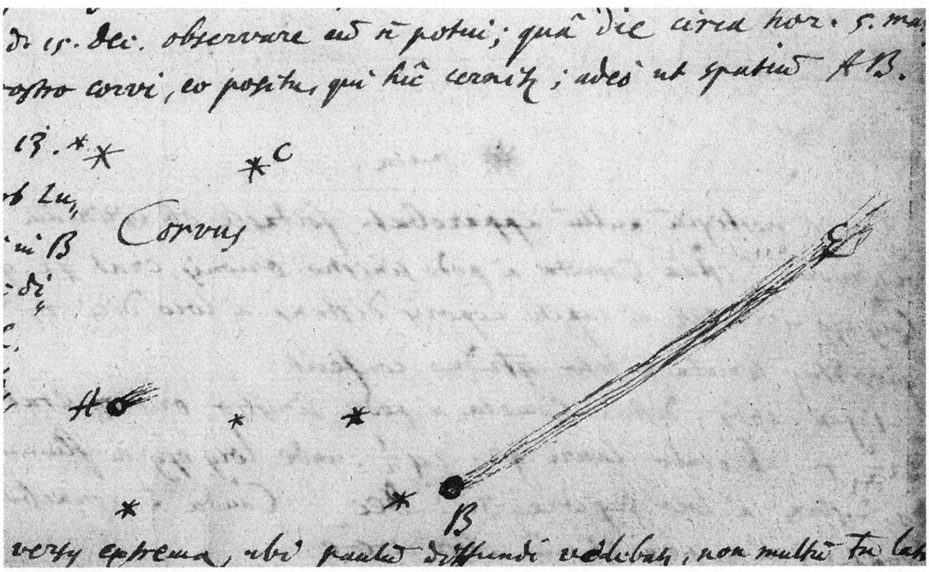

FIGURE 1.3 * Christiaan Huygens's graphic rendering of the 1664–65 comet from January 2, 1665; Royal Society Early Letters, vol. H1, fol. 43. © The Royal Society. Reproduced by permission.

sets of drawings—sets both depicting comets, made in pen and ink, apparently by the same agent in the same city less than two decades apart? Any satisfactory explanation would need to attend to a range of factors surrounding their production and reception. We would want to note how Robert Hooke had been charged by the Royal Society in the winter of 1664–65 to make observations of comets "with all diligence and exactness."[8] Not yet thirty years old when he received this commission, Hooke was then "on loan" to the Royal Society from chemist Robert Boyle and angling to have his appointment as the institution's Curator of Experiments made permanent.[9] Given what recent scholars have seen as a taste for minute pictorial finish and graphic precision in the early Royal Society, the publicly exhibited 1664 comet drawings might be cast as hewing close to his patrons' visual fancy—at a moment in which Hooke's courtship of elite taste was keenly opportune.[10] Even though Hooke likely did not execute these 1664 drawings, we could also stress their strong continuities with the program of draftsmanship he was then theorizing and exhibiting to the Royal Society.[11] By contrast, this explanation might well continue, the 1680s drawings were produced at a moment when Hooke occupied an international profile in the experimental-philosophical community and a range of prominent positions in Restoration London; among other titles, Hooke was then Secretary of the Royal Society, Professor of Geometry at Gresham College, and leading collaborator with Christopher Wren in the architectural rebuilding of fire-ravaged London.[12] Made by a powerful philosopher who no longer had to impress patrons, we could conclude by emphasizing how Hooke's late drawings apparently had an audience of one. As his period editor claimed at least, they were produced "only as Helps to his own Memory."[13]

The concerns of this chapter lie less in elaborating an explanation like the one I have just sketched than with the methodological contours framing such an interpretation. When plotted as a chronological displacement of the closed, distinct, and certain by profusions of ruptured, ephemeral forms—when framed as a classic art-historical problem—these drawings steer interpretation directly to the heart of matters that have informed the understanding of Hooke's visual project for three decades and implicitly implicated it for a century. That is, the conception of the seventeenth century as a period defined by a fundamental shift in ways of seeing registering in visual artifacts was one staked out by no less a figure than Heinrich Wölfflin. From "the perception of the object by its tangible character—in outlines and surfaces," so Wölfflin claimed in his seminal studies of baroque art, the seventeenth century marked a turn away from the Renaissance's privilege of the static and the certain toward a desire for movement, instantaneity, and, above all, an embrace of the experience of the eye: "a perception which is by way of surrendering itself to the mere visual appearance" (eine Auffassung, die dem bloßen optischen Schein sich zu überlassen imstande ist).[14] A century old though Wölfflin's reading is, the "surrender to the eye" he diagnosed has continued to inflect a range of important interpretations. Where Svetlana Alpers influentially positioned Hooke as exemplary of the acquiescent reception of optical appearances in seventeenth-century northern European visual culture, interpreters like Stuart Clark have more recently explored the vertiginous anxieties that such reliance on the visual provoked among early modern intellectuals—what Clark has called the period's "loss of optical nerve."[15] From crisp articulation in the 1660s to smudgy, blobby blurs two decades later, we might thus say, Hooke's shifting graphic style materializes a final collapse of the Renaissance's tactile linearity and a triumph of baroque, painterly sight.

The appeal of plotting experimental draftsmanship in the terms of this organizing art-historical narrative is that it actually seems to speak to the observational concerns voiced by Robert Hooke and his colleagues. Philosophy's impending experimental reformation depended, so Hooke declared contemporaneously with the mid-1660s comet sightings, on "a *sincere Hand*, and a *faithful Eye* to examine, and to record, the things themselves as they appear."[16] By the 1680s, it was as if that impulse to reconcile optical data into graspable, haptic form had been abandoned entirely, sacrificed to smudgy evocations of retinal events. Aided then by his weird, halting drawings, Hooke had foresworn his theories about comets in favor of optical experience alone. The comet drawings were made, he claimed, as he threw "aside all manner of Hypotheses concerning them . . . to observe them as if there never had been any such Appearance before, and to attend wholly to what the Appearances themselves could teach me."[17] Framing the drawings in art history's treasured narrative forms thus ends up producing a story eminently conducive to histories of the humble, prolix nature of English empirical science.[18]

Yet concealed by this attractive, interdisciplinary synthesis that has dominated recent work on the Royal Society is the devilishly clever and surprisingly successful bricolage at the heart of Hooke's 1680s comet drawings. To make that ingenuity comprehensible, this chapter aims to subtly short-circuit the interpretive desire that would pitch Hooke's celebrated graphic work from the mid-1660s as a classic apex and its strange, hybrid sequels as only so many surrenders. Beginning with the totemic *Micrographia* (1665), I trace the intersections between conceptions of drawing and experimental-philosophical identity contemporaneously worked out by Hooke, Christopher Wren, and other prominent figures of the early Royal Society. If instrumentally aided observation was to be a heroic task of discovery that figured the experimentalist as a "new Columbus," then drawing would edify beholders by presenting certified sights through techniques of graphic enclosure theorized by and practiced in the early modern artistic tradition. But what if that target of observation changed every time it was seen, rupturing any graphic boundaries by which it would be enclosed? How could the experimentalist retool his eyes to see entities that corroded not only his theories but those very seen bodies themselves? These are questions that Robert Hooke came to ask when looking at comets in the early 1680s. What makes Hooke's late, lambasted drawings so fascinating, then, is how they reveal a philosophical observer's recursive, self-conscious negotiation between the guidance of artistic perception and places where other lessons had to be learned. Likewise, wrestling with Hooke's experimental draftsmanship and the conceptions of teaching theorized through it provides an opening in which to explore what art-historical narration reveals and conceals, where our eyes too must turn outward.

Not So Much a Teaching as an Entangling

That Hooke would position his 1680s drawings at a juncture with teaching gestures toward the theme of educational reform which runs like a red thread through early modern natural philosophy. Rather than following the canonical authority of ancient thinkers like Aristotle or Galen, so men and women across Europe (and beyond it) increasingly argued, natural knowledge would best advance by integrating new technologies, giving new credence to the evidence of observation, and profiting from an expanded field of printed sources. This work often found encouragement at courts and urban centers, in spaces where university-educated scholars were gaining new contact with and respect for the skilled manipulations of matter achieved by artists and craftsmen.[19] Commensurate with an increasingly global traffic in material goods, mathematical academies emerged in trading cities such as Florence, Venice, Seville, and London, teaching skills to merchants and navigators that were beyond the curriculum of the great medieval universities.[20] Drawn together with a range of esoteric sources, these urban, courtly modes of natural knowledge claimed an ambition long asserted by the magical tradition. Not only would the investigator understand nature, but she would change it. The aim of the "new sciences," so Francis Bacon claimed, was not learned debate but "command of nature in action."[21]

Many among the philosophers in seventeenth-century England who sought to study nature by experiment knew the universities and their learning were not the often-claimed bastions of reactionary conservativism.[22] In the volatile atmosphere of the English Civil Wars (circa 1642–51), reformist currents had substantially transformed the teaching, the demographics, and even the educational intent of England's universities. At Cambridge, innovations ran the gamut from the embrace of metaphysical Cartesianism in the Platonist circles of Henry More to the intensive botanical studies pursued by John Ray, the likely draftsman of the 1664 comet drawings. The "Philosophical Club" catalyzed by John Wilkins at Oxford's Wadham College in the later 1640s built on the physiological tradition brought to the university by William Harvey (1578–1657), encouraging a strand of mechanical experimentalism later transmitted to the Royal Society by Robert Boyle, Christopher Wren, Robert Hooke, and others. Infused with strains of millenarian Protestantism, the reforms effected by this new learning were seen by many in experimentalist circles as so many symptoms of much broader transformations. Rather than training the sons of the elite for service to crown and gown, universalizing education would flow out among the people, unleashing English economic power even as it effected the topsy-turvy conditions of the end of days.[23]

Whatever else it may have been, *Micrographia*, the magnificent illustrated volume of optically assisted research that Hooke published in 1665 under the Royal Society's imprimatur, aimed to speak with that reformist, collectivist project.[24] Individually, so Hooke claims in the preface, human knowledge is extremely limited; the reasoning subject relies on ideas formed from sensory impressions and stored in the faculty of memory. Because the production and management of ideas are so routinely flawed, thinking readily deteriorates into little more than guessing. "We often take the *shadow* of things," Hooke cautions, "for the *substance*, small *appearances* for good *similitudes*, *similitudes* for *definitions*; and even many of those, which we think to be the most solid definitions, are rather expressions of our own misguided apprehensions then of the true nature of things themselves."[25] For Hooke and his colleagues, the collaborative project of experimental philosophy offered a crucial remedy to these individual faults. By enlisting masses of participants to perceive nature and commit their observations to public record, the grounds of reasoning could be shifted from the unreliable "Brain and the Fancy" of discrete individuals to broader, agreed standards. In *Micrographia*, Hooke modeled this process through Harvey's crucial and then still-recent insights on the circulation of the blood. Beginning in the hands and eyes, knowledge would pass into the bodily organs. "And so, by a *continual passage round* from one faculty to another," Hooke argues, "it is to be maintained in life and strength, as much as the body of man is by the *circulation* of the blood through the several parts of the body, the Arms, the Fat, the Lungs, the Heart, and the Head."[26] As a healthy body sustains vitality by communicating nourishment between its various faculties, so would experimental philosophy perfect natural knowledge by gathering, testing, and integrating data from collaborators of diverse geographical, confessional, and other stations.[27]

By design, this theory introduced an important asymmetry between the ambition of the philosophical collective and its individual contributor. If the collective enterprise aimed at nothing less than a reformation of philosophy, the individual's contribution was all the more valuable for being humble. Framing a frequently cited passage from *Micrographia*, Hooke stakes out the terms for individual entry into experimental philosophy this way: "There is not so much requir'd towards it, any strength of *Imagination*, or exactness of *Method*, or depth of *Contemplation* (though the addition of these, where they can be had, must needs produce a much more perfect composure) as a *sincere Hand*, and a *faithful Eye*, to examine, and to record, the things themselves as they appear."[28] The discrete, philosophical contributor should trust to the larger experimental body the provision of refinements like imaginative analysis and methodical exactness; what the individual could meaningfully supply were reliable data gathered from tactile and optical inputs. So, Hooke claims, it is this that *Micrographia* aimed to deliver.[29]

Yet any hopes that an innocent eye and hand could easily traverse the microworld are leveled by the assault of *Micrographia*'s first plates. The gnarled and blasted triangular form at the head of the volume's second plate, we are told, depicts nothing less than the magnified "top of a small and very sharp needle"[30] (fig. 1.4). As this ferociously clever, programmatic plate visualizes, the gap between what the hand perceives and what the aided eye can see quickly expands into a broader lesson, one that implicates the beholder and the very ontology of *Micrographia* itself.[31] Consider the monstrous, mottled blowfish-like spot along the left edge of the plate, which represents the magnified form of the encircled dot directly above it. Labeled *A*, this dot both is and represents the printed impression of a grammatical full stop.[32] Appearing "like a great splatch of London dirt" under the microscope, Hooke explains, irregularities such as these are congenital to prints by their facture. Flaws follow inevitably from "the uneven surface of the paper, which at best appears no smother than a very course piece of shagd cloth, next the irregularity of the Type or Ingraving, and a third is the rough Daubing of the Printing-Ink that lies upon the instrument that makes the Impression."[33]

Proffering both instrumental and unaided views, Hooke's plate does two inimical things simultaneously. By delivering printed images of instrumental revelations, the plate alerts the beholder to the disconnect between the ideas formed about objects from the body's unaided sensations and what becomes apparent of those objects from even slight instrumental enhancement. Yet by exemplifying printed form (the period) and then depicting its deterioration under only a modest degree of magnification, *Micrographia* also enjoins the beholder to take a critical relation to the very printed impressions through which these microscopic disclosures have been delivered. Even if the spurious printers' sallies that Hooke would call "Mr. Engraver's Fancy" could be eradicated, the volume shows us that through their faulty, human

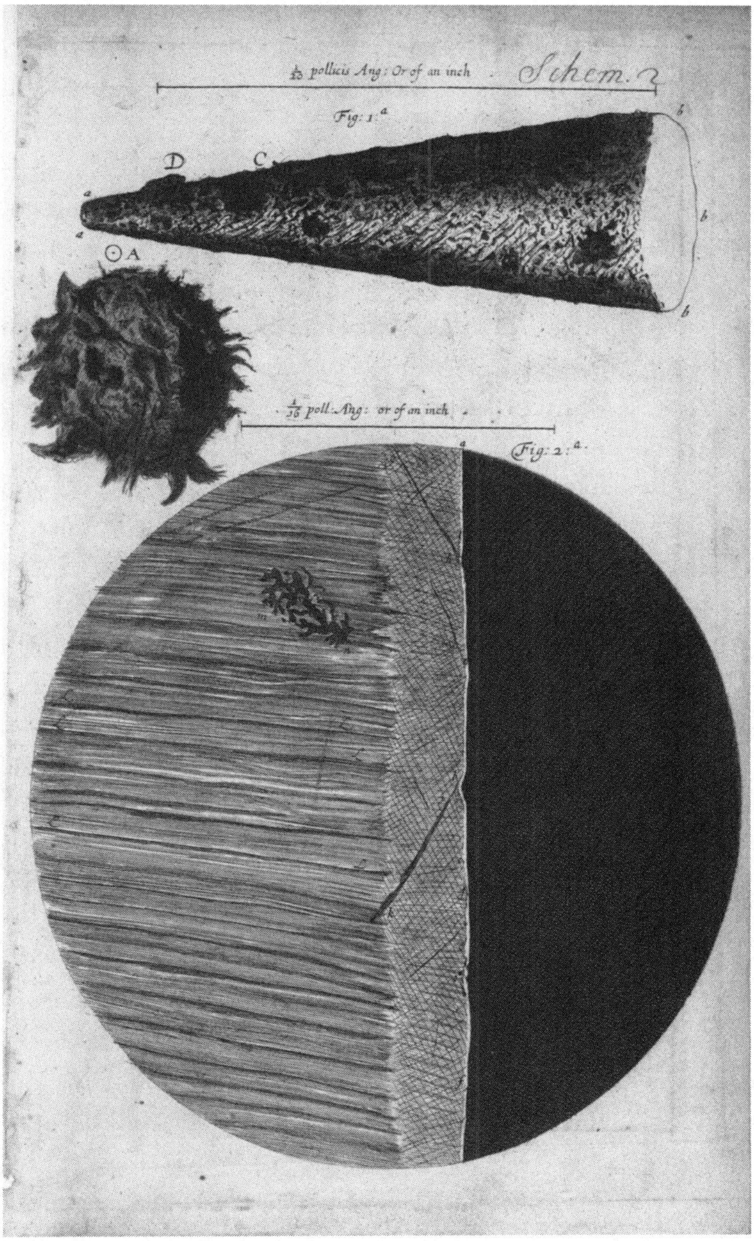

FIGURE 1.4 * Print depicting the magnified tip of a needle and the blade of a razor with an enlarged and exemplified grammatical point; from Robert Hooke's *Micrographia; or, Some Physiological Descriptions of Minute Bodies made by Magnifying Glass* (London: John Martyn and James Allestry, 1665), scheme 2. © The Royal Society. Reproduced by permission.

facture, artifacts such as printed impressions would always be—could *only* be—a precarious calculus of material, accident, and luck.[34] Hooke's prints work to breed their own doubt, implicating themselves in the treacherous epistemological terrain they iconoclastically visualize.

Guiding and cautioning, the didactic dimensions of *Micrographia* are embedded in its form and content alike. The volume arrays a revised hierarchy of making, beginning with lowly human artifacts, progressing through the mineral, vegetable, and animal kingdoms, before ascending to the celestial studies with which it concludes.[35] What *Micrographia* repeatedly sets out to teach, though, is a new protocol of observation by which "the things themselves as they appear" can be apprehended. First, the microscopist has to learn to consult many different instruments to determine the veracity of his observations.[36] With these tools, Hooke outlines a range of standards that he had used in his observational practice; he describes working in a room with one south-facing window, placing the microscope on a table three or four feet away from that window.[37] He teaches the reader how to measure the degree of an instrument's magnification.[38] He relates his own method for eliminating the inevitable variations of daylight when drawing; he tells us that he draws at night with a lamp equipped with a brine-filled, light-amplifying globe.[39] If microscopy would seem to privilege the sense of sight, Hooke schools the user in how to integrate qualifying information from other sensory inputs. Analyzing mold found growing on a leather bookbinding, Hooke tests the tactile properties of its spores by probing them with a pin. He even submits them to the olfactory and gustatory senses. "Both their smell and taste, which are active enough to make a sensible impression upon those organs," he reports, "are unpleasant and noisome."[40]

Most important, Hooke points the microscopic user toward the range of tacit skills necessary for navigating amid sophisticated optical instruments and the unfamiliar microworlds they disclose. Aside from then-insurmountable challenges such as correcting lenses' spherical and chromatic aberrations, *Micrographia* indicates the substantial skill required to light, focus, and sight a target object with a seventeenth-century microscope.[41] Once such practical fluency with these expensive, finicky instruments had been acquired, Hooke narrates the patient posture of inquiry requisite to interpreting the visual data delivered by them. Lustrous to the microscopic percipient as it is to the unaided eye, a peacock feather, for example, is a slippery target. As Hooke cautions, "Every new position of it to the light makes it perfectly seem of another form and shape, and nothing what it appear'd a little before."[42] So episodes like this disclose only too plainly, attending to the phenomenology of microscopic appearances turned out to be an extremely complicated matter. "Nay," Hooke continues of the peacock feather, "it appear'd very differing often-times from so seemingly inconsiderable a circumstance, that the interposing of ones hand between the light and it, makes a very great change, and the opening or shutting a Casement and the like, very much diversifies the appearance."[43] As any would-be investigator needed to learn, the minute targets of microscopic observation responded to the hand, eye, and body of the percipient in surprising ways, which could be signaled on paper but mastered only through practice.

Like any "explanation" of virtuosic skill, we might ask whether Hooke's narrated procedures truly open them to the reader or, rather, create conditions under which their results appear all the more spectacular.[44] Whatever answer we give, it is clear that Hooke had the technical chops to support, if not outstrip, the practices he reported. Recent scholarship has done much to illuminate the complexity of the dialogue connecting Hooke and Christopher Wren to London's most skilled instrument makers as they designed, built, and reworked their instrumental devices.[45] But nowhere in seventeenth-century English microscopy is the showing and telling of how to plumb these depths of microscopic appearances made more a matter of virtuosic display than in Hooke's study of the pellucid eyes of the gray drone fly (fig. 1.5). As we see in *Micrographia*'s incredible plate, the fly faces outward with its bat-like nostrils, twin projecting proboscises, and prickly orchid of insect tongue sprouting from the wiry tufts skirting its bearded visage. These features are dwarfed by the massive optical banks that frame its severed head like a displaced rib cage. Split down their iridescent center and aligned with the vertical axis of the page itself, the eyes are subdivided into ever-smaller ranges of fully functional optical organs.[46] As powerful as this image is, Hooke makes even more of the observations that had brought the reality of the fly's eyes into visibility. "It is exceedingly difficult in some Objects," he claims, "to distinguish between a prominency and a depression, between a shadow and a black stain, and a reflection and a whiteness in the colour. Besides, the transparency of most Objects renders them yet more difficult then if they were opacous."[47]

Unfamiliar terrain such as this was not for the faint of heart and had even led the learned badly astray. Noting that "in one kind of light," flies' eyes appear like a punctured lattice, Hooke explains that this "probably may be the Reason, why the ingenious Dr. Power seems to suppose them such"—a telling dig at a rival microscopist.[48] To prove himself no comparable dupe to appearances, Hooke then supplies a list of the ostensible, but incorrect, faces that flies' eyes can present: "In the Sunshine they look like a Surface cover'd with golden Nails; in another posture, like a Surface cover'd with Pyramids; in another with Cones; and in other postures of quite other shapes."[49] What any aspiring phenomenologist of microscopic appearances would need, then, was the very protocol of observational variation, rigor, and instrumental inventiveness that Hooke had prescribed to sound such murky waters.

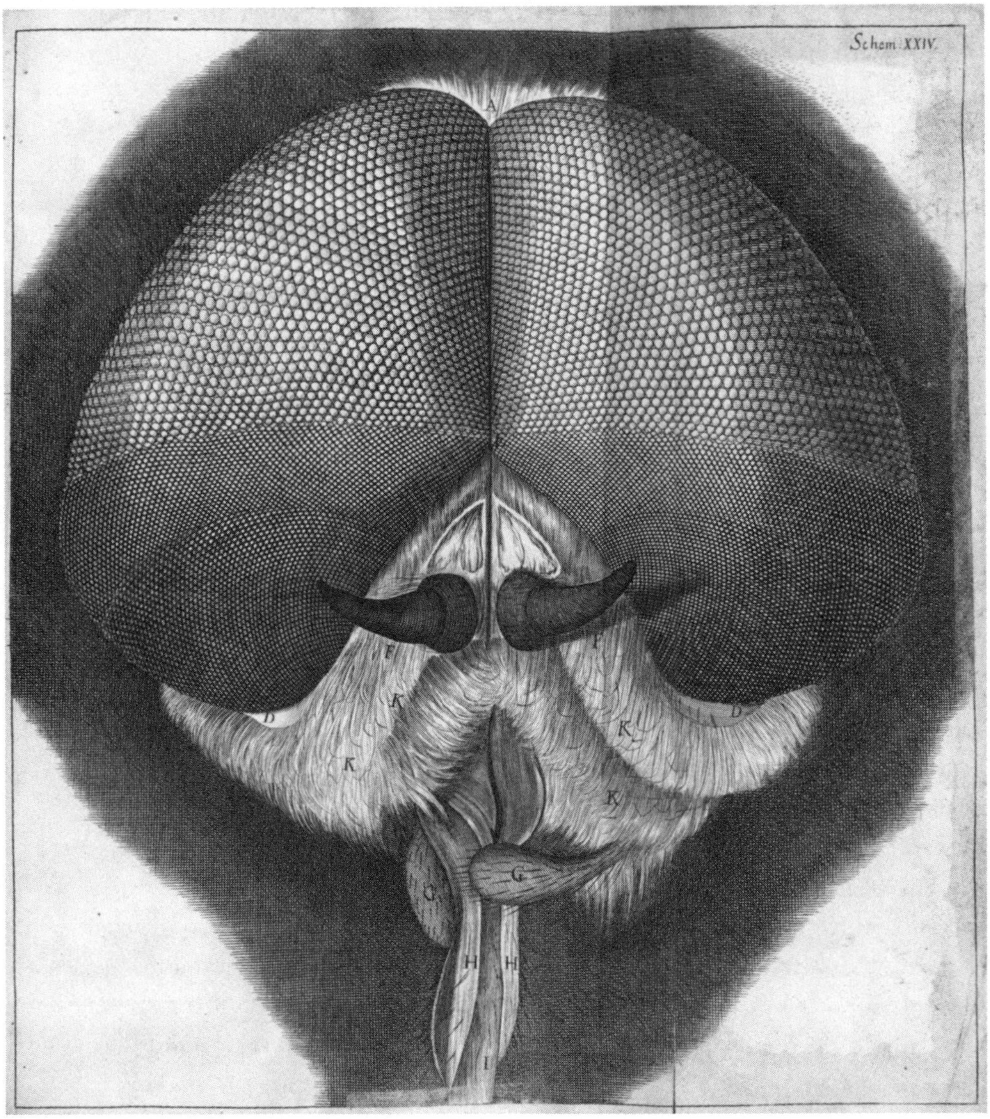

FIGURE 1.5 * Print of the magnified head of a gray drone fly as published in Robert Hooke, *Micrographia; or, Some Physiological Descriptions of Minute Bodies made by Magnifying Glass* (London: John Martyn and James Allestry, 1665), scheme 24. © The Royal Society. Reproduced by permission.

Critically, the act of drawing was to visualize less this temporal sequence of observation than what it had reliably established. In *Micrographia*, Hooke forcefully stages a clear, hierarchical demarcation between observation and graphic rendering. As he puts it, "I endeavored (as far as I was able) to discover the true appearance, and next to make a plain representation of it."[50] An experimentalist had to complete an exhaustive project of perceptual discovery before any actual picture-making could begin. Hooke's articulation of this point through the case of fly eyes deserves close attention:

> Of these kind of Objects there is much more difficulty to discover the true shape, then of those visible to the naked eye, the same Object seeming quite differing in one position to the Light, from what it really is, and may be discover'd in another. And therefore I never began to make any draught before by many examinations in several Lights, and in several positions to those Lights, I had discover'd the true form.[51]

The variety of drawing to which the microscopist should aspire would thus aim to represent the bedrock quiddity of the target—to make visible its "true form" as it hides beneath and amid deceptive appearances. Drawing must be retrospective and synthetic, giving lucid exposition to the results of scrutinizing discovery.

Hooke effectively embeds in *Micrographia*'s plates intermediary visualizations of how this synthetic pictorial realization could be accomplished. In a detail inserted into the volume's twenty-third plate, we see are shown a cluster of nineteen forms shaded with looping hatches and darkened at the interstices that join them together (fig. 1.6). Fictively projecting upward as convex hemispheres, each of these bulbs betrays the reflections of two illuminated windows cut with a transom and mullion, all distorted by the individual globule's curvature. These clustered forms denote discrete groupings of eye pearls that Hooke had observed while studying the gray drone fly's head. In these languid orbs, he had been able to see not only "a Landscape of those things which lay before my window," but also "the Image of the two windows of my Chamber."[52]

Observing the pattern of the windows' distorted reflection on the surface of fly eyes was certainly useful to Hooke. It aided him in distinguishing the true convex, hemispherical structure of the eyes from their likeness to nails, pyramids, and cones or the other shapes proposed by rival microscopists such as Power. From small samples of optical surface like these, Hooke then made inferences about the general shape of all the gray drone fly's thousands of eyes.[53] He offers no claim to have observed—or that an aspiring microscopist should try to see—each and every

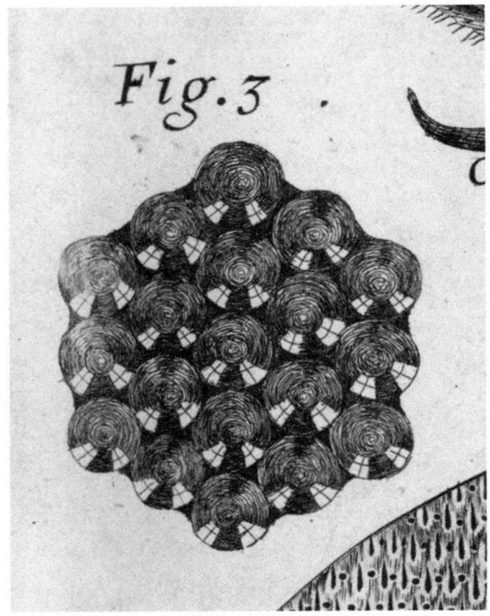

FIGURE 1.6 * Printed sample of fly eyes with distorted reflections of windows as published in Robert Hooke, *Micrographia; or, Some Physiological Descriptions of Minute Bodies made by Magnifying Glass* (London: John Martyn and James Allestry, 1665), scheme 23 (detail). © The Royal Society. Reproduced by permission.

one of the fly's fourteen thousand eyes. Instead, Hooke creates a mathematical projection based on counting "certain rows of them several ways, and casting up the whole content."[54] To this project, appearances like windows' distorted reflections on optical surfaces were artifacts of particular observational conditions; they could appear when fly eyes were inspected during the day with a microscope in a room with two windows. But appearances such as these were precisely the kinds of accidental effects that Hooke aimed to penetrate beyond in his protocols of observational discovery. Tellingly, his synthetic depiction of the drone fly's head has integrated the convex, hemispherical eye forms discovered through these window reflections, but eliminated the ephemeral appearance of the reflections themselves.

What I mean to suggest is that in moving between Hooke's fragmentary fly-eye globules and the resplendent drone fly head, we gain an unnerving sense of how the instrumentally aided draftsman was to relate to "the things themselves as they appear" in the mid-1660s. Beneath the objective lenses of the sophisticated optical instruments the investigator had to learn to command, things appeared often opposite to expectation and laden with accidental trappings. Microscopic things disclosed themselves partially, in time and in fragmentary aspect—and this only when parsed with observational vigilance, technical skill, and a patient, flexible posture of inquiry. Discovering what Hooke called "true form" was not only to be a matter of processing and synthesizing perceptions; it would also require making inferences from sensory data to project the nature of targets that could never actually be seen in full. The implication of this is that the true forms the draftsman aimed to represent were not really *seen* forms at all.[55] Instead, what would be drawn were markings that could effectively denote a reliable calculus of contours, dimensions, and features ascertained from a target by instrumental means. What we see as the etched and engraved presentation of the magnified drone fly's head is a master class in this project. Hooke's print is a concretion of data collected from discrete, observational acts conducted in time, then integrated, projected, abstracted. Plotted against a darkened field and accentuated with a streak of undulating shadow that rakes from snout to beard to eye, the startling, dazzling image makes visible the fly's most compelling features in their brazen glory.

Outlining

Daunting a "textbook" as it is, minor circumstantial evidence suggests that segments of Robert Hooke's audience actually did try to learn microscopic seeing and drawing from works like *Micrographia*. The virtuoso Samuel Pepys famously recorded in his diary how he had attempted to parlay his reading of Henry Power's text on microscopy into practice. "[I] read a little in Dr. Power's book of discovery by the Microscope," Pepys reports, "to enable me a little how to use and what to expect from my glasse."[56] Hooke's friend William Cole, a collector and aspiring naturalist from Bristol who we will meet again in chapter 4, certainly knew *Micrographia* and had begun microscopic drawing by 1669. Although his tutelage from *Micrographia* is unknowable, Cole promised Hooke that he would "draw the figures of them [ferns] all as they appear by the microscope, together with their seeds, and to add the description of all considerable circumstances, and join them to the rest of my draughts of that kind."[57] It is also known that a young Isaac Newton made drawings from *Micrographia*.[58] But more compelling than this fragmentary evidence is the concordance between the conception of instrumental draftsmanship articulated in *Micrographia* and the mechanical devices by which Royal Society Fellows were contemporaneously teaching drawing.

If the things themselves had appeared to *Micrographia*'s trained investigator in time, then the draftsmanship taught among London's experimentalists was an intensively spatial enterprise. We have no clearer instance of this conception of drawing than the "perspectograph" that Christopher Wren designed as an aid to the novice draftsman. To make a drawing, the user would place the device on a tabletop, as appears in the print published by Royal Society Secretary, Henry Oldenburg, in 1669[59] (fig. 1.7). Sighting the target of depiction through a pinhole attached to an adjustable arm at left, the user manipulates a counterbalanced beam fitted with a metal index at its left tip and a graphic stylus at its center. Holding

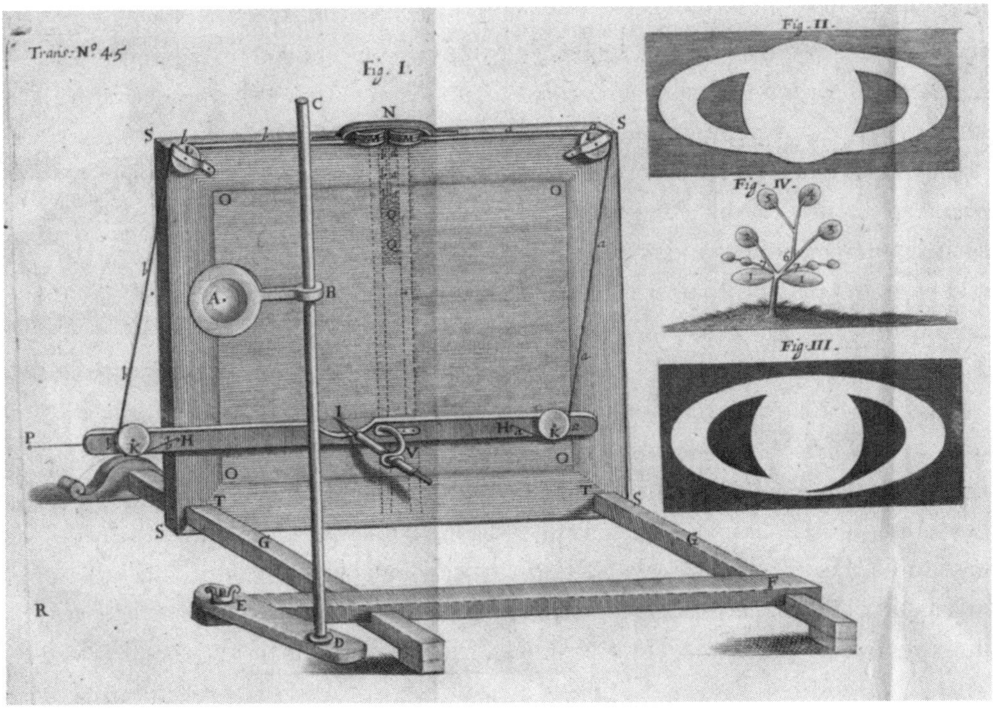

FIGURE 1.7 * Engraving of Christopher Wren's "perspectograph" as published in *Philosophical Transactions of the Royal Society of London* 4 (1669). © The Royal Society. Reproduced by permission.

this stylus against the sheet of paper tacked to the device's vertical backboard, the draftsman brings the metal index into view and drags it around the contours of the target objects visible within the field fixed by the sight. The user would need to exert a carefully calibrated muscular force to guide the pin around the boundaries of each seen feature while simultaneously maintaining enough stress on the stylus to register its path on the graphic surface. By design, the machine envisions drawing as a displaced registration of the haptic action of running a finger around entities in the visual field—what Oldenburg glosses as "an instrument . . . for drawing the out-lines of any object in perspective." Wren's device grounds drawing in the demarcation of seen bodies' spatial limits.

Art historian Svetlana Alpers has influentially juxtaposed the conception of depiction embraced in northern European visual culture to that of Italian Renaissance theorist Leon Battista Alberti.[60] What I would stress is that in its approach to pictorial facture, Wren's device actually hews closely to Alberti's vigorous theorization of the outline and to the famous artistic aid he invented to promulgate that theory.

In his revolutionary *De pictura* (1435), Alberti had articulated a crucial division in the nature of seen bodies. Visible entities possess two permanent properties that concern the painter: the body's surface (which may be flat, concave, convex, or otherwise) and its boundary—what Alberti calls "the brim, or the fringe."[61] Drawing on medieval traditions of mathematicized, perspectival optics, Alberti points the painter to the keen importance of the extrinsic rays which propagate to the eye from these objective boundaries. These rays "hold on like teeth to the whole of the outline, [and] form an enclosure around the entire surface like a cage."[62] When a painter begins her pictorial work, it is first to the reliable data of extrinsic rays that she should turn her attention as she inscribes the spatial boundaries of objects. Nothing is more helpful in this fundamental project of circumscribing extrinsic radiation, Alberti claims, than the innovative grid of threads that he calls "the veil." Introducing this grid at the intersection of the visual pyramid that the artist would paint, the view is fixed and flattened as "the position of the outlines and the boundaries of the surfaces can easily be established on the painting panel."[63] If stabilizing both hand (with counterbalanced weights) and eye (with pinhole), Wren's instrument similarly assists the delineation of extrinsic rays propagating between the seen object and a lone, seeing eye at a fixed intersection of the visual pyramid.

As circumscription or the correct delineation of outlines had stood for Alberti as the sine qua non of good painting, so a sequential, graphic progression from articulating limits of represented bodies to suggesting ephemeral effects of light and shadow became standard pedagogical precepts of early modern Europe's art academies. Symptomatic of this approach are the directions for draftsmanship set out by G. B. Armenini, a cleric and connoisseur of painting in the Northern Italian courts, in his 1586 *On the True Precepts of the Art of Painting.*[64] Therein, Armenini encourages the student to "begin very lightly to form all the profiles and contours distinctly according to the model [that is, a pictorial exemplar], but without sketching the hatchings which serve for shadows."[65] Moving from generality to specifics, from the outside of the depicted form inward, "beginners will have the courage to take the pen and delineate the sketch very lightly until all the contours are finished. The hatching has to be done in the same way with the pen."[66] Denotation of transient effects of light and shade would be added into the pictorial matrix only once the spatial boundaries of bodies had been ascertained and plotted.

Even in a culture as artistically stunted as Restoration England is often supposed to have been, this organizing privilege of the outline was well known. Taking his cue from artistic treatises, John Evelyn counseled graphically inclined Royal Society Fellows in 1662 to "first draw the exact lineaments, and proportion of the subject you would expresse in *profile*, *Contours* and single lines only; and afterwards by more frequent and tender hatches in the lighter places, strong, bold, or cross in the deeper."[67] By 1669, enterprising artists such as Alexander Browne were showing English amateurs how to construct depictions of human bodies from the schematic step-by-step aggregations of outlines pioneered by circles surrounding the Bolognese academy of the Carracci brothers[68] (fig. 1.8). Most important, Italianate conceptions of drawing as proceeding from defining contour to accidental, shadowed effect were being disseminated through London's new academies of life-drawing such as that associated with Peter Lely, the painter to whom Hooke was apprenticed.[69] At the heart of Lely's practice was one of the greatest collections of drawings in later seventeenth-century Europe, its strength concentrated in sixteenth-century Italian figural drawings.[70] Fully mingling teaching and being taught, Lely saw his collection of Italian drawings, so one leading scholar has put it, "as a varied source of potential ideas, postures and compositional groupings, to be adapted and interspersed with his own preparatory drawings. In other words, a pattern book . . . for his assistants and a fund of potential ideas for his own work."[71]

No more direct evidence of this attitude do we have than a bound folio of drawings now known as the Dulwich College Album.[72] In addition to copies after reproductive prints of Mediterranean antiquities and pounced reproductions of Lely's own designs, the album contains a range of life studies associated with Lely's pioneering "Academy of 1673."[73] Conventionally the most advanced stage of the student's training, drawing from the life model was often conducted with pupils gathered in a semicircle around the posing model.[74] Newly imported to later seventeenth-century England from the Continent, these conditions can be inferred from a pair of life studies in the Dulwich Album (fig. 1.9 *a*, *b*). In each drawing, a muscular figure leans forward like a boxer, clenched right fist poised to deliver a blow. Consistent in size and pose, the two drawings have been rendered from slightly different locations: whereas the ink sketch positions the beholder almost directly behind the model, the red chalk drawing views the subject from a slightly more obtuse angle, revealing more of the jutting head, coiled right shoulder, and perspectival recession of the extended right foot. Where the young colleague standing to his right had confined himself to red chalk, the draftsman of figure 1.9*b* has added exaggerating retouches. With a thin wash and small repertoire of curling, inked strokes—like that rampant cobra denoting the musculature of the right buttock—this draftsman has made volumetric suggestion of light and shade in a highly stylized way. Up from outlines, we see the rudimentary chiaroscuro depiction being built.

That Robert Hooke knew these basic pictorial precepts is demonstrated by the mechanical improvements he offered to novice draftsmen. A lecture from 1694 describes how a camera obscura could be adapted so that it would allow even the

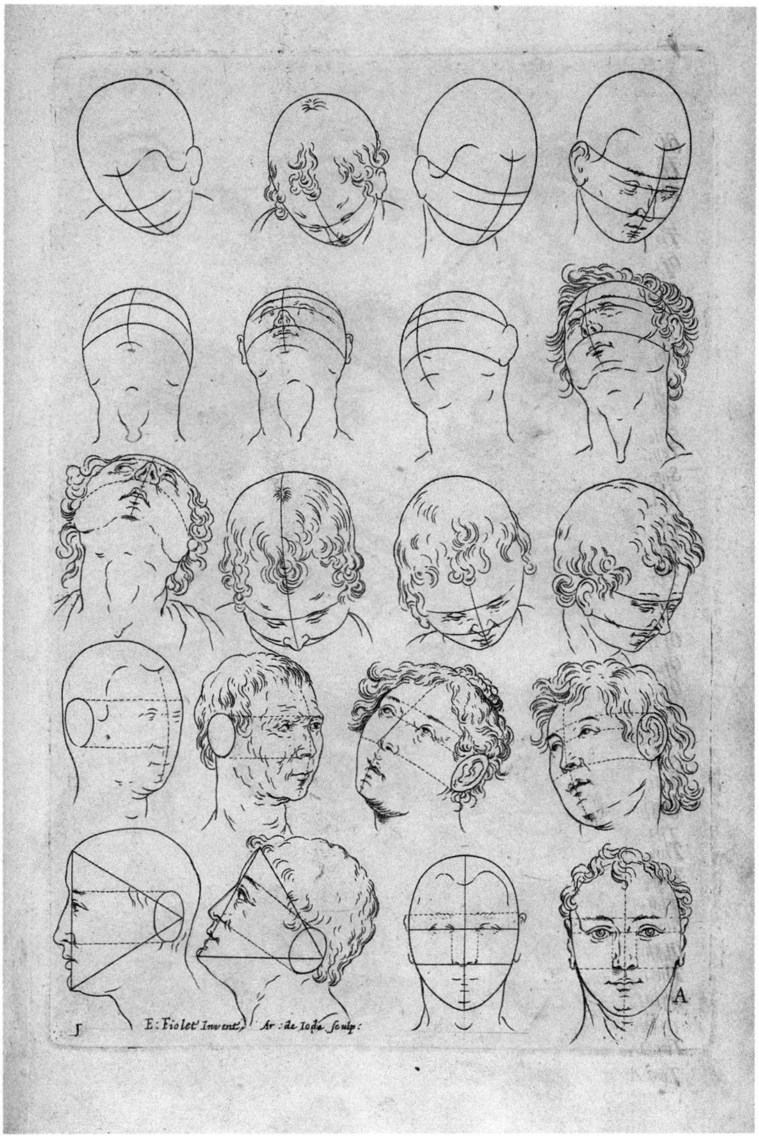

FIGURE 1.8 * Schematized outlines for drawing human heads; based on Odoardo Fialetti's *Il vero modo et ordiner per dissegnar tutte le parti et membra del corpo humano* (Venice: Sadeler, 1608) as published in Alexander Browne's *Ars Pictoria: An Academy. . . .* (London: A. Tooker, 1675), plate 1; Huntington Library, RB 351942. Reproduced by permission of The Huntington Library, San Marino, California.

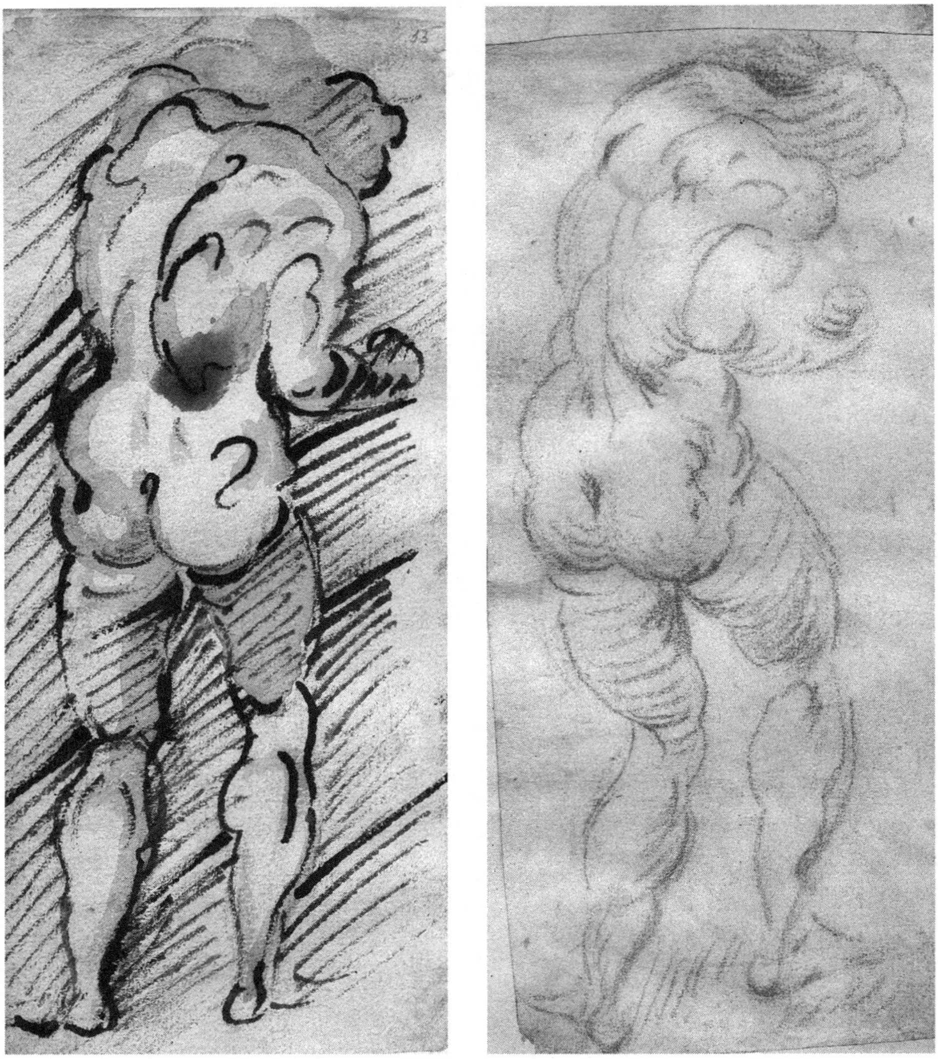

FIGURE 1.9*a*, *b* * Anonymous figure drawings by students in Peter Lely's "Academy of 1673"; from Dulwich College Album, fols. 18v–19; Dulwich College, London. Reproduced with kind permission of the Governors of Dulwich College.

most unskilled observer to make scientifically useful pictures. In a scheme that he proposed to the administration of the Royal Mathematical School in London "for the Use of the Children," Hooke enthused about how such machine-assisted drawing would demand only an ability to "nimbly run over, with his pen, the boundaries, or out-lines of the thing to be represented; which being once truly taken, 'twill not at all be difficult to add the proper shadows and light pertinent thereunto."[75] Three years later, he again extolled the virtues of the camera obscura to the Royal Society. His optically enhanced method would provide "a way to give the true outlines of an object by the shadow of it and that so exactly represented that the least hair will be truly Represented or traced on the table where it is to be drawn by the shadow of it."[76] In typically iconoclastic fashion, Hooke asserted that "all the Skill and Dexterity of the Best Painter Limner or Drawer in the world" could not compete with the image contrived by plotting out the contours projected by such a mechanical aid. Deprecating toward artistic capacities though Hooke is here, his sequence of instrumentally aided graphic facture is effectively that stipulated by Alberti. Successful drawing needed to proceed by enclosing the targets of representation on paper with and within reliable outlines. "We must first endeavour to make *letters*, and draw *single* strokes true," so Hooke noted in *Micrographia*, "before we venture to write whole *Sentences*, or to draw large *Pictures*."[77]

Micrographic Facture, Terrestrial Theory

Apprenticed in a milieu where circumscription mattered and well versed in its practical application, Hooke took strategies of spatial enclosure still further in the facture and philosophical self-styling of instrumental draftsmanship during the mid-1660s. When studying the formal patterns of crystallization in snowflakes in late 1662, Hooke quickly found himself at an impasse. "Such an infinite variety of curiously figur'd" forms presented themselves, so he phrased it, that "it would be as impossible to draw the Figure and shape of every one of them, as to imitate exactly the curious and Geometrical *Mechanisme* of Nature in any one."[78] Aided by optical instruments, however, his analysis revealed telling regularities within this baffling profusion. Each falling crystal possessed "six principal branches, all of equal length, shape and make, from the center, being each of them inclin'd to either of the next branches on either side of it, by an angle of sixty degrees."[79] These are not merely generalizations from empirical observation but a veritable recipe for the graphic means by which Hooke set out to deliver snowflakes as pictures (fig. 1.10). As demonstrated by a surviving drawing from December 1662, he marked off fourteen circles with a compass, using a straightedge to subdivide each encircled form into six equal segments. On this skeletal ground plan, he then superimposed a range of differentiating, symmetrical markings painted in a thin ink wash.

Blocky and stylized, these drawings suggest that such a rough method of incised preparation would be more appropriate for an exercise in modeling the principles of crystallization—what Hooke called nature's "curious and Geometrical *Mechanisme*"—than for depicting specific entities. All the more surprising, then, is the

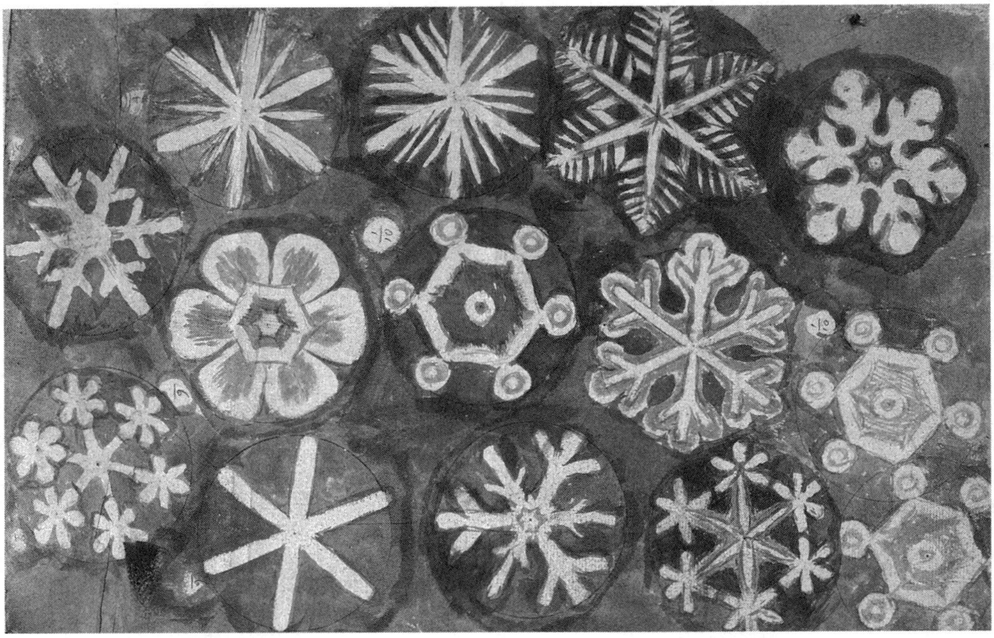

FIGURE 1.10 * Robert Hooke's pen-and-ink drawing *Figures Observ'd in Snow* (December 1662); from Register Book (Original) 2:62. © The Royal Society. Reproduced by permission.

force with which this hexangular armature acts in his most elaborate surviving preparatory drawing for *Micrographia*: a study of frozen urine crystals also prepared in late 1662. (See figure 0.2.) In strikingly authoritative, inked strokes, Hooke has rendered frozen and magnified urine crystals as a feathery network of herringbone hatches. Enclosed by wavy, curvilinear flourishes, his frothy, bulbous fields of gossamer pen work gradually coalesce into the six characteristic segments formed by the vertebral intersection of each crystal's stem-like spines. Although it is compelling to imagine that Hooke must have used some kind of optical projection technique to produce this drawing, close inspection reveals the incised path of a compass and straightedge (imperceptible in photographic reproduction) deployed as the pen's preliminary, imposing and indenting the crystal's hexangular form directly into the page.[80]

This is a draftsmanship of assurance, one where the materiality of graphic process is ingeniously harnessed into the spatial ordering of perceptual observations uncovered and ascertained in time. It is also a visualization of shrewdness. Bearing the trace of compass and rule, these drawings manifest to their intended audience of influential Royal Society elites how Hooke could reduce nature's profusion to attractive graphic form through geometrical devices—this at precisely the moment in the mid-1660s when he was angling for a professorship of geometry at London's Gresham College.[81] Yet suggestively, it is Hooke himself who sets out a compelling model for how we might apprehend this approach to draftsmanship and the mode of philosophical self-styling materialized in it. In a major theoretical text from the later 1660s, he identified an ideal embodiment of the principles that should guide experimental discovery in the most famous explorer of the early modern world. Like Christopher Columbus, Hooke proposes in *A General Scheme, or Idea of the Present State of Natural Philosophy* (circa 1666–68), this investigator needs to possess deft skill allowing penetration into nature's workings, an unending commitment to purpose, and a willing capacity to share collected intelligence with the learned world.[82] Specifically, the experimental philosopher should develop a solid plan of investigation "as *Columbus* did in the discovery of the New World."[83] Should he not have sufficient means to conduct necessary research, this philosopher should "here also imitate *Columbus*, [and] endeavor to be provided with Ships, and Men, and Money, and all those Assistances he finds requisite for the thorough Prosecution of this Discovery."[84] Once into a program of research, the investigator "ought also to proceed with the greatest Circumspection and Diligence to find out such things, as are Indications of what he seeks, and from those to take Incouragement to prosecute his Intentions, as *Columbus* did from the decreasing Depth of the Sea, the Drift of Weeds on the Surface of the Water, and the White Clouds that appeared near the Horizon, and the like to incourage and direct him in his Course."[85] Guided by his intellectual insights into natural phenomena, the Columbus-like philosopher would remain true to his convictions until his vision of the world is shown to all.

Robert Hooke's modeling of the enterprise of experimental philosophy on that of Columbus is intriguing, as it departs in important ways from the humble, modest, and pious modes of selfhood often associated with the early Royal Society.[86] What makes this modeling particularly interesting, though, is the way it envisions the project of experimental drawing. At once, Hooke claims, the Columbus-like experimentalist needs to "be able to design and draw very well, thereby to be able both to express his own Ideas better to himself, to enable him to examine them and ratiocinate upon them himself, and also for the better informing and instructing of others."[87] This mediating function for design-inflected draftsmanship is effectively materialized in *Micrographia*'s reflection-dotted swatch of fly eyes. (See figure 1.7.) Through the acts of perceptual engagement of which this image is the result, Hooke had learned something new about a specific natural target—he learned that discrete specimens of drone fly eye have particular, determinable structures. By studying these data, however, the philosophical investigator would also then need to reformulate his ideas about the broader structure of insect heads, the optical organs of submicroscopic creatures, and even, as we will see in chapter 3, the nature of nonhuman consciousness. Drawing guides and clarifies the ideas that lead to discoveries.

Columbus also provided Hooke with a broader example of how the experimental philosopher should become a teacher of his insights:

> He ought, as *Columbus* did, freely and impartially to discover what he finds. . . . And whatsoever he registers, he ought to do it in the plainest, shortest, and most significant Description, the Matter is capable of, and in such a Method as may neither cause Repetition of History, in more places than one, nor the Rejection of some others, because it fits not punctually to his Method.[88]

Though not specifically designated here, the act of drawing theorized and practiced in the early Royal Society fits well with the general category of what Hooke calls "registration": the economical, synthetic work done once the heroic project of observational discovery has been completed.[89] Like the visualization of the gray drone fly's "true form," the discoveries registered by Hooke's Columbus stand at a certain distance from sight. In step with contemporaneous colonial arguments, discovery demanded more than just bare beholding of a previously unknown entity. According to an influential theorist such as Hugo Grotius, after all, a claimant to discovered land had "not merely 'to seize it with the eyes [*oculis usurpare*] but to apprehend it.'"[90] Early Royal Society Fellows pursued such apprehending depictions through the spatial enclosure of discovered bodies. Based on the optically reliable evidence of extrinsic radiation, these practices of graphic outlining could be taught to amateurs by mechanical aides, elaborated through engagement with artistic traditions, and inscribed into the very fabric of their graphic work. Whether depicting flies, crystals, or snowflakes, the graphic enterprise endorsed and performed by Hooke for the Royal Society in the mid-1660s thus aimed to teach what experimental investigation had discovered in powerful, concise form. Following Columbus, these images would deliver new worlds to their patrons in the "shortest, plainest, and most significant description, the Matter is capable of." So why in the early 1680s would Hooke have ruptured that happy aesthetico-epistemological union of artistic tradition and scientific synthesis "to attend wholly to what the Appearances themselves could teach me"?

Drawing, Desperately

It is ironic that the first comet in recorded history to be discovered by a telescopic observer was the massive comet of 1680.[91] Visible for over four months, this striking object elicited an incredible range of apocalyptic responses from popular commentators and philosophical observers alike. Traces of a venerable tradition of cometary prognosis certainly permeated the account given by Royal Society Fellow John Evelyn, who described the object as "an obscure bright colour, very much in shape like the blade of a sword. . . . What this may portend, God only knows."[92] Even the notoriously freethinking Edmond Halley read dire tidings in the blazing path of this new star. So close had the comet come to the earth, Halley would subsequently claim, that if it had arrived a month later, "it would have produced some change in the situation and species of the Earth's Orbit, and in the length of the year."[93]

Often interwoven with these prognostic inflections, seventeenth-century philosophical observers asked three major questions of comets: their physical form, their orbital patterns, and their possible periodicity.[94] While calculating orbits and determining periods would become central topics of advanced mathematical research after the publication of Isaac Newton's *Principia* in 1687, Hooke had long taken a different approach. Looking askance upon talk about comets as "Targets, Shields, Spears, and Daggers, with Hands, &c. of Dragons and Serpents, and such like" that his colleagues circulated or, worse, visualized in their philosophical publications, he even doubted the parallax measurements with which his contemporaries were estimating comets' speed and distance from earth.[95] Consequently, Hooke vowed to make observations "only for enquirng into the constitution of these wonderful bodies."[96]

According to the elegant physical theory Hooke worked out by the late 1670s, comets originate from semisolid, spheroid bodies that have catastrophically decomposed through massive internal instabilities. As volcanic eruptions and earthquakes reveal the inner volatility of the earth, Hooke reasoned, so proto-cometary bodies are beset by internal agitations exacerbated by corrosive contact with surrounding aether. Gradually disintegrating, these vulnerable bodies lose mass and attractive force on their component particles, which then trail behind as the characteristic antisolar tail. As its head begins to destabilize, the comet's magnetic and gravitational relations are disturbed, throwing it from a stable orbit. Projecting tangentially in a rectilinear path according to René Descartes's laws of motion, the comet enters into the attractive fields of other bodies in its new path—encounters that only intensify its disintegration and lengthen its tail.[97] Published in his *Cometa* (1678), Hooke's theory offered a template for explaining the observed form and unusual trajectories of comets while elucidating their genesis from the deterioration of stable celestial bodies.[98] Less successful was his explanation of what he took to be comets' generation of light, a shortcoming he tried to remedy by appealing to a range of ad hoc models. Comets ignite like torches in aether, he speculated, while their luminous form might be explained by imagining that flaming fireballs are being projected at equal velocities into parabolic paths from every point of the comet's solid core.[99]

It is precisely this kind of hypothesizing that seems to be targeted for destruction by the drawings of the early 1680s. To see comets, Hooke again mandated a rigorous observational regimen whereby the telescopic observer would subject sightings to a range of optical instruments and a battery of tests. "I was not assured by the appearance of one or a few observations," he insists, "but by the repetition of them some hundreds of times, and by changing the position of the tube, apertures, eye glasses, posture of my head, and the viewing of them with the right eye, and sometimes with the left, that if those appearances had been caused by anything peculiar either in the glasses or the eyes, I might have found them out."[100] Whereas his microscopic targets had registered to the human senses of sight, touch, smell, taste, and potentially hearing, data from celestial phenomena such as comets could only possibly be detected by vision and touch. After offering a series of instrumental measures by which to amplify the radiant heat emanating from the moon and the stars, Hooke concludes that these effects were altogether "very insensible and inconsiderable" to terrestrial detection.[101] Conceding the reliance of human perception of comets on optical data alone and acknowledging the flights of fancy that such inscrutable entities had long encouraged among beholders, Hooke thus offers his crucial renunciation. He would abandon "all manner of Hypotheses concerning them. and to observe them as if there never had been any such Appearance before, and to attend wholly to what the Appearances themselves would teach me."[102]

Making drawings stands in this sacrificial project to beat back the idols of eye and mind. Anticipating a skeptical reception for his claims, Hooke avers, "I do verily believe that there was no kind of Fallacy in them, but that any other Persons might have seen the same, had they heedfully attended what they saw; which to do, nothing is more advantageous than the present designing and drawing what is seen, and writing a Description thereof."[103] No longer is drawing to be an ex post facto summation of discovered entities, nor will it even serve as an intermediary supplement to the Columbus-like philosopher's cogitations. Instead, graphic action is a vital engine for the investigator's deliberative attention whose products can uniquely cut through the received wisdom of perception and tradition—even the observer's own treasured conceptions. Outfitted from this graphic arsenal, Hooke declares, "Tho' I have already published my Conjectures [on comets] . . . I did not at all confine myself to be of that Opinion, or not to seek farther to inform my Judgment by other Appearances I should happen to observe in my future Trials."[104] Through drawing, Hooke would abdicate his de facto role as teacher, foregoing the theoretical claims about comets that he himself had defended in public, and be taught by "Appearances themselves."

What has happened here? Instead of consolidating a conception of the instrumentally aided draftsman as a Columbus-like explorer who confidently discovers and maps new worlds, it is as if Hooke's writings of the early 1680s had positioned that heroic conception of the self as one of drawing's casualties. Categorically indifferent to theories or their partisans, data detected in graphically aided observation would mercilessly pierce any attempt to order nature to some subjective conceit. If anything, the graphic forms in which Hooke made his observations would seem to materialize these radically corrosive objectives. Gone from the 1680s drafts are the single-page vistas wherein mid-1660s observers had reduced confusing celestial data to the lucid exposition of a particular nocturnal event. In their place, Hooke's notations for a single night from the 1680s often sprawl out over several pages,[105] split out into multiple representations of the target,[106] or, worse, map several observational events onto the same graphic matrix.[107] Nor were the visual languages of celestial representation immune to challenge. Where a system for denoting the "fixed stars" as six-pointed asterisks and the comet as an outlined dot with a stippled tail had dominated drafts of the mid-1660s, any number of smudges, stipples, streaks, and combinations of pen and pencil were deployed (often in superimposition) in the 1680s. It is as if the kind of authoritative, didactic, outline-based draftsmanship commanded by the heroic, Columbus-like experimentalist of the 1660s had come to connote the "high piece of Arrogance and over valuing ones own Judgment" that the halting, hesitant forms of the 1680s mercilessly undermined.[108]

But if we needed any further proof that these late drawings constitute a retreat—perhaps even a Wölfflinian optical surrender—we might well find it in the views of their sole period commentator. Understudy and occasional draftsman to Hooke, Richard Waller assembled his mentor's papers after Hooke's intestate death in 1703 and published *The Posthumous Works of Dr. Robert Hooke* from them in 1705. Introducing Hooke's 1682 comet lectures therein, Waller wrote the following about his encounter with the drawings:

> Though indeed these Figures were all very rudely designed, only as Helps to his own Memory, which the Author himself could much better have fitted for the Graver; yet, I have endeavoured to supply this Defect as well as I was able, and I hope the Reader will pardon the Failures. I think they pretty well answer his Descriptions, those being my chief Directors in perfecting the Draughts.[109]

Designing, so Hooke had claimed in the mid-1660s, was a crucial skill through which the Columbus-like experimentalist reformed and economically communicated his ideas. Defective and "very rudely designed" as they appeared to Waller, Hooke's strange images of the 1680s would seem only to betray more systemic faults of the reasoning that memory would serve. Chastened away from his heroic configuration of seeing over drawing, Hooke's confused smudges might be said to index a reversion to more familiar modes of Restoration philosophical identity and an acquiescence to the passive reception of optical appearances. Less than comets, we could conclude, what these drawings show is the policing of the self requisite to the formation of the valid scientific image.[110]

Drawing as an Education

Attractive as it may be, I think that such a reading works to obscure much of the audacious, mercenary cleverness of Hooke's 1680s comet drawings and, with it, the kinds of drawing skills actually required for experimental-philosophical observation. To get at these points and, perhaps shockingly, what Hooke's comet drawings seem to have gotten right, we need to look again at the facture of the 1680s drawings and the juxtaposition they offer to the more comfortable aesthetico-epistemological terrain mapped by *Micrographia*.

And to do so, we need to look down. It is important to recognize that Hooke's notations of the early 1680s are incomprehensible if separated from the text to which they repeatedly refer and in which they would ultimately be discovered. Scrawled in pen across the coarse paper in which the drawings are now bound is the following inscription: "Several Papers relating to Comets wrote by Dr. Hook. I found them in Baeyer's Uranometria which I bought at his auction."[111] Hooke's editor Waller affirms that these papers were returned to him for inclusion in the *Posthumous Works*, having been recovered by Dr. John Woodward in a volume purchased at the 1703 auction of Hooke's possessions.[112] "Baeyer's Uranometria" refers to a central work of seventeenth-century astronomy of which Hooke owned several copies.[113] First published by Augsburg jurist Johann Bayer in 1603, the lavishly illustrated *Uranometria* is a star chart noted by historians of science for its collation of the positions of the stars in the Northern Hemisphere contemporaneously calculated by Tycho Brahe, with new information about the southern constellations freshly delivered by Dutch navigators. Most important, *Uranometria* is recognized as the work that codified the astronomical convention of designating each star in the Ptolemaic constellations according to their magnitude of brightness.[114] Descending from *alpha*, Greek letters came to denote the brightest stars seen by seventeenth-century Europe's astronomers, while descending Roman letters would indicate the fainter.

Important as Bayer's work would be to establishing the lingua franca of modern astronomy, the frequent reprinting of *Uranometria* throughout the seventeenth

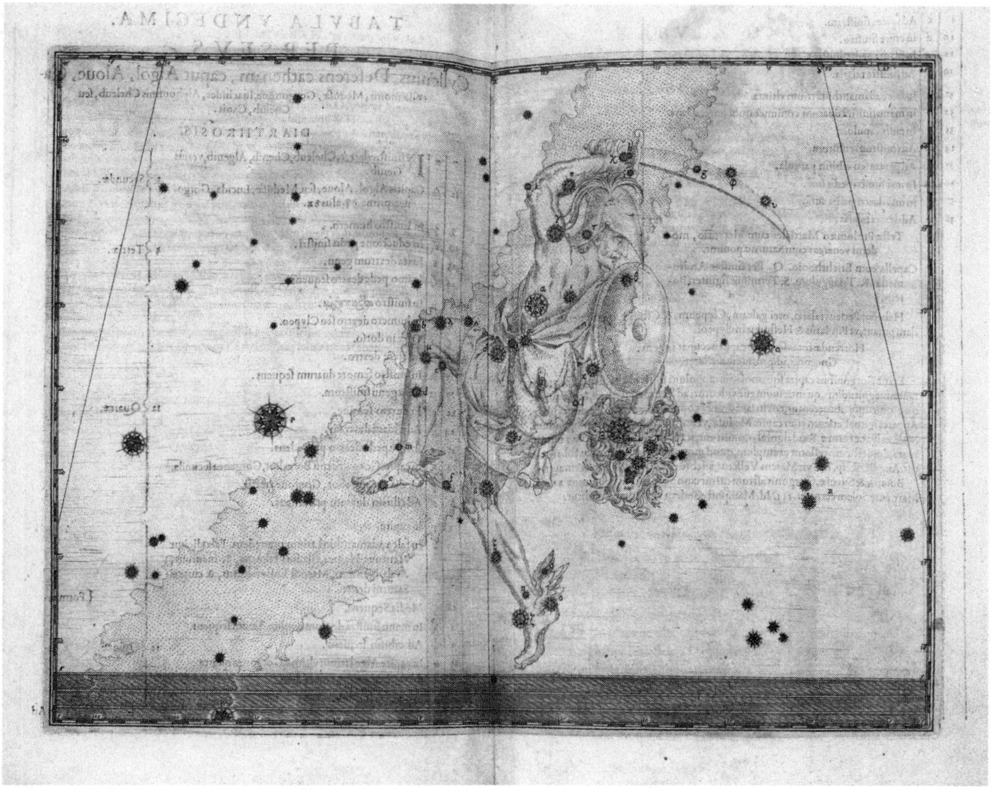

FIGURE 1.11 * Alexander Mair's print of the constellation Perseus as published in Johann Bayer, *Uranometria* (Augsburg, 1603), n.p. © The Royal Society. Reproduced by permission.

century was due in large part to its striking copperplate engravings by Alexander Mair[115] (fig. 1.11). Mair's folio-sized prints represent the heavens in a cunning mixture of the seen, the signaled, and the unseeable. First, each plate maps the celestial positions of a given constellation's stars in accordance with Tycho's observations. Mair denotes those verified star positions on the page with a quasi-naturalistic symbol: a six-pointed asterisk form, which varies in size and internal complexity to designate distinct classes of astral brightness. Because these symbols remain standard throughout the volume, a beholder can potentially read off values of astral brightness by visual comparison within and between plates. But Mair's grades of star character lack clear syntactical differentiation; since modest increments of graphic size and complexity matter to this loosely regimented formula, it is easy to confuse the class difference signified by, say, the star in Perseus's right shoulder from that at his right wrist (fig. 1.12). Consequently, Mair's symbols are further distinguished by the Greek and Roman alphabetic letters of Bayer's innovative system, which have been inserted

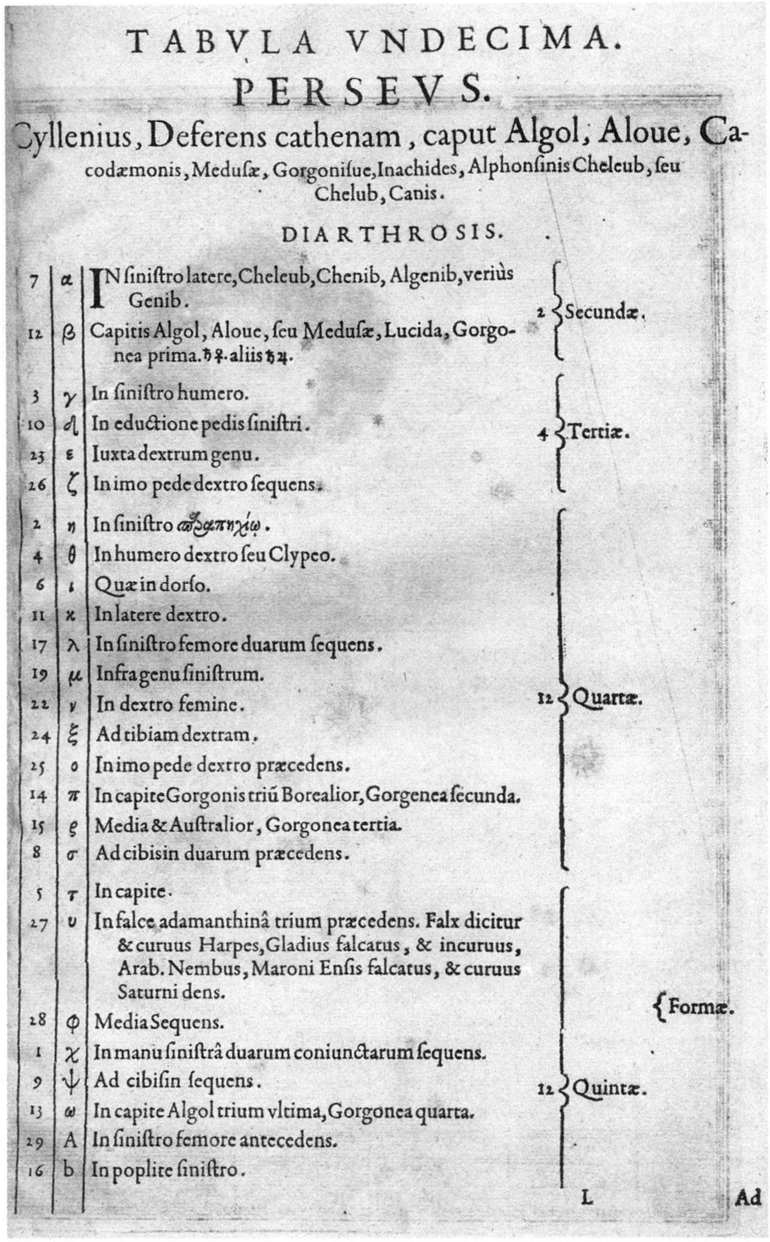

TABVLA VNDECIMA.

PERSEVS.

Cyllenius, Deferens cathenam, caput Algol, Aloue, Cacodæmonis, Medusæ, Gorgonisue, Inachides, Alphonsinis Cheleub, seu Chelub, Canis.

DIARTHROSIS.

7	α	IN sinistro latere, Cheleub, Chenib, Algenib, veriùs Genib.	2 Secundæ.
12	β	Capitis Algol, Aloue, seu Medusæ, Lucida, Gorgonea prima. ♄☿. aliis ♄♃.	
3	γ	In sinistro humero.	4 Tertiæ.
10	δ	In eductione pedis sinistri.	
23	ε	Iuxta dextrum genu.	
26	ζ	In imo pede dextro sequens.	
2	η	In sinistro ἀκροχειρίῳ.	12 Quartæ.
4	θ	In humero dextro seu Clypeo.	
6	ι	Quæ in dorso.	
11	κ	In latere dextro.	
17	λ	In sinistro femore duarum sequens.	
19	μ	Infra genu sinistrum.	
22	ν	In dextro femine.	
24	ξ	Ad tibiam dextram.	
25	ο	In imo pede dextro præcedens.	
14	π	In capite Gorgonis triũ Borealior, Gorgonea secunda.	
15	ρ	Media & Australior, Gorgonea tertia.	
8	σ	Ad cibisin duarum præcedens.	
5	τ	In capite.	12 Quintæ.
27	υ	In falce adamanthinâ trium præcedens. Falx dicitur & curuus Harpes, Gladius falcatus, & incuruus, Arab. Nembus, Maroni Ensis falcatus, & curuus Saturni dens.	
28	φ	Media Sequens.	
1	χ	In manu sinistrâ duarum coniunctarum sequens.	
9	ψ	Ad cibisin sequens.	
13	ω	In capite Algol trium vltima, Gorgonea quarta.	
29	A	In sinistro femore antecedens.	
16	b	In poplite sinistro.	

{Formæ.

L Ad

FIGURE 1.12 * Table ranking and locating stars in the constellation Perseus; from Johann Bayer's *Uranometria* (Augsburg, 1603), n.p. © The Royal Society. Reproduced by permission.

into the plates and amplified in its adjoining tables. Finally, Mair has embedded both his star symbols and reference letters into glorious pictures envisioning the Ptolemaic constellations as mythological beings in the robust, classicizing style of northern mannerism.[116] (See figure 1.11.) These figures are what philosopher Nelson Goodman calls "unicorn-pictures"; they are denoting symbols of fictional beings, even though there are no such entities to denote.[117] Indeed, *Uranometria*'s plates insist on visualizing these distinctions between the verified and the imagined. Using clustered billows of stipples, Mair has suggested the evanescent contours of those mythological beings imaginatively projected into the heavens while contrasting them with the hard-edged symbols that designate positions and gradations of real fixed stars. Part celestial map, part astral notation, and part figurative meditation on the palimpsestic being of constellations, *Uranometria*'s plates thus plot cutting-edge, certified coordinates into ancient, useful fictions.

To seventeenth-century philosophical observers such as Robert Hooke and Isaac Newton, Bayer's volume was a constant point of reference as they pursued the comet through the skies above England during the winter of 1680–81.[118] What becomes clear from following their progression, though, is how Hooke's drawings variously deployed and fragmented Bayer's plates to train deliberative attention on the comet. Consider the drafts made on January 7, 1681. In the upper right-hand corner of the page, Hooke depicts and labels the celestial event as it appeared "to the naked eye" at eleven o'clock at night. Adjacent to a chiasmic form *1*, the protracted dribble of the comet *C* stands between asterisks designating the delta and alpha stars in the constellation Andromeda. So the legend explains, the comet had appeared to Hooke's unaided eye between Andromeda's imagined head and chest, the region of sky occupied by the wavy locks falling onto the chained maiden's right shoulder in Bayer's plate (fig. 1.13). On the left side of the page, Hooke denotes this celestial event, now optically inverted and reversed, as seen minutes earlier through his telescope. The comet is the spreading fountain plume at left, graphically darkest (and so, optically brightest[119]) at its curved head and fanning broadly outward in a series of loose strokes.

Hooke's crisply drawn, rectilinear geometrical forms framing the comet's head are, however, not visible entities at all. They are invented, heuristic armatures expressing the site of the comet's appearance in relation to fixed positions of the "telescopical starrs," enumerated here by points 2 through 8. By constructing these firm guides, Hooke could reliably locate the position of the comet's celestial appearance night after night, thus expanding his time spent studying the object's features. Having meticulously plotted the site of the comet's emergence for weeks, so Hooke noted on the evening of January 16, 1681, deceptive effects of bad weather proved no obstacle. "The moon being now in the first quarter shone bright and besides the air

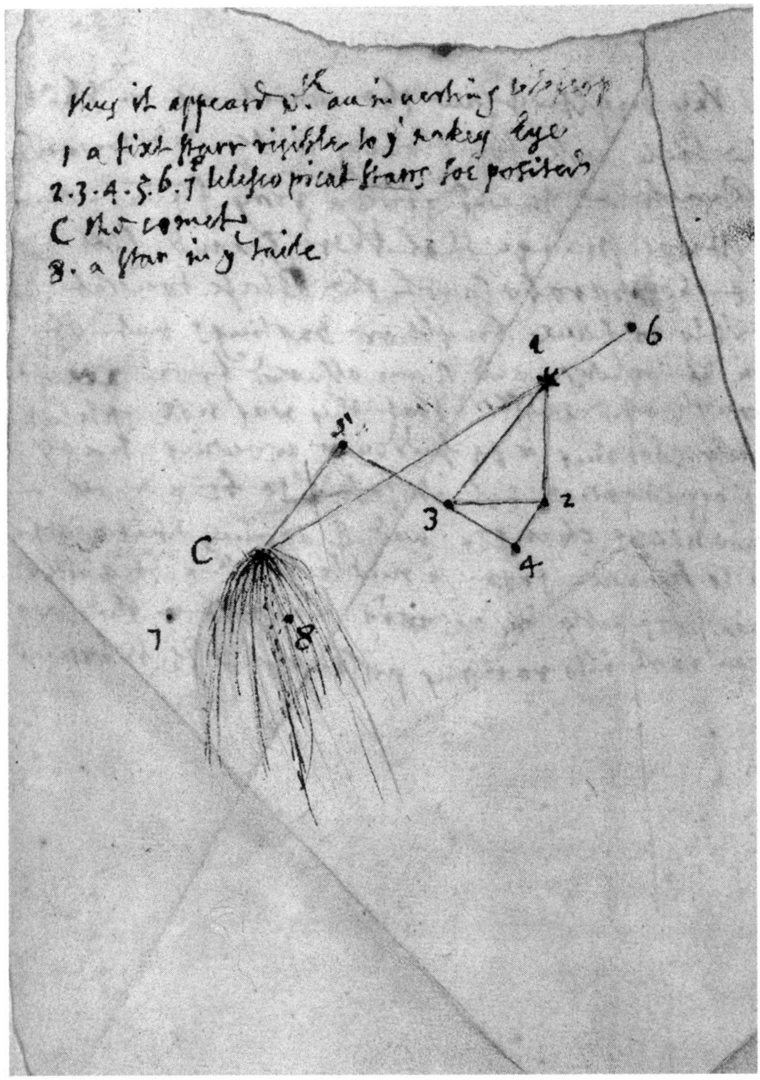

FIGURE 1.13 * Detail of Robert Hooke's telescopically enhanced comet drawing and notations from January 7, 1681; Royal Society Classified Papers, vol. 24, fol. 287. © The Royal Society. Reproduced by permission.

at the time I lookd was a little hazy," he reported, "but knowing whereabout to look for it I quickly found it by the glass and took its position."[120] For an astronomical draftsman like Hooke, Bayer's *Uranometria* provided not only the names for the stars (according to their valuation in the Greek and Roman alphabetic notational system) but a reliable standard against which observations of novel or mobile entities could be compared.

Lexicon and guide, Bayer's *Uranometria* also became a keen representational resource that Hooke activated in his drawings of the early 1680s. If Mair had plotted Tycho's "fixed stars" as outlined asterisks—graded to designate astral brightness, labeled with text—embedded into stippled depictions of anthropomorphized Ptolemaic constellations, Hooke's drawings effectively disassembled those stacked modes of reference to bring the target of his observation into focus. Exemplary here are the drawings Hooke made on the night of January 30, 1681 (fig. 1.14). Framed in the box at the base of the page, the comet visible "to the naked eye" is a circumscribed, wing-like entity sailing below the constellations Andromeda and Triangulum. In the data supplied by his fourteen-foot reflecting telescope, though, Hooke could find no stable, bounded target in the flaring, effervescing object suitable for his long-vaunted project of pictorial circumscription (fig. 1.15). Instead, through six sequential views on the opposite side of the sheet, he shows how the body of the comet appeared to be in significant flux even as he watched it:

> I observed very often and very plainly this night as I had done divers time before that the appearance of it was a perfect flame but very thin and that it waxed and undulated to and fro and did sometimes burn cleerer and sometimes less. Sometimes on the one side and then on the other and now and then in the middle which was more notable a little on the side next the sun. It would sometimes appear with little or noe blaze but a small stream very faint issuing from it as at C.; and at other times for the twinkling of an eye I could see a very small bright point in the middle of it as at C which appeared no bigger than 2 but then was immediately covered with the flamule about it, which flammal would sometimes be considerably bigger and brighter than the other and sometimes seem to have several brighter parts in it as at F.[121]

In Hooke's drawings, dots and asterisks had signified the fixed positions of labeled stars, while rectilinear strokes heuristically plotted the location of the comet's appearance among them. Yet the mobile, mutable head of the comet is here depicted (and depicted repeatedly) as a cluster of stipples opening into very different postures. Projecting a great plume at *A*, the comet's nucleus appears so greatly dispersed at *E* that it more resembles a spray of water than a solid entity.

Nothing in either Hooke's prior observation or his elegant comet theory could possibly answer to the violent, projecting transformations visualized in these drawings. As he had written in 1678, "I could not perceive any kind of motion in the parts of it, such as is observable in flame, smoke, or other steams rising from a burning or hot body."[122] Equally, if he had then supplemented his theory of internal

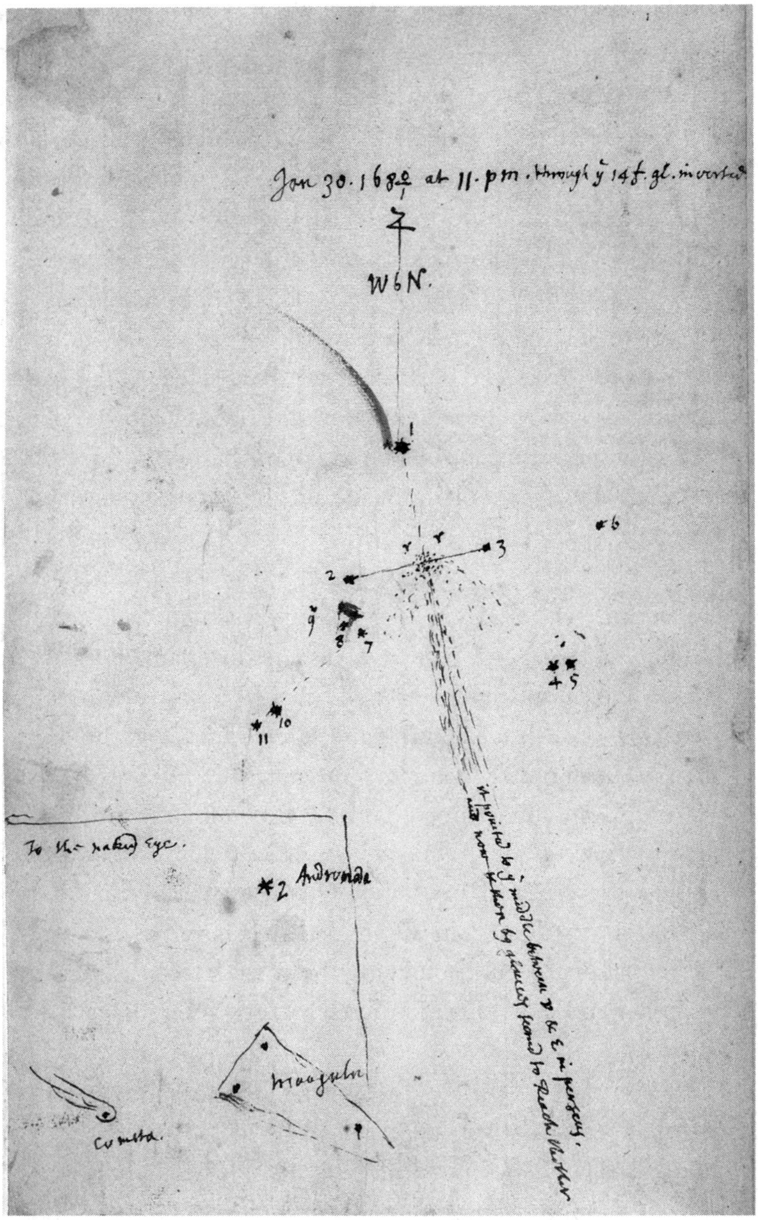

FIGURE 1.14 * Robert Hooke's naked-eye and telescopically aided drawings of comets from January 30, 1681; Royal Society Classified Papers, vol. 24, fol. 297. © The Royal Society. Reproduced by permission.

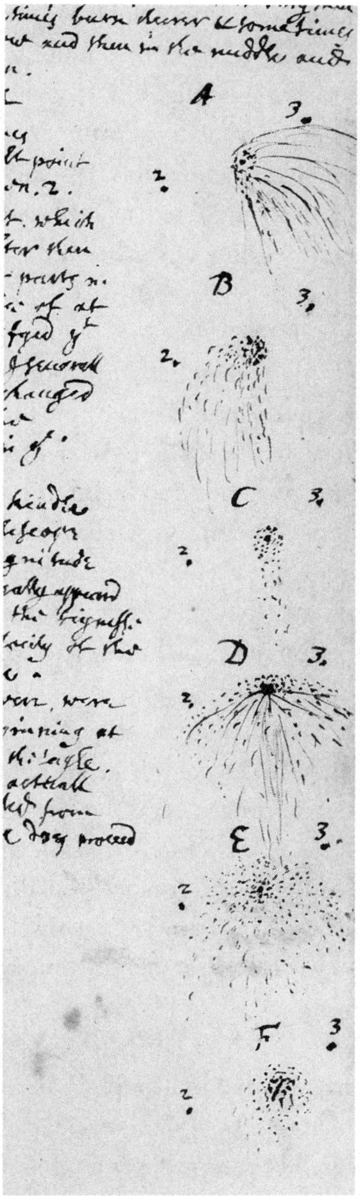

FIGURE 1.15 * Details of Hooke's drawings of events within the rupturing nucleus of the comet from January 30, 1681; Royal Society Classified Papers, vol. 24, fol. 296. © The Royal Society. Reproduced by permission.

disintegration accelerated by external corrosion to explain comets' luminosity through models of torches or projectiles, such ad hoc analogies were completely unequipped to handle the new comet's eruptive ejections of matter "in the manner of a sudden Spouting of Water out of an Engine."[123] Worse, while a knowable core of thing—"true form"—had been amenable to instrumental discovery in the 1660s, Hooke's telescopic draftsmanship of the early 1680s was made all the more intractable by this fluxing, ontological lability of the very target of observation. Where he had earlier sought to model comets' gradual decomposition and response to gravity with a solid, steel-coated ball evolving in a beaker of sulfuric acid, Hooke's drawings struggled now to visualize the massive, lacerating volatility within this exploding celestial body.[124] Its target disintegrating violently and visibly, Hooke's graphic work aimed to bring serious attention to a decaying object as it blazed and flashed in "Mutations and strange Appearances, possibly never heeded by any before."[125]

This is what we see cunningly visualized in Robert Hooke's drawings—images that repurpose *Uranometria* to stage a departure from his own model of artistically informed graphic practice. Although his visual languages do indeed differ from the familiar vocabulary of the mid-1660s, Hooke's is no surrender. Where Alexander Mair had graphically juxtaposed the fixed stars' stable positions and hierarchical order to the ephemeral ontology of constellations, Hooke has appropriated *Uranometria*'s serene, stippled forms to encode barely perceptible events within distant, moving, unstable bodies. Exploded and multiple though they may appear, these strange images struggle to make visible features that no one—including Hooke himself—had previously been able to see.[126]

Strange as it appears, this ingenuity has not been lost to all interpretation. At the return of Halley's comet in 1835, astronomers in England and Germany began to note a range of odd phenomena in the head of the comet as it approached perihelion. As editor James William Grant wrote in the *Monthly Notices of the Royal Astronomical Society*, "These consisted of irregular emanations of a luminous substance from the nucleus. . . . No parallel to these singular phenomena could be found in the observations of any comet of recent times."[127] Perusing the work of seventeenth-century philosopher Hooke, Grant made a surprising discovery:

> I was astonished to find in the "Posthumous Works" of that philosopher, an account of his observations of the comets of 1680 and 1682, which, in so far as regards the physical changes above referred to, tallies completely with the observations of the comet of 1744 . . . and with numerous observations of Halley's comet.[128]

In early eighteenth-century London, however, the "visual desperation" required for such prescient perception did not make for easy looking.[129] Preparing the plates for his posthumous volume of Hooke's work, editor Richard Waller effected a massive formal revision upon this collection of drawings (fig. 1.16). The luminous, udder-like comet forms apparent in the prints Waller subsequently published are much closer

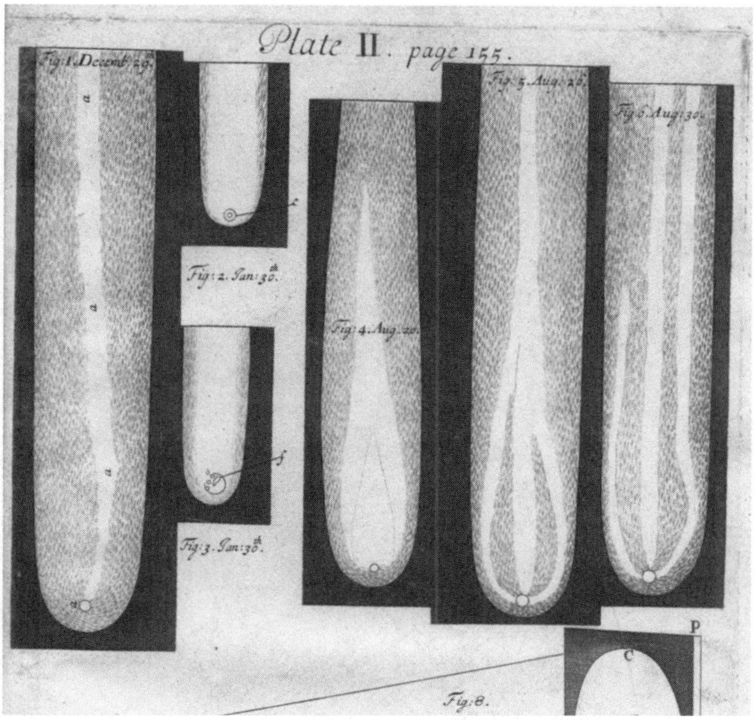

FIGURE 1.16 * Engravings of comets from the early 1680s; published in *The Posthumous Works of Robert Hooke*, ed. Richard Waller (London: Sam. Smith and Benj. Walford, 1705), plate 2. © The Royal Society. Reproduced by permission.

to the visual vocabulary of Hooke's 1678 *Cometa* than to the complex notations devised in his actual drawings (fig.1.17). Rather than attempting to replicate these radically corrosive, ingeniously perceptive drawings in print, Waller had defaced them, dismissing them as but mnemonic aids that failed in their supposed attempt to replicate Hooke's earlier aesthetico-epistemological positions.

If it is unlikely that Waller would have been alone then or now in preferring his printed forms to Hooke's "very rudely designed" comet drawings, perhaps the violence effected by his engravings is only an appropriate response to Hooke's iconoclastic project. After all, the posture of learning diligently from what nature can teach is surely one of Renaissance art's most subversive legacies. "The painter will produce pictures of little excellence if he takes other painters as his authority," so an exemplary figure such as Leonardo da Vinci would claim, "but if he learns from natural things he will bear good fruit."[130] Recent studies have reminded us of the anarchic impulse implicit in such appeals to the authority of nature, their repudiation of the hierarchical systems of apprenticeship to particular masters and of the

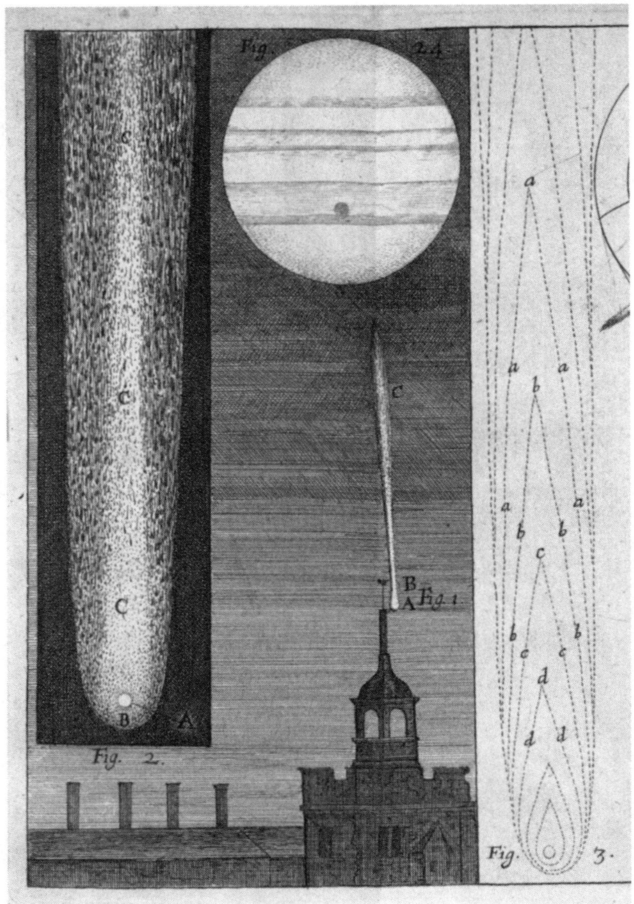

FIGURE 1.17 * Etching and engraving of the 1677 comet as published in Robert Hooke's *Cometa* (London: J. Martyn, 1678). © The Royal Society. Reproduced by permission.

broader aesthetico-epistemological orders that structured early modern practice.[131] So, perhaps, might we read Hooke's turning to "what the Appearances themselves could teach me" as a heroic staging of this ultimately Oedipal move. After all, his telescopic discoveries of the early 1680s were graphically accomplished by renouncing, not only much of what he had learned from his artistic training with Peter Lely, but the very theories and practices he would go on endorsing in the teaching of aspiring experimental draftsmen. Hooke's surrender to Wölfflinian opticality would, by this account, amount to aesthetic and epistemological insurrection.

Yet what I think is more telling to a reading of the cunning, furtive intelligence of visual practice in Restoration London is the way in which Robert Hooke's small graphic revolution was achieved. He did not invent his strange marking system ex nihilo, let alone did he appeal to some "copying" of natural appearances. Instead, he recruited the systems of alphanumeric notation and the valuations of outlined fixity and stippled ephemerality that he found in Bayer's *Uranometria*, the book (itself shot through with artistic schemata) that guided his nightly observations. He used those repurposed tools to craft highly artificial visual languages by which he could sustain observational focus on an elusive, nearly invisible target of investigation. Whatever aesthetic judgment Hooke's difficult drawings might warrant, their vitality to the study of wicked intelligence lies precisely here: in disclosing how the conventionalized schemes of the early modern artistic tradition were being tortured, ruptured, and recombined for incredibly clever purposes at and beyond that tradition's borders. Let us now dive deeper into the heart of experimental draftsmanship's philosophical workings.

CHAPTER TWO

Knives Out

Thinking On, With, Through, and Against Paper in the Mid-1660s

IN THE SPRING OF 1663, fishermen pulled a massive herring from the frigid Atlantic waters off Thurso, the northernmost town in the cold, northerly land of Scotland. Dietary staple of early modern northern Europe as herring was, this singular, monstrous example quickly became an object of philosophical reflection through a paper proxy dispatched to the Royal Society of London[1] (fig. 2.1). Looking at that artifact, we see the herring in profile as if splayed for preparation on a fishmonger's block, a surface that has effectively impressed itself into pictorial service. This image has not been drawn in any conventional sense. It has been cut from two sheets of writing paper glued together to accommodate the fish's substantial, nineteen-inch length. Down from the projecting lower lip, we can imagine a picture maker cleaving, scalloping out the forward forms of the pectoral, pelvic, and anal fins in one long, undulating slice. Reversing this stream, oblique wedges would then have been snipped to denote the characteristic fork of the herring's tail and the interstices where coarse, net-like fin membrane rejoined the sleek contour of iridescent, scaly body.

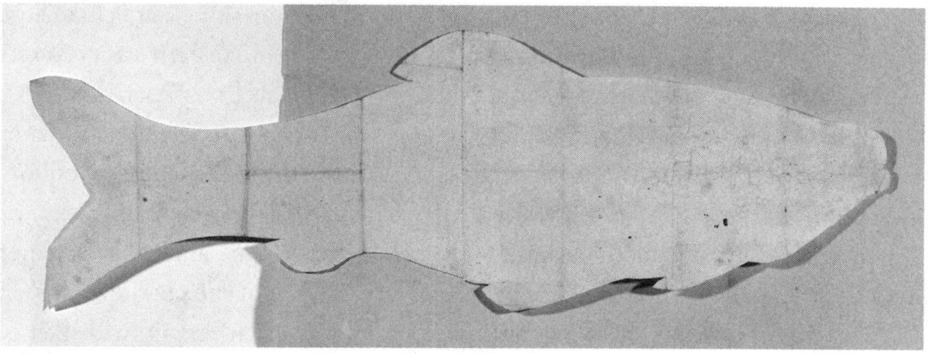

FIGURE 2.1 * Anonymous cut-and-pasted paper herring sent to the Royal Society in spring 1663; Royal Society Classified Papers 13:1. © The Royal Society. Reproduced by permission.

Brilliant though the herring's superficial features may have been, we can learn nothing about them from this image. Only a small, circular denotation of eye and two halting, pen-stroke gills inform us about its physiognomy. Boundaries and not the contents of the herring's body are what matters here. Measurable, made through physical contact between piscine body and cut paper, the image has perhaps been proffered as a reliable means of communicating that most dubious of claims: the capture of a truly massive fish. So its inscription declares, the image possesses "the exact proportion of an hering taken in the water of Thurso May 1663." Sliced down to size on the light, dry medium of paper, the herring has acquired, we might say, an abstracted, mathematical intelligibility that was sorely needed for studying those slippery denizens of the cold, inky depths. After all, as this paper herring was arriving in early Restoration London, experimentalist and future architect Christopher Wren was making "Schemes of several Fishes dissected, in which the Fabrick of the Parts appear'd very often irregular, and differing much both from Brutes, and one another."[2]

Cut, not drawn; outlined, not rendered, both referent to and bearer of the monstrous dimensions it isomorphically exemplifies, the paper fish would seem to be more an exception to the rule of scientific draftsmanship than a meaningful specimen of it—more red herring than *Clupea harengus*. Yet outlandish as this fish might appear, a growing body of scholarship has begun to make a broader case for the intellectual force of early modern "paperwork." Acts of cutting paper apart and pasting it back together were, it is claimed, central to the ways in which men and women in early modern Europe read, traveled through space, integrated information, produced their books, and understood their drawings.[3] As historians of science now study the agency of "paper-tools" in Enlightenment-era chemistry, theorists such as Bruno Latour locate the defining power of Western science in its peculiar alchemy: "the transformation of rats and chemicals into paper."[4]

Guided by the totemic herring, this chapter tells the tale of a single cut-paper object that came into existence at a nexus of manual skill, technological competition, and frankly wild speculation among leading French and English agents of experimental philosophy during the tumultuous years between 1664 and 1668. A paper model of a telescopic micrometer (fig. 2.2*a*, *b*), the object at the chapter's center, was designed and drawn by Robert Hooke, working in coordination with Henry Oldenburg, Royal Society Secretary and publisher of the *Philosophical Transactions of the Royal Society of London*; an unknown engraver; and John Martyn, London bookseller and the Royal Society's official printer. Although only a few examples of this assembled object survive, any talk of rarity is misleading.[5] Hooke's paper model was made as a multiple. It was designed to be printed, cut, pasted, and sold by Martyn from his shop at The Bell in the broken shadow of St. Paul's Cathedral destroyed by the Great Fire of 1666.[6] Narrating this artifact's birth as a picture and its maturation as an object in the hurly-burly of 1660s experimental life, I seek to elaborate both the ingenious problem-solving Hooke's paper micrometer performed and the broader varieties of conceptual shape-shifting it enabled as prompter to experimental intelligence.

Intensifying the focus on Hooke's draftsmanship begun in chapter 1, then, this chapter's aim is to show how the philosophical stakes of experimental graphic practices become truly perceptible only as we dive with their makers down into the cut, pasted, pitted, stained, and ultimately rapturous mess of material facture. My argument has three key parts. After sketching its origins amid the vicissitudes of precision instrumentation in the mid-1660s, I read the paper micrometer that Hooke fashioned in the fall of 1667 as a puzzle. I tell this story by reconstructing the experimental, visual, and ethical conflicts in which the paper model emerged so as to illuminate its elegant, puzzle-solving wit. The second part of the argument then turns the situation around, exploring relations between this pragmatic, cut-paper artifact and what Hooke and his milieu called "art." As opposed to the devious artistic entrapments he was then envisioning, I will propose that Hooke's light, labile object can best be seen as valuable to the experimental intelligence insofar as it could both solve puzzles *and also* undermine its own authoritative structure, prompting and stimulating imaginings anew. The chapter concludes by suggesting how Hooke himself attempted to conceptualize the strange ontology and curious epistemological function of such experimental artifacts by redeploying nothing

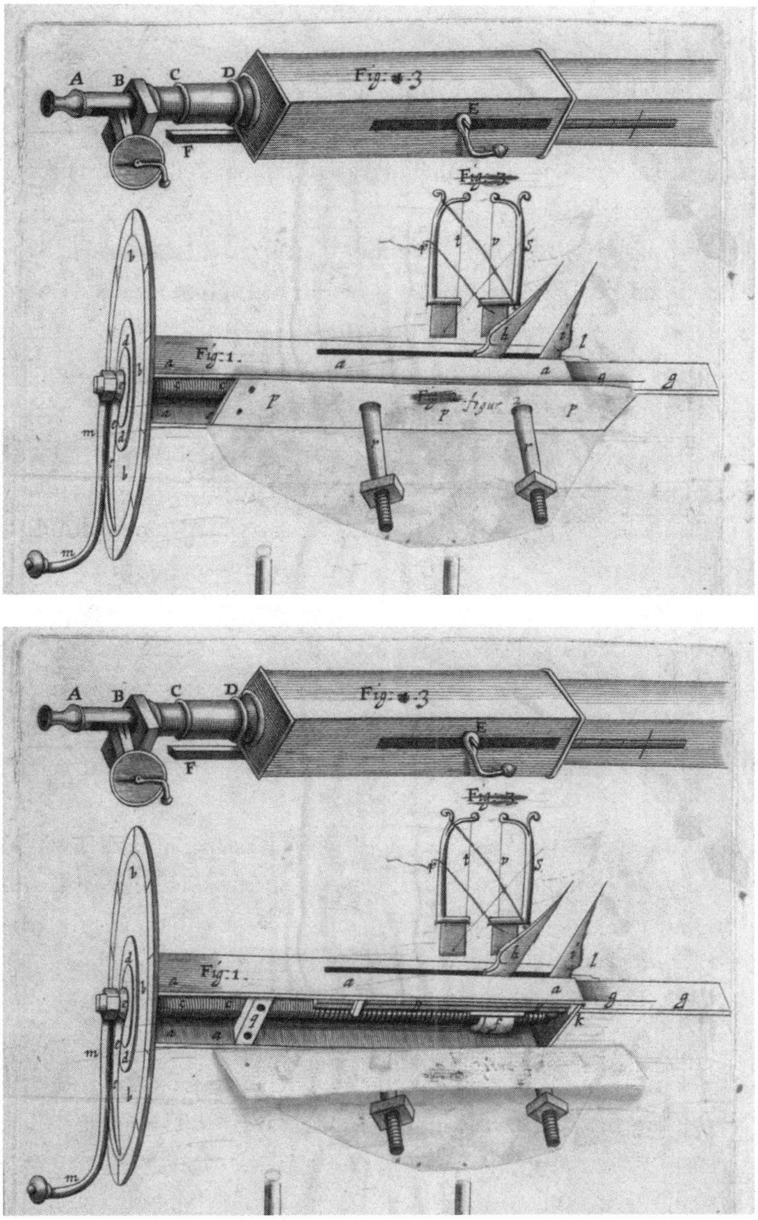

FIGURE 2.2*a*, *b* * Print designed by Robert Hooke with paper patch showing Richard Towneley's micrometer closed and opened; offprint of illustrative material for *Philosophical Transactions* 29 (November 11, 1667), now bound into Royal Society Register Book (Original), vol. 3, fols. 227–30. © The Royal Society. Reproduced by permission.

less than his own pivotal work on gravitation and attraction at a distance, lines of thinking also first publicly presented in the mid-1660s. Neither a comprehensive history of scientific drawing in the early Restoration nor even a systematic study of Hooke's draftsmanship, I see this chapter as a set piece—a framed construct through which to analyze the violent unfolding of experimental intelligence rippling outward though graphic practice.

But before descending into the rabbit hole that is this chapter, a note is warranted on the mechanical device at its center. Although it too is an instrument of precision optics, a screw micrometer promises much less to visual perception than does a telescope or microscope.[7] If the wondrously strange sights glimpsed from above and below precision-ground, objective lenses privileged by chapter 1 can signal the "new worlds" awaiting the seventeenth-century experimentalist's enhanced eye, measuring devices such as the micrometer would seem positively designed to expose and rectify the congenital frailties any aspiring Columbus trucked along with him into the ocular field. Instructively, Hooke implicated his audience in that very problematic with a self-performing image. Promoting the use of mensurated telescopic sights in astronomy, Hooke enjoined the reader of his 1674 *Animadversions* to "take a sheet of white Paper, and thereon draw two parallel Lines . . . at four or five inches distance, then draw as many other small lines between them at right angles to them, and parallel one with another, as he thinks convenient."[8]

Darkening in alternate spaces to form black-and-white bands, the reader was to hang this paper on a well-illuminated wall and walk as far back as he could until the striped markings became indistinguishable. By measuring the distance from the paper to that point of optical failure, the observer could calculate the smallest angle at which his eye could still make differentiations—a sobering "Standard, by which he is able to limit the bigness and exactness of his Instruments, beyond which all magnitude and curiosity is not only useless, but of much detriment."[9] If micrometric devices suggest a commensurate prick to the telescope's plentiful visual reveries, this chapter proposes instead that attention to the rich materiality of Hooke's paper micrometer can actually reveal the cosmological scale it imagines and the wicked intelligence flowing through it. Binding near to far, inside to outside, beast to machine and mind to body, we will see how this humble paper artifact effectively outwits the art objects by which Hooke himself was bewitched, just as it takes its epistemological function and ontological identity from a "System of the World" pitched against and interwoven with that of Isaac Newton. Not only is the paper micrometer filled with humans and animals like a good Latourian hybrid; but, by thinking with it, we might exclaim with David Bowman, "Oh my God!—*it's full of stars!*"[10]

The Micrometer

In late December 1666, Adrien Auzout, astronomer and member of Paris's newly constituted Académie royale des sciences, wrote to the Royal Society of London, announcing his new method for dividing "one foot into 24000. or 30000. parts."[11] Auzout knew that London's experimental philosophers would quickly recognize the keen advantages to astronomical and terrestrial observation entailed by such a device. It would allow an observer to measure the apparent size of objects sighted through a telescope at incredible precision, thus increasing the exactitude with which astronomical phenomena or distances of military targets could be calculated.[12] Within months, Auzout's priority and proclaimed technical supremacy were indeed challenged by English astronomer Richard Towneley (1629–1707). From Halifax in West Yorkshire, Towneley had read Auzout's claims in the pages of the *Philosophical Transactions of the Royal Society*, the seminal scientific journal published by Henry Oldenburg. By Towneley's reckoning, national pride was at stake. In the wake of London's devastating plague of 1665 and fire of 1666—and on the eve of England's ignominious defeat in the Second Dutch War—Towneley proclaimed that he would be "look't upon as a great Wronger of our Nation" were he not to contest French priority by advancing his own achievement.[13]

Using instrumental techniques developed by fellow Yorkshire mathematician William Gascoigne in the late 1630s, Richard Towneley had crafted a telescopic micrometer with a precision superior to that asserted by his French rival. "It is small," he explained, "not exceeding in weight, nor much in bigness, an ordinary pocket-watch, exactly marking above 40,000 divisions in a foot, by the help of two indexes."[14] As one historian of science elaborates, Towneley built on Gascoigne's recognition that "by introducing two fine-pitched screws in the eyepiece . . . one could 'enclose' the lunar or solar image produced telescopically, and knowing the exact focal length of the object glass and the linear pitch of the screw, calculate the angle subtended to a hitherto unobtainable degree of precision."[15] As visualized in the folio-size ink drawing he sent to the Royal Society in the summer of 1667 (possibly following the instrument itself), Towneley represented less the design of his mechanism than the micrometer put to use (fig. 2.3). From the darkened inlet of the hourglass-shaped eyepiece at left (labeled *A*), Towneley depicts a telescope spanning the horizontal axis of the page, progressing from a sequence of movable, circular stems to the square, adjustable shaft of the focal plane. The outward-turned face of the micrometer appears just below the letter *B*. Rupturing the image's perspectival construction in its emphatic frontal address to the picture plane, this minutely calibrated face and the upward-slanting, hatched casement are all Towneley shows

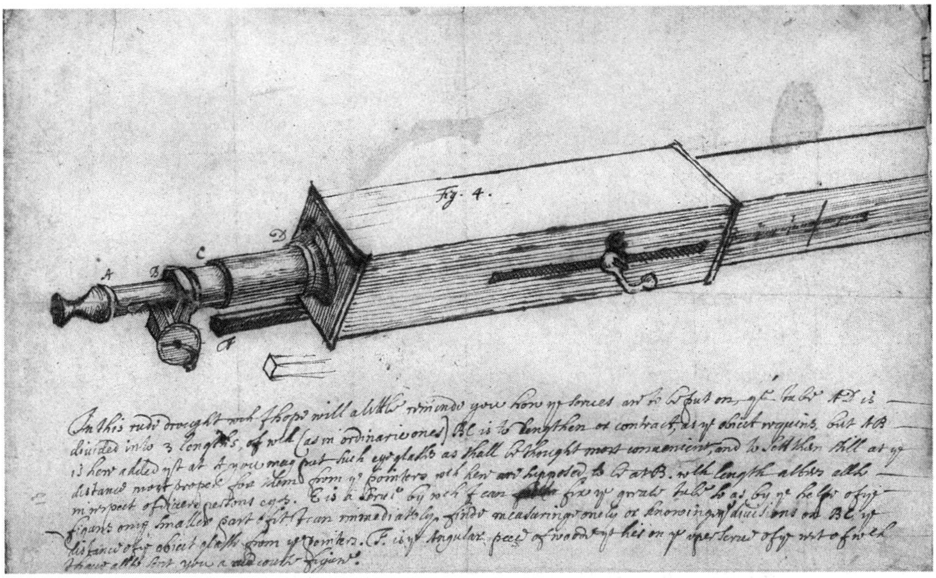

FIGURE 2.3 * Ink drawing of a telescope installed with Richard Towneley's micrometer sent to the Royal Society in 1667; Royal Society Classified Papers 2:13, fol. 1.

us of his micrometer; as would have been required for its use, the device's intricate pointers have disappeared into the telescope's focal field. Accompanied by a detailed inscription and an additional drawing of an instrument stand, Towneley presents the micrometer as one component among several broader improvements to telescopic technique.[16]

Profiting from collaboration yet keen to assert priority; contesting international rivalry while advancing personal interest; depicting his instrument without divulging its essential design: Richard Towneley's representation of his telescopic micrometer registers the tectonic pressures weighing on authorship in the early experimental community.[17] In his case, Towneley's bet-hedging precautions were entirely justified. Already preempted by Auzout, Towneley found a new claimant on his micrometer in the person of Robert Hooke, a keen inventor of precision mensuration instruments through which he hoped to claim the longitude prize among other emoluments.[18] From London, Hooke quickly shadowed the instrument's development. When note of Towneley's device was first published in the *Philosophical Transactions* in May 1667, Hooke responded with a short piece in the same volume entitled, baldly, "More Wayes For the Same Purpose."[19] As it appeared in mid-November 1667 in the Royal Society's Register Book (the institution's central mechanism for recording important

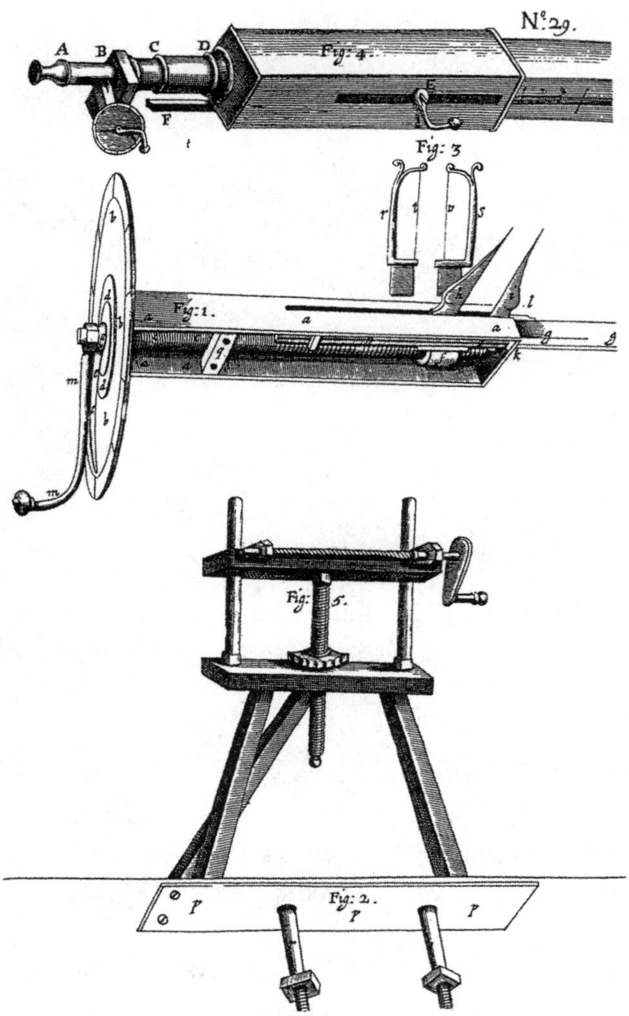

FIGURE 2.4 * Printed plate showing Robert Hooke's design for the representation of Richard Towneley's micrometer, with the "uncut" patch (see fig. 2) at the plate's base, as printed in *Philosophical Transactions* 29, no. 2 (November 11, 1667), frontispiece.

research and allocating inventive priority), the micrometer was likewise noted as "being contrived and sent by Mr. Richard Towneley to the R. Society, and this following description of it being made by Mr. Hook was read before the Society."[20] And when printer John Martyn assembled illustrations of the micrometer for the *Philosophical Transactions* of November 11, 1667 (fig. 2.4), Oldenburg credited "the ingenuity of Mr. Hook"—not only for negotiating between the far-flung Towneley and local engravers but for making "the draught of the figures, representing the

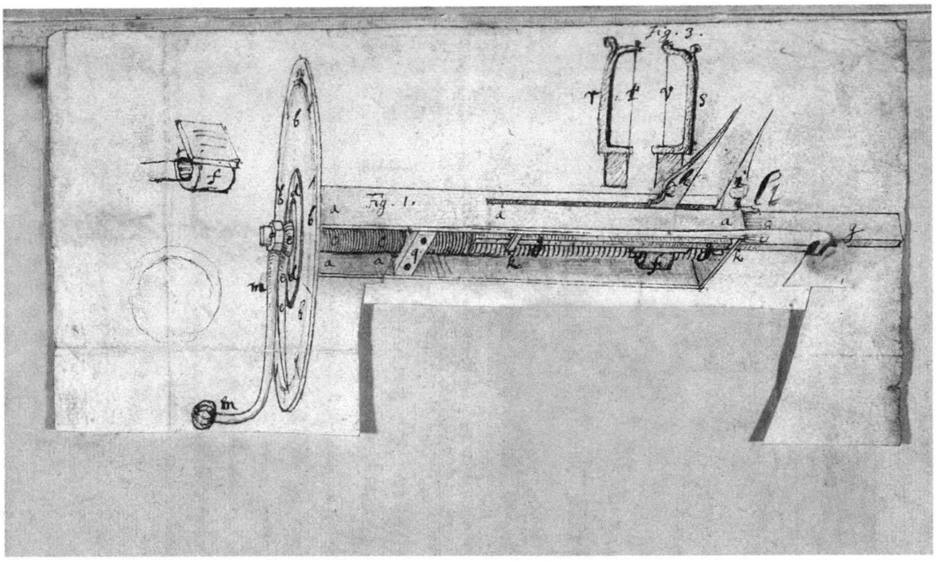

FIGURE 2.5 * Robert Hooke's ink-and-pencil drawing of Towneley's micrometer (autumn 1667); Royal Society Classified Papers 2:13, fol. 3. © The Royal Society. Reproduced by permission.

new instrument it self, and the description of the same."[21] While it is clear that two of the designs engraved for Oldenburg—and contemporaneously copied into the Royal Society's register—derive directly from the drawings Towneley had sent to London, the image at the center of the print is that which sprang from Mr. Hooke's aforesaid "ingenuity."[22]

In the surviving pencil-and-ink draft he made for this engraved figure, Hooke's cunning can be observed as if in slow motion (fig. 2.5). As in the subsequent *Philosophical Transactions* print, Hooke's drawing has fundamentally altered our point of view from Towneley's transmitted sketch to show the micrometer from the side, its face appearing like a flywheel in oblique perspective. Rotating the handle at extreme left, we can infer, turns the axle-like screw at the instrument's core, moving the pointer threaded to it. Protruding upward through an incision in the micrometer's casement, this index would have been visible through the telescope's focal plane, where it could measure features of observed targets relative to its stationary twin at right. Yet Hooke's drawing not only alters the pictorial schemata and point of view through which we see Towneley's device; it manifests a restlessness with existing form, a desire to mutate the object itself. Note, for example, the small design labeled *fig. 3* directly above the micrometer's sights. As Hooke explains, this figure denotes an alteration proposed by "some ingenious men" to enhance the micrometer's functionality: "two sights fitted with hairs . . . may be conveniently used in the place of the solid edges of the sights."[23] Omitted from the version of his drawing copied into the Royal Society's official register, Hooke's pictorial additions betray his regard for the Towneley micrometer as an evolving design rather than as a static, authorized object—a view clearly not shared within the Royal Society.[24]

Far more radical volatilities were subtending Robert Hooke's engagement with the micrometer, however. As the material facture of his preparatory drawing demonstrates, Hooke was deeply ambivalent about how to represent Towneley's precision instrument. Should his depiction confine itself to detailed articulation of the instrument's exterior features? Or should he show by pictorial means those delicate, movable mechanics that would have been concealed from view when the instrument was in use? At the outset, Hooke clearly preferred the first option; he began to depict the micrometer from the side, with its lower casement panel and threaded attachment screws projecting perspectivally outward toward the beholder, much as the device appears in the "enhanced" state of the *Philosophical Transactions* plate. (See figure 2.2*a*.) But Hooke revised this approach in media res. Using a razor or knife, he cut away the portion of his drawing representing the lower casement and screws. The damage inflicted on the existing page is substantial; nearly a quarter of its paper surface has been removed. Next, with red wax (material that remains visible at the far right corner of the incision), he attached a patch to the back of his dissected drawing (fig. 2.6).[25] On the recto of this thin strip, Hooke drew a detailed representation of the micrometer's interior mechanics, drafted so as to appear continuous with the existing image. As its surviving form suggests, the pictorial continuity of front and back, wounded page and patch, is nearly seamless.

Perhaps it could be objected that Hooke's incision into his micrometer draft was just a practical remedy to a graphic error—a *repentance*, as it is known to connoisseurs of drawing.[26] After all, the image was to be seen only by the engravers, a pragmatic lot for whom the excess expenditure of labor required to redraw fine details of micrometer innards would have been unnecessary. Yet when preparing his print for the *Philosophical Transactions* in mid-November 1667, Hooke effectively reversed the procedure he had earlier performed on his own drawing, insisting on the beholder's haptic encounter with the representation. He devised an engraved, pictorial patch showing the outer casement and screws of the micrometer that was to be cut and pasted on top of the printed image of the micrometer's interior. (See figure 2.4.) Through this printed, pasted addendum, the beholder was to toggle between views of the instrument's interior or exterior. In an explanatory text

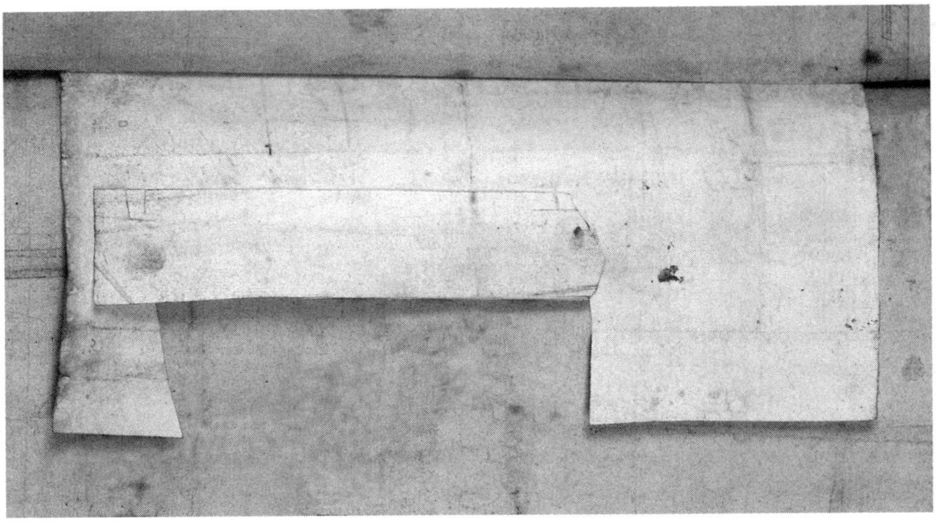

FIGURE 2.6 * Verso of Robert Hooke's drawing of Towneley's micrometer revealing where and how it has been cut and pasted back together (autumn 1667); Royal Society Classified Papers 2:13, fol. 3v. © The Royal Society. Reproduced by permission.

printed in the *Philosophical Transactions*, Hooke stipulated that "Fig. 2. represents the moveable cover containing the screws, to be by the bookseller cut off by the pricked line (x x x) from the paper, and to be fitly placed on Figure 1."[27] (See figure 2.2*a*, *b*) Consequently, the depicted micrometer *qua* paper model could be seen in two distinct ways: "By the taking off, as it were, or folding up of this cover, the inward contrivance of the screws and sights may appear."[28] Exterior appearance and interior systems would be made visible effortlessly, at will.

Never again would Hooke appeal to this strategy of flap-happy, fugitive sheet representation. His closest approximation of that technique can be found nearly fifteen years later, in the second number of his short-lived scientific journal, *Philosophical Collections*. (See figures 4.6–7.) There, as we will see in chapter 4, Hooke instructed his reader to slice up the illustrative plate to create a "paper instrument" of Dr. Robert Wood's device for rectifying solar and lunar calendars. So, if it is effectively a pure unicum, the question stands to be asked: why, in the fall of 1667, would Hooke have gone to all the trouble and expense to fashion such a weird, useless object and then insist upon its mechanical replication? And what could this monstrous stillbirth possibly reveal about experimental intelligence? By tracing the origin of this object back through the interfaces of machine and living bodies in 1660s experimental culture, I want to begin to answer these questions by pulling forth the object's positive, puzzle-like dimensions.

Morbid Vision

By the late summer of 1662, several different programs of physiological research rooted in the dissection-based tradition of William Harvey were moving closer together, converging on a workable account of respiration.[29] From the days of their celebrated collaborative experiments in late 1650s Oxford, Robert Hooke and Robert Boyle had played a key role in that conversation. Through experimental evidence of birds and small mammals that sickened or died in the vacuum chamber of the air-pump Hooke built for him, Boyle explained the influx of air into the lungs through the decrease in pressure when the bellows-like construction of the chest dilated.[30] As Boyle collaborated with Hooke to identify chemical components in air essential to vitality, physicians such as William Croone in London and Walter Needham in Oxford were codeveloping a powerful program of anatomical investigation. Building on Harvey's experimental framework, their trials demonstrated that withholding air from a creature would cause its vital signs to diminish rapidly. But should air be promptly pumped back into its lungs, these physicians found, the asphyxiated beast could often be resuscitated and restored to full health.[31] What these overlapping chemical, anatomical, and experimental studies were seeking was as much an understanding of respiration as of observable relationships between systems—how the respiratory and circulatory systems interacted with each other and, specifically, whether any "visible passage" could be identified between them.[32]

Responsive to this optical search for cardiopulmonary connections, the program of anatomical investigation that Hooke perfected in London between autumn 1664 and summer 1668 amounted to a radical visualization of the symbioses sought by his colleagues.[33] His program of experimental vivisection was a simple as it was brutal. Removing the animal-subject's rib cage and the tissue encasing its heart, Hooke would intentionally sever the windpipe, replacing natural respiration with a supply of air delivered into the lungs by a tube connected to bellows. Mechanically controlling the influx of air into the animal's body, he could directly observe effects caused by privation or artificial provision of air on vital organs. This is how Henry Oldenburg reported Hooke's paradigmatic trial in early November 1664:

> By means of a pair of bellows (when the thorax was laid quite open, and ye whole *venter infirmus* also) and a certain cane thrust into the wind pipe of ye Animal, ye heart continued beating for a long while, at the least an houre, even after the diaphragme had been cutt away in great part, and ye pericardium removed from the heart. And from several trials it seemd very probable, yt this motion might have been continued as long, as there had been any blood left

within ye vessels of ye dog: for ye motion of ye heart seemed very brisk and lively, after an houres time from ye first displaying ye diaphragm; and upon removing ye bellows, ye lunges would presently begin to sink, and ye heart begin to have irregular, thick, and convulsive motions: but upon renewing ye motions of ye bellows, ye heart recovered its former motion, and the irregular one ceased.[34]

Here, if anywhere, we are truly in the province of the historically novel "Baconian" sense of experiment.[35] No longer could experiment be understood as an illustration or exposition of a priori principles; it had become the torturous "vexations of art" by which philosophers would produce new facts about nature.[36]

Sanctioned by anatomical tradition, productive of new data, and oddly fascinating to boot, these experiments profoundly revolted Hooke. In a letter written to Boyle soon after the first trials in November 1664, he outlined a powerful conflict between the desire to see at the heart of his research program and his disgust at its violence.[37] The letter opens with Hooke describing the progress he had made studying light's passage through various media, promising to analyze "refractions also of the parts of the eye & of each of which as fast as I can make them."[38] From this dry, "ballistic" account of vision, he quickly turns to the uncomfortably liquid body of the dog rendered by his anatomical experiment.[39] He writes, "The other experiment (which I shall hardly confess make again because it was cruel), was with a dog which by means of a pair of bellows wherewith I filled his lungs and suffered them to empty again."[40] Whereas Oldenburg had built his narrative of this event around passive-voiced constructions, Hooke is unequivocal in stating his intervention. "I was able," he claims, "to preserve [the dog] alive as long as I could desire, after I had wholy open'd the thorax and cutt off all the ribs and open'd the belly nay I kept him alive above an houre after I had cutt off the pericardium and the mediastinum."[41] Nor was Hooke ambiguous about the pain and suffering involved. "I shall hardly be induced to make any further tryalls of this kind," so he pledged to Boyle in conclusion, "because of the torture of the creature."[42]

Hooke's recognition of wickedness—of "cruelty," "torture"—in his treatment of animals is striking.[43] A sympathetic reader of René Descartes and a noted practitioner of mechanical philosophy, he might well have been expected to understand animals as machines[44] (fig. 2.7). Indeed, commenting on the sole target of his 1660s vivisection experiments that he actually drew, Hooke deployed that machine analogy as a term of praise. "The contrivance for erecting and retracting, or sheathing the teeth" disclosed in this English viper, Hooke wrote, was "very pretty, and like all other articular motions of the body very mechanical."[45] Because he possessed this

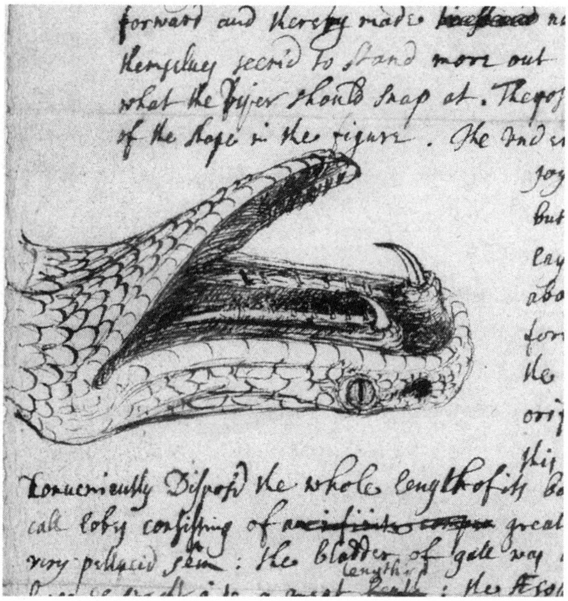

FIGURE 2.7 * Robert Hooke's ink drawing of the head of a vivisected snake from November 23, 1664; Royal Society Register Book (Original) vol. 3, fol. 65. © The Royal Society. Reproduced by permission.

conceptual framework for understanding beasts in terms of machines, it is tempting to think that any qualms induced by his experiments' violent literalizations—those concrete fusions of beast-machines with mechanical devices like bellows—could have been peremptorily dismissed.[46]

Yet as Boyle had earlier done, Hooke steadily developed a range of prophylactic measures in the mid-1660s by which to shear his visual excavations from their hideous violence.[47] Contemporaneously in *Micrographia*, he was advocating for the microscope specifically on the grounds that it enabled the philosopher to see deeply and invasively into bodies without destroying the target of inspection. The microscope proves miraculous, Hooke writes, "for the discovery of Nature's course in the operations perform'd in Animal bodies, by which we have the opportunity of observing her through these delicate and pellucid teguments of the bodies of Insects acting according to her usual course."[48] Compared with vivisection, the microscope's advantage was undeniable to any student of Lady Nature:

> Whereas, when we endeavour to pry into her secrets by breaking open the doors upon her, and dissecting and mangling creatures whil'st there is life yet within them, we find her indeed at work, but put into such disorder by the violence offer'd, as it may easily be imagin'd, how differing a thing we should find, if we could, as we can with a *Microscope*, in these smaller creatures, quietly peep in at the windows, without frighting her out of her usual bias.[49]

Because of the transparency of insects' protective exoskeletons, microscopic observation could provide visual access to interacting bodily systems in ways utterly foreign to the violent methods of the gross anatomist. And although no comparable optical technique could then allow the philosopher to "quietly peep in" to the bodies of the larger beasts of his experimental trials, Hooke was eager to defray the cruel cost of their visualization (as we will see in chapter 3) by employing anesthetic techniques then being developed by colleagues such as Christopher Wren.[50] "Certainly," Hooke noted of his ongoing vivisections, "the inquiry would be very noble if we could any way find a way to so stupify the creature as that it might not be sensible wch I fear there is hardly any opiate will performe."[51]

Scholars of early modern visuality have often asserted a delight taken in probing, dissecting, and otherwise anatomizing targets of observation, especially among artists and experimentalists of northern Europe. In classic studies, Svetlana Alpers has emphasized how depicted objects were "exposed to the probing eye . . . by the technique of flaying them," while Barbara Maria Stafford narrates an Enlightenment visuality where "digging knives, invading scissors, sharp scalpels . . . stood for an investigative intellectual method that uncovered the duplicity of the world."[52] What I would emphasize here is the *extremity* of Hooke's struggle to stabilize such rupturing vision by means of instrumental and psychotropic supports. On the one hand, he had argued for microscopic seeing against gross anatomy on the grounds that it enabled the experimental beholder to look inside organic bodies without either ruining them or implicating the investigator in the disfiguring scene of violent observation. Conversely, he had proposed, if an animal subject could be sufficiently "stupefied," or anesthetized, then the experimental vivisectionist should not only be freed from implication in torture, but the enterprise could be elevated to true nobility. By transporting the animal into an opiate-induced torpor, the probing investigator would spare the creature sentience of its physical violation and nullify the charge of cruelty against himself. Prying into interacting organs of living animals or peering through their window-like bodies were ignoble enterprises if, and only if, the target beast was sensible of its inspection.

Whatever we make of Hooke's tortured ethical reasonings, the crucial point is this: by the early years of the Restoration, experimentalists in Oxford and London had come abreast of a conflict between a desire to see into the very depths of organic life and a repulsion at the stain caused by indulgence of such morbid vision. In the first instance, I think the peculiar intelligence materialized in the paper micrometer that Hooke crafted during the late fall of 1667—a moment chronologically coincident *precisely* with the resumption of his experimental vivisection trials after the disruptions caused by London's plague and fire—derives from its ability to address that conflict, to transform its menacing dynamics into a puzzle.[53]

Puzzling

Like many acts of cunning, Hooke's cut-and-pasted paper micrometer came together through some quick-witted opportunism.[54] On October 24, 1667, approximately three weeks before the patched print went to press, John Collins, mathematician and then-recent Fellow of the Royal Society, donated a collection of books to the institution's library. Among this donation (and signed by Collins himself) was an incredible specimen of anatomical representation: Johannes Remellin's *Catoptrum Microscopium* (1639).[55] In a series of prints of almost incomprehensible complexity, Remellin's engraver, Stepan Michelspacher, had depicted full-length male and female bodies as figures that a beholder could open in a "paper dissection" to reveal upward of eleven layers of printed cut-and-pasted viscera[56] (fig. 2.8*a*, *b*). Given the highly public nature of Collins's donation and the bibliophilic Hooke's intimate connection to the Royal Society library, it is impossible to escape the conclusion that Hooke would have become aware of Remellin's work in late October 1667.[57]

Before exploring exactly how Hooke put Remellin to work, the immediate chronology of events needs to be recapitulated. It reads this way: in the spring of 1667, Richard Towneley sends word and then images of his newly revamped micrometer to London, contesting the claims of French rival Adrien Auzout. Hooke takes keen interest in Towneley's device; by the early fall of 1667, he produces a pencil-and-ink drawing depicting the micrometer at close range with its interior mechanics concealed, likely based on firsthand knowledge of the instrument. With the onset of the cooler autumn weather preferred across Europe for anatomies, Hooke simultaneously recommences a program of fascinating but repulsive vivisection experiments, just as Johannes Remellin's illustrated volume is publicly donated to the Royal Society. Because Hooke is already conceptually equipped to think of beasts as machines, the arrival of Remellin's foldout depictions of human anatomy in late October 1667 suggests a way to turn those terms around, to represent machines after the manner of organic bodies.[58] By the second week of November 1667, then, Hooke designs a foldout print inspired by Remellin, having already cut apart his existing machine drawing and repaired it with a pictorial patch. Not only did this resulting paper object show Towneley's device open and closed at the user's pleasure, but it could, as it were, nobly gratify a morbid, penetrative desire to see. If this is a desire that Hooke possibly shared with contemporary painters in the northern European

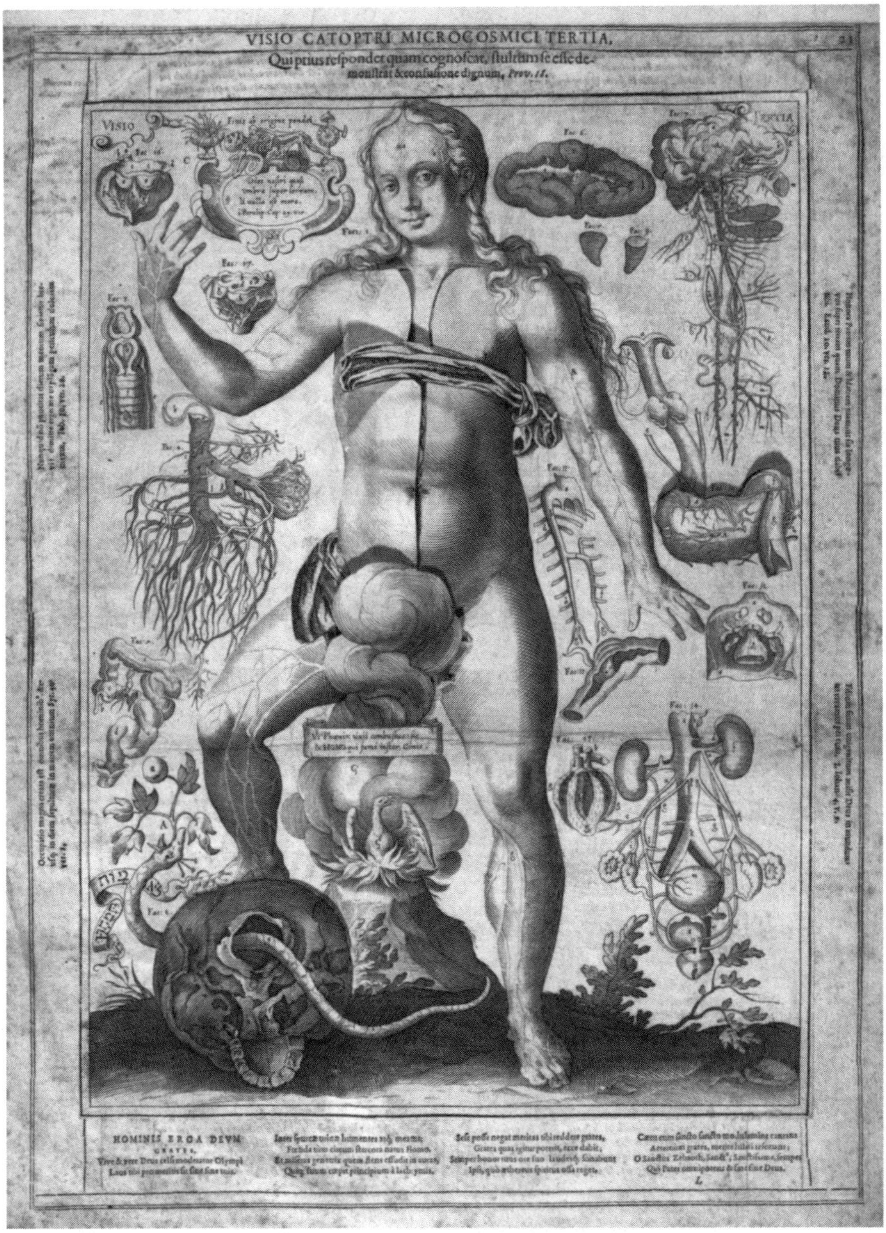

FIGURE 2.8*a* ∗ Stephan Michelspacher's engraving of female anatomy; in Johannes Remellin, *Catoptrum Microscopium* (Ulm: Johannis Gorlini, 1639), n.p. © The Royal Society. Reproduced by permission.

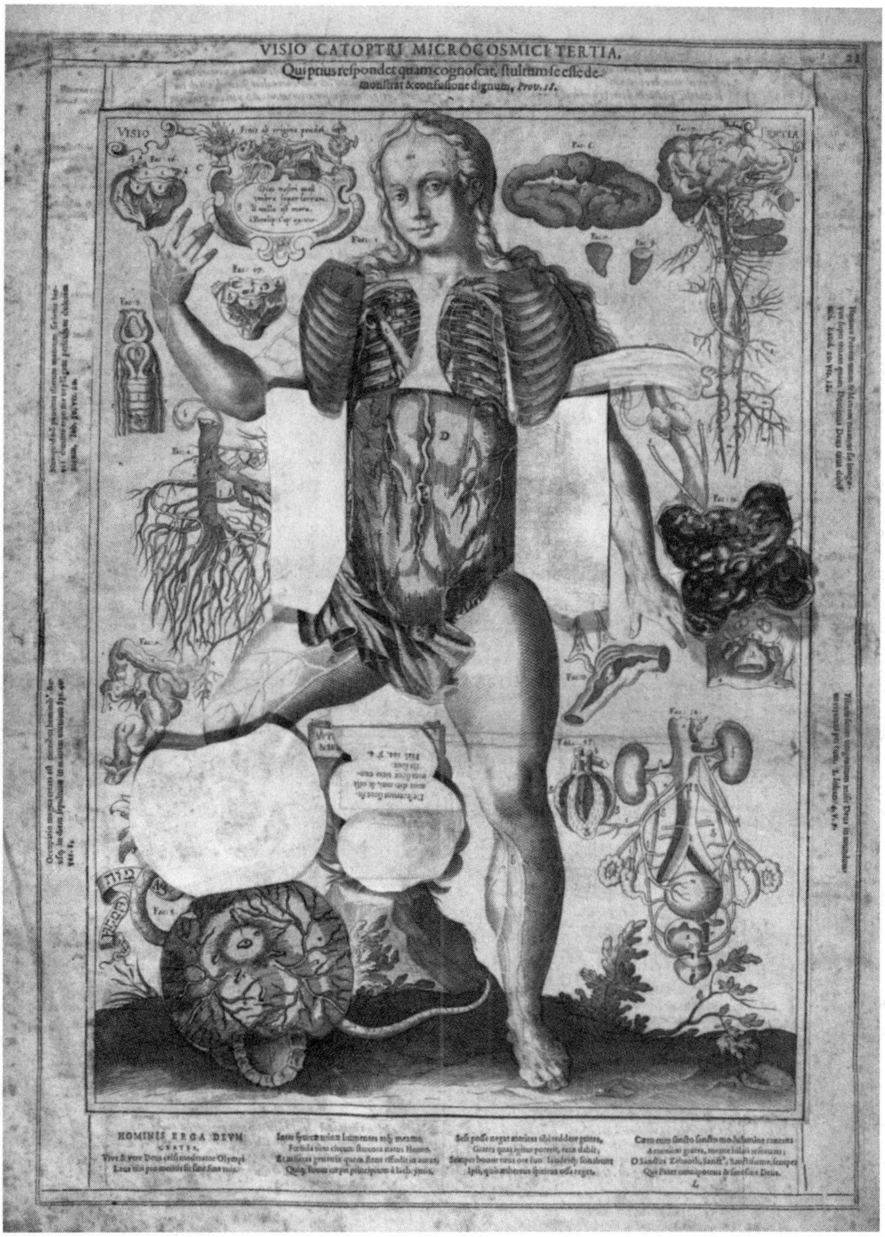

FIGURE 2.8*b* * Michelspacher's engraving of female anatomy opened; in Johannes Remellin, *Catoptrum Microscopium* (Ulm: Johannis Gorlini, 1639), n.p. © The Royal Society. Reproduced by permission.

tradition, it is one that had *explicitly* haunted his colleagues' experimental work since the later 1650s. Responsive to broader artistic and experimental currents yet reducible to neither, the ingenuity of Hooke's object devolves, in this first instance, on its ability to liquidate visual and ethical conflicts through opportunistic, conceptual flux—to solve a puzzle. Pilfering shamelessly from Remellin and sacrificing the physical integrity of his own micrometer drawing, Hooke crafts an object that could simultaneously represent its mechanical target while also delivering the provocative visual fantasy of peering inside bodies, albeit transferred into nonthreatening, paper form.

Now, as theorists of various intellectual stripes remind us, modern science—that enterprise at least traditionally seen as inaugurated by Robert Hooke and his colleagues—is veritably defined by its ability to reduce the buzzing, blooming chaos of the natural world to radically simplified forms.[59] The efficacy of "normal" or paradigmatic science is contingent, so Thomas Kuhn famously put it, precisely on its targeting of only those problems reducible to the conditions of the puzzle. Because solutions to such problems are guaranteed to exist, the investigator can confidently proceed in the conviction that "if only he is skilful enough, he will succeed in solving a puzzle that no one before has solved or solved so well."[60] Hooke's paper object is puzzle-like in this sense.[61] It is innovative neither in its conceptualization of the machine/animal analogy nor in the practical terms of how to represent interior and exterior views of a target simultaneously; those solutions were secured by the standard works of Descartes and Remellin, respectively. Yet Hooke's paper micrometer concretizes those conceptual and practical models in an elegant and highly efficient way such that what had been unmanageable and obstructive in two discrete domains becomes luciferous when combined. As with a Kuhnian puzzle or other pursuit where the existence of a successful resolution is guaranteed from the outset, moreover, the criteria for assessing these intentional, "positive" ends of Hooke's object would effectively be aesthetic. What matters is neither the intrinsic interest of the question to be solved nor its originality as a problem, but the economy, ingenuity, and even grace with which the solution is achieved in practice.[62]

Noting these elegant means to puzzle-solving ends and, at least as presented thus far, the object's sheer uselessness for ameliorating the experimental situation, might we not simply think of Hooke's paper micrometer as an *artistic* work of protoparadigmatic science?[63] After all, Hooke's pictorial training can surely be witnessed in the drafting and fabricating of his paper micrometer, and no shortage of critical commentary has asserted an aesthetics of "plain style" as a crucial desideratum shared between experimental philosophy and the arts of seventeenth-century England.[64] Albeit in different terms, categorizing Hooke's paper micrometer as an art object will indeed find still further support once we take into account the varieties of conceptual play to which it quickly gave rise. Nonetheless, as I will argue in the next section, what is most interesting—and most instructive to the student of wicked intelligence—about Hooke's cut-paper creation is the traffic between puzzle-solving and imaginative prompting enabled through it. And when weighted against the formidable power that Hooke and his contemporaries attributed to artists and their works, such a light, labile object and the thoughts it generated will need to be apprehended by altogether different criteria.

Aesthetics of Entrapment and the Artist's Intelligence

If a puzzle-solving capacity of Hooke's paper micrometer may be allowed, attributing actual intelligence to such an artifact seems an entirely more precarious business. In the mid-1660s, Hooke himself was using relations between animals and machines to warn against that very move. "When we find Flies swarming, about any piece of flesh," he observed, we should not explain their behavior by positing any insect rationality. Instead, we should credit their ability to detect food or identify a suitable habitat to what Hooke called "the excellent contrivance of their machine, to excite and force them to act after such and such a manner."[65] Citing the cunning ostensibly possessed by a huntsman's trap, Hooke elucidated the point this way:

> In the contrivance killing a Fox or Wolf with a Gun, the moving of a string, is the death of the Animal; for the Beast, by moving the flesh that is laid to entrap him, pulls the string which moves the trigger, and that lets go the Cock which on the steel strikes certain sparks of fire which kindle the powder in the pann, and that presently flies into the barrel, where the powder catching fire rarifies and drives out the bullet which kills the Animal; in all which actions, there is nothing of intention or ratiocination to be ascrib'd either to the Animal or the Engine, but all to the ingeniousness of the contriver.[66]

We mistake effect for cause, patient for agent, when we assign intelligence to the trap rather than to the trapper, to the beast-machine rather than the divine creator who contrived it. Yet few places in Hooke's work frame the stakes of such slippery entanglements more potently than his discussions of artists and their works. This nexus is important for the conceptual concerns I want to elaborate, as the figure of the trap has loomed large both in recent writing on the technologized image and in studies of early modern art objects.[67] In introducing the views of Hooke and his colleagues, we can see how their readings of the trap-like products of early modern artistic intelligence possess a hold on mind and body instructively *different* from that of the paper micrometer. If, as Hooke seems to suggest, art is a trap, the objects made for and by the experimental intelligence always retain a trapdoor.

Cognizant as only a true iconophile can be of the power of images, Hooke frequently cautioned his colleagues about art's dangerous capacities to ensnare and to mislead the active imagination.[68] "Instead of giving us a true Idea," he warned in a lecture from 1694, prints in travel books were all too frequently the baseless products of what he called "Mr. Engraver's Fancy."[69] These bogus images "misguide our Imagination, and lead us into Error, by obtruding upon us the Imaginations of a Person, possibly, more ignorant than our selves."[70] In a telling registration of what one scholar has called the special "authority of print" in early modern Europe, Hooke also extended his reservations to works of art more broadly conceived.[71] "Pictures of things which only serve for Ornament or Pleasure," he adjures in the *General Scheme* (circa 1666–68), "is rather noxious than useful, and serves to divert and disturb the Mind, and sways it with a kind of Partiality or Respect."[72] Even the comparatively staid visual experience of literary art presented a menacing aspect to the experimental imagination.[73] As Hooke implores in the preface to a history of Ceylon penned by a former prisoner in South Asia, "Read the Book it self, and you will find your self taken Captive indeed, but used more kindly by the Author, than he himself was by the Natives."[74] Where contemporary Fellows of the Royal Society such as John Evelyn could celebrate graphic art's optical deceptions as "the *Magic*, and innocent *Witch-craft* of lights and shades," Hooke perceived crafty manipulations that were far less innocent, far more capable of deluding and ensnaring the experimentalist.[75]

Yet, as with the devious tale of painterly entrapment with which he framed his own life narrative (so we saw in the introduction), Hooke possessed a curious, persistent fascination with artistic power. Opening an undated lecture on the effects of music, he sets out a beguiling tale of artistic enchantment. "A certaine excellent & skilfull Musitian," we read, had circulated tell of his musical powers far and wide. When summoned to perform before the Danish king, "the artist . . . did by various notes, Strains, & moods, soe excite, <forme> & change the passions not only of the King, but of all the auditors that they were sufficiently satisfied both of the Power of Musick, and of this artist's skill."[76] Not content at simply performing his ability, the musician uses his royal audience to demonstrate the true potency of his art. Music and musician "soe powerfully wrought upon the attentive king, that it put him into a violent Phrensy; Wherein he not only fell upon his dear freinds, & councellors, beating and kicking them, but went on to kill severall of them."[77]

Nearly contemporaneous with the paper micrometer, Hooke was attributing similarly efficacious powers to visual artists and their techniques of optical projection. Through a magic lantern-like contrivance, he warns, priests and other shady dealers could summon "Apparitions of Angels, or Devils, Inscriptions and Oracles on Walls; the Prospect of Countryes, Cities, Houses, Navies, Armies; the Actions and Motions of Men, Beasts, Birds, &c. the vanishing of them in a cloud, and their appearing no more after the cloud is vanisht."[78] Seeing such wondrous displays would be no less overwhelming than hearing the Danish musician's bewitching performance. "Spectators, not well versed in Opticks," Hooke observes, ". . . would readily believe them to be supernatural and miraculous, and would as easily be affected with all those passions of Love, Fear, Reverence, Honor, and Astonishment, that are the natural consequences of such belief."[79] With passions stirred irresistibly, the spectator of this contrivance—what one French reader of Hooke's text called "l'art de fair des apparitions peintes"—would be entirely entrapped, raw putty to the artist's wicked manipulations.[80]

Robert Hooke was, of course, hardly alone in seeing early modern artists and art objects as capable of wielding formidable, malevolent power over the mind and body of the beholder.[81] Artists themselves delighted in trafficking such sentiments. Witnessing copyists torturing themselves below his Sistine ceiling, Michelangelo reputedly exclaimed, "Oh, how many men this work of mine wishes to destroy."[82] What is more, a recent, influential tradition of art-historical interpretation has done much to emphasize how the experience of beholding early modern art constituted a violent assault by which the stupefied observer would be utterly transfixed. Benvenuto Cellini's *Perseus and Medusa* (circa 1545–54, Piazza della Signoria, Florence), by Michael Cole's reading, crafts a force field in which beholders and works by rival artists were petrified.[83] Louis Marin casts this "Medusa effect" as a general symptomatology of Caravaggio's blinding art, while Michael Fried has plotted the origins of his absorptive tradition around that "poke in the eye" from which a stupefied beholder would be incapable of looking away.[84]

Hooke's ambivalence toward such boggling works of stupefying artistic power could perhaps be seen as manifesting professionalization anxieties as experimental philosophers arrogated and repressed their "artisanal" roots.[85] But I think these overwhelming aesthetic entrapments need to be seen as fundamentally different from the ontology and epistemic function of an artifact like the paper micrometer Hooke constructed and deconstructed in 1667–68. Crucial to the student of wicked intelligence is reckoning with the relative weakness—better, the *lightness*—of Hooke's cut-paper object and the cognitive advantages its unsublimated fragility could offer to experimental philosophy's devious cleverness.[86] Bringing that epistemological slipperiness now to the fore, I will conclude by suggesting how Hooke himself seems to have attempted to theorize that objecthood and light cognitive force through nothing less than universal gravitation.

Prompting, Paper

Throughout Hooke's writing, bare beholding is repeatedly staged as debased brutality. Human beings are elevated above all other creatures, so we read in the opening lines of *Micrographia*, because "we are not only able to *behold* the works of Nature . . . but we have also the power of *considering*, *comparing*, *altering*, *assisting*, and *improving* them to various uses."[87] If this bricolage or "thinkering" with materials is clearly instantiated in the paper micrometer's cut-and-pasted form, a more capacious approach is needed to elucidate the radical availability of such an artifact to the labile reimaginings that Hooke saw as a uniquely human prerogative.[88] To this project, the "pretense theory" of representation advanced by philosopher Kendall Walton is especially useful. In Walton's terms, the paper object made by Hooke might be understood as a representation insofar as it prescribes a game of make-believe.[89] As a "prop" that possesses a "principle of generation," the paper micrometer authorizes a public and socially comprehensible game in which beholders pretend that inked markings on cut-and-pasted paper are Richard Towneley's telescopic micrometer.[90] Privately (and perhaps unconsciously) for Hooke and his close affiliates, I have suggested, the paper micrometer also functioned as a game of make-believe in which opening and closing a haptic representation solved by displacement the puzzle of how to manage the morbid, conflicted desire to see systemic interactions inside living bodies. But while his appeal to game-playing felicitously overlaps with these concerns for puzzle-solving, Walton's approach is particularly instructive for understanding the range of ad hoc imaginings by which Hooke's authorized object was quickly swamped.

What I mean is this: in late February 1666, Hooke had presented a "very pretty" lecture to the Royal Society on the interrelated trades of papermaking and felting.[91] Assisted by his own heavily annotated drawing of a felter's workshop (fig. 2.9), he explained how workers in both trades processed and laboriously refined fermented cloth rags.[92] Depicted in procedural sequence from left to right, Hooke's drawing shows a craftsman oscillating raw lamb's wool with a bowstring on a slotted table called a hurdle.[93] As the artisan in the central foreground mans the basoning table, where thick batts of wool are compressed together and heated from below by a pan of burning charcoal, a workman at center right squeezes excess urine and wine lees from hot, matted wool on the "waking plank." A comrade in the distance fits the felt to a hat mold, while we see the hat brim beaten on "their Block which they call

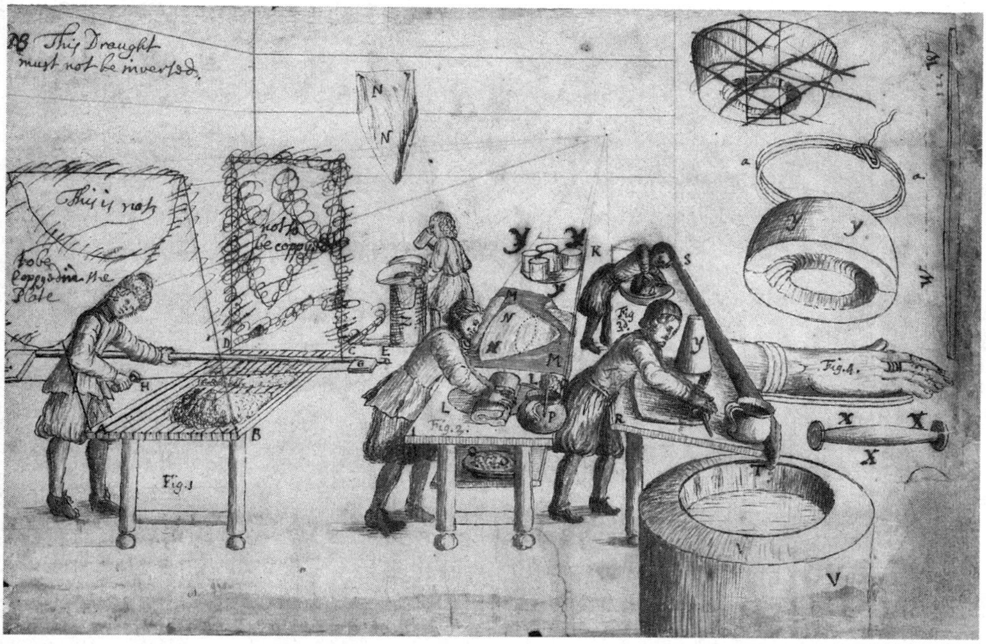

FIGURE 2.9 * Robert Hooke's ink drawing representing processes of felt hat-making (circa February 1666); Royal Society Classified Papers 20:96. © The Royal Society. Reproduced by permission.

a fryer" at far center. Signaled by the vanishing point at upper right, Hooke has clearly tried to integrate this scene into a singular, perspectival picture evocative of Abraham du Bosse's contemporaneous workshop views.[94] Yet the copious density of artisanal information he had amassed ruptures spatial form, spilling outward into a sequence of disproportionate details at right and mandating extensive corrections within the pictorial world of the workshop drawing at left.[95]

Hooke was clearly struggling to bring this plenum of artisanal knowledge about felt and paper to order *on* paper in early 1666. But the condensation of animal and machine bodies at work with the cut-and-pasted micrometer in late 1667 soon made the object and substrate of his studies newly available as a prompter for "toying with, exploring, trying out new and sometimes farfetched ideas."[96] In a portion of a major theoretical text likely composed circa 1668, Hooke encouraged his reader to indulge the following piece of make-believe: "Suppose we compare Paper or Hats with the Skin of an Animal."[97] Stipulating shared properties between the two systems, Hooke then uses exploration of the papermaking process he had studied and depicted so intensively in 1666 to illuminate an obscure, taboo target of his long-standing fascination. In making both paper and felt, he writes:

> We find that the Artists endeavor first to cut, grind, or beat into small parts the Materials they are to use, then to make them soft, light, and supple with Water or other Liquor, then to dispose, place, or put it into such Forms or Moulds as they may shape it into what Form they design it, then they there so work and order it, that the Moisture may by degrees waste, and the solid Parts unite more closely together, adding to it such glutinous Matter as may make it fit to stick, hold and grow stiff, firm, and strong together; and lastly, by several ways they smooth and color or beautify the Surface. . . . Now Nature in preparing the Matter that does repair the Skin or Flesh of the Body, seems to proceed much after the same method, the food is chosen, then ground or chewed, then digested or brought into the Form of a Milk or Froth, then dispersed over the Body, by degrees thickened, and lastly polished.[98]

Students of technology have often aligned "dreaming with" materials and making references external to the practiced craft as damning hallmarks of the amateur.[99] To Hooke's way of thinking, though, paper was not only a useful surface on which to prop public, authorized games of make-believe and to resolve semiprivate visual/ethical puzzles; it was simultaneously a made thing whose production history, physical properties, and even poetic dimensions could prompt philosophically generative, ad hoc imaginings of their own.[100] Reciprocally, by taking his graphic material back to its dark, wet roots, paper had also enabled Hooke to *change the question*. Instead of acting only as dry, white support to pictorial visualization of animal/machine interior systems, paper-as-made-thing had now become a conceptual model for exploring how organic tissue comes into being in the first place. By repurposing the material constraints from which his picture-puzzle of late 1667 had been fashioned, the binding snare had been slipped. A trapdoor had opened in the liquidated floor of the trap.

I think that precisely this persistent suggestiveness of paper's materiality and the consequent drift of active, experimentalist imagination away from the central, authorized representational games played on it need to be seen as potentially advantageous epistemic features of Robert Hooke's paper object. One way to press this point might be through appeal to recent work in analytic philosophy of science on the "autonomy" and nonmimetic, fictionalized dimensions of scientific models. Indifferent to their putative creators' intentions, this literature argues, models are subservient to neither theory nor observation, but gain epistemic value by resistant mediation between them.[101] "Model systems are interesting," writes philosopher of science Roman Frigg, "exactly because more is true of them than what the initial description specifies; no one would spend time studying models if all there was to know about them was the explicit content of the initial description."[102] Or, as Tarja Knuuttila and Atro Voutilainen have pithily put it, "A model has an existence of its own. For this reason we cannot be totally in charge of it, however purposefully fabricated it may be."[103]

Now, exactly *how* it is that scientists learn from these fabricated imaginings and make those unintended discoveries is currently a matter of lively debate in the philosophy of science.[104] But the analytic tools forged in those debates can help us to reconsider the peculiar nature of historical artifacts such as the paper herring and Hooke's model micrometer, as well as the species of intelligence they encourage. First, unlike overpowering artistic works that would mystify their material means and entrap stupefied beholders, these humble paper objects command a fundamentally different kind of hold on the eye, mind and body of the experimental beholder. After all, what had struck an influential early modern theorist such as Giorgio Vasari as "absolutely astonishing" in Michelangelo's marble sculpture was that "a stone, formless in the beginning, could ever have been brought to the state of perfection which Nature habitually struggles to create in the flesh."[105] Puzzling through technical competition, visual desire, and ethical conflict, the transformation of materials enacted in the game of make-believe sustained by Hooke's micrometer was of an altogether lighter kind. Never fully transformed like that massy miracle of representational semblance Vasari saw in Michelangelo's marble, the micrometer's paper remained available both for ongoing physical manipulations by cutting and pasting, as well as for guiding ad hoc, imaginative promptings. Intended publicly as a representation of a machine while deployed semicovertly as a substitute visualization of animal bodies, Hooke's object opened a cognitive space where the materials for graphic representation became available as resources for remodeling the targets *of* that representation—where thinking on paper could be pursued by thinking with paper, while those materials of thought subtly altered the shape of the philosophical questions being asked.

But how, it might here be asked, can we possibly hope to reconcile procedures like these with what we know of Restoration experimentalism? Can they be squared with experimentalists' proclaimed methods of collective induction?[106] Should thoughts like these be placed at the very forefront of a rigorous mechanizing of philosophy?[107] Or need we see them as lingering residue of modes of "analogical" reasoning rooted in an *episteme* of Renaissance magic?[108] What I propose in conclusion is that Hooke himself sketched outlines for the most compelling answer, guided by his own innovative thinking on celestial mechanics and attraction at a distance.

"I have often wondered," Hooke announced to the Royal Society in a lecture from late May 1666, "why the planets should move about the sun according to Copernicus's supposition, being not included in any solid orbs (which the antients possibly for this reason might embrace) nor tied to it, as their center, by any visible strings."[109] Hooke's question presumed Descartes's second law of nature, whereby all bodies move in rectilinear paths such that rotating bodies will always tend to recede from an orbital center.[110] But to yield the curved paths of planets orbiting in the Copernican system, Hooke continued, rectilinear motion must be inflected by "an attractive property of the body placed in the center; whereby it continually endeavours to attract or draw it to itself."[111] To elucidate that claim, he proposed a material model with his aforementioned "visible strings" very much in play. Hanging a wooden ball attached to a conical pendulum from the ceiling of the Royal Society's meeting room at Gresham College, he demonstrated how the swung bob produced different curved motions depending on the tangential velocity at which it was released. In this model scenario, the swinging ball was to represent an orbiting planet, the action of the earth's gravitational force would constitute the attractive influence of the Copernican sun, and the whole enterprise was to be mathematically reconciled with general mechanical principles—although, as Hooke put it mildly, "all the appearances of it are not exactly the same."[112]

Contemporaneous with his work toward the paper micrometer, this lecture and trial of orbital dynamics are well known to historians of science insofar as they constitute key moments in Hooke's contribution to the foundations of Classical mechanics. Following on from those essays of the mid-1660s, he advanced the rudiments of a "System of the World" in lectures from 1670 (published in 1674 as *An Attempt to Prove the Motion of the Earth by Observations*). Conceiving all celestial bodies as exerting mutually attracting force on themselves and other entities within the "sphere of their Activity," Hooke stipulated that bodies will move in inertial rectilinear motion until deflected into curves by another body's influence.[113] That attractive force acts not at a constant rate, as contemporaries such as physicist and physiologist Giovanni Alfonso Borelli had contended, but as a function of "how much the nearer the body wrought upon is to their own Centers."[114] Publishing in the year after Christiaan Huygens had made stipulation of the outward-receding tendency he called *vis centrifuga*, Hooke was unable to articulate the rate at which his centripetal, attractive force would vary with distance, claiming that it was "not yet experimentally verified."[115] Yet if he likely possessed that relation by the spring of 1676, Hooke clearly knew it by the first week of January 1680. It was then that he famously reported to Isaac Newton, "The Attraction always is in a duplicate proportion to the Distance from the Center Reciprocall."[116]

How exactly did Hooke acquire knowledge of this principle? When did he learn that gravitational attraction operates at "a duplicate proportion to the Distance from the Center Reciprocall"? And had he, in that fateful letter of early January 1680, inadvertently given Newton a tool crucial for understanding celestial mechanics and universal gravitation? An erudite literature in the history of science has explored these questions in detail, seeking to adjudicate between the rival claims advanced in the furious priority dispute that erupted between Newton and Hooke on the eve of the publication of Newton's *Principia* in 1687.[117] Current views are divided as ever. Far from supplying Newton with a key mechanism for conceptualizing curved paths outside the supposedly obfuscating notion of centrifugal force, Domenico Bertoloni Meli contends, Hooke's thinking on curvilinear motion circa 1680 was actually *guided* by Newton. By contrast, Michael Nauenberg asserts, Hooke presciently and legitimately arrived at the inverse-square relation through imaginative strategies of model-based reasoning, strategies perhaps not unlike his adventures with the paper micrometer.[118]

However these debates are understood, my proposal is that by following Robert Hooke's own ingenious deployment of mutually attractive action at a distance, we can gain unique insight into the workings of experimental intelligence and the agency of artifacts such as the paper micrometer to it. In a crucial lecture from the early summer of 1682 that will be considered at length in chapter 5, Hooke articulated an elaborate theory of memory and cognition. Seated deep within the fleshy folds of the brain that had been detailed so lovingly by Christopher Wren in the mid-1660s (see fig. 0.7), Hooke imagined the human soul as the "*primum movens*, the self-moving principle" of the body.[119] The soul fabricates representational artifacts based on sensory information and stores them for its own cognitive use; "idea" is Hooke's name for this category of fabricated, epistemological tool.[120] "The *Supellex* of the Soul," as he puts it, ideas are "the Instruments it makes use of in the apprehending of things or Actions past; and by these it becomes sensible of all that it really knows."[121] How, though, could the soul meaningfully know through making and using its representational idea-artifacts?

To solve this problem, Hooke models that experimental intelligence by appealing to nothing less than orbital dynamics and the inverse-square law of gravitational attraction.[122] Envisioning the soul and its orbiting ideas as the sun and planets in the Copernican solar system, he puts it this way:

> The Radiation of the Soul is more powerful upon Ideas at a nearer than at a further Distance; and their Reaction is also more powerful back again, and that in a duplicate proportion to their Distance reciprocal. . . . And thence it is, that the Memory of things long since done is for the most part very faint.[123]

Though we will return to this strange and fundamentally important lecture, Hooke's basic contention might be summarized as follows. The soul knows by creating, storing, and then radiating energy upon physical artifacts called ideas. As the sun is to its planets, so the soul influences and illuminates most immediately those ideas that are spatiotemporally closest to it. Made and altered by the soul's pulsating radiations, ideas can nonetheless exert a mutual, attractive force *back upon the soul*—and do so at a "duplicate proportion to their Distance reciprocal." What Hooke is proposing here, then, is an account of how those material artifacts fashioned and used intentionally by the experimental intelligence also act necessarily and reciprocally back upon the intellect in unforeseen, unexpected ways. Far from being some extraneous, marginal concern to experimental philosophy, moreover, Hooke has deployed a cutting-edge component of celestial mechanics to explain the physical basis of this cognitive drift and philosophical generativity prompted by artifacts.

These are the terms under which we should see the paper micrometer, I think. The paper micrometer is an idea. It is a fragile, material artifact fashioned by an active intellect to represent visual and tactile sensory reports about a mechanical device. Informed by seeing frightful anatomical sights and ingenious foldout depictions, this printed, cut-up, and pasted-together idea has also been physically revised, thought over, reasoned with, or, as we will see Hooke calling it, "compounded."[124] Yet, made from a lithe, physical substance under extensive influence by an operating intellect, that idea also wields its own weak, reciprocal influence. Governed by a universal attractive force, this decaying thing fabricated from knowable craft processes can alter the direction of the operating experimental intelligence *slightly*. That is, prompted by this idea and its material facture, Hooke had remapped his hard-won puzzle-solution of seeing animal/machine interiority back through the laborious cutting, grinding, compressing, and drying of paper he had studied the previous year to conceptualize the generation of animal tissue. Thus, where the wolf trap in *Micrographia* was only a mechanical extension of the hunter's bloody intent—and where art had been but a devious, transformative trick by which the artist stupefied his boggled beholder-victim—the ideational trapdoor that is the paper micrometer unexpectedly redirects the flow of its maker's intellection as it binds the interior, mental artifacts of experimental thought to the very physics of the stars.

I hope these lines provoke some trepidation in the reader. This unease should stem not from the precarious-sounding connections between celestial mechanics, gravitation, and that diabolical paper herring we have followed into such deep waters of experimental philosophy—as historically we are on solid ground. After all, an oft-told tale in the history of science reminds us how the publication of the period's great statement on gravity (Isaac Newton's *Principia*) was nearly sunk by the monstrous paper fish of Francis Willughby's *Historia Piscium*, a floundering volume then paid to Newton's bested rival in gravitation (Hooke) in lieu of his annual salary.[125] Instead, I think our worries should be directed first toward the model of experimental intelligence we have seen Hooke constructing around and through his paper micrometer. How would Hooke's ideas not become the very Baconian *idols* that the much-vaunted collaborative, inductive empiricism of the Royal Society was reportedly designed to eradicate? "Every one is more prone to exercise Fancy in building paper Theories," so Christopher Wren was then warning after all, "than patient to first pile the unsure Foundation and hew solid Materials out of the History of Nature."[126] These concerns will figure centrally in chapters 4 through 6.

Yet more problematic, I hope the reader will find, is the rapidly vanishing prospect of a meaningful rapprochement between visual art and experimental philosophy. If, as we have seen, Hooke was both deploying and decrying the artistic skills through which he crafted his crafty micrometer, then I have almost perversely drawn from art-historical scholarship to positively distance that object's ontology and epistemic function from art. Treated to such poaching, the art-historical reader might well feel like John Flamsteed, who around 1686 wrote to none other than micrometer maker Richard Towneley in outrage at the creative "borrowing" practiced by his former friend Edmond Halley. The true culprit was not hard to find. Halley, so Flamsteed noted, "is got into Mr. Hooke's acquaintance, has been his long intimate, and from him he has learnt these and some other disingenuous tricks."[127] Taken out to these distant bounds of experimental graphic practice by Hooke's dirty tricks, what can an art-historical reader possibly hope to learn from his wicked intelligence? It is time to stare directly into the eye of that question.

CHAPTER THREE

Pictorial Intelligence

Peter Lely, Experimental Culture, and the Parameters of Painting

DURING THE AFTERNOON meeting of the Royal Society of London on December 19, 1667, founding Fellow Thomas Povey delivered a discourse entitled "A Secret in the Use of Painting."[1] A colonial administrator and courtier, Povey (1614–circa 1705) was also a respected connoisseur of visual art. Through the mid-1660s, virtuosi had flocked to his fashionable townhouse at Lincoln's Inn Fields to admire what John Evelyn called his "vases in imitation of porphyry . . . the inlaying of his closet; above all, his pretty cellar and ranging of his wine-bottles."[2] Povey in fact remains well known among art historians for his important patronage of Samuel van Hoogstraten, a Netherlandish painter of trompe l'oeil wizardry then resident in London.[3] So powerful were Hoogstraten's "strange things" on the minds of contemporaries that Povey's fellow naval administrator Samuel Pepys could but marvel at "how they do delude one's eye . . . methinks it would make a man doubtful of swearing that he ever saw any thing."[4]

Given these artistic contacts and wondrous collections, it is curious to note that the so-called secret Povey delivered to the Royal Society in December 1667 was hardly news at all. In a city positively itching for news and novel intelligence, his "Mystery worthy the Inquisition of a Philosopher"[5] was nothing but the preparation of egg tempera as a painting medium—a technique known to classical antiquity and thoroughly disseminated across Europe since the fifteenth century.[6]

Povey's philosophical defense of egg tempera makes for strange, even uncanny, reading. His narrative pulls against familiar stories about relations between visual art and experimental science until they begin to unravel. At once, Povey lauds the unusual intimacy created between beholder and target of representation by egg tempera's matte surface. "It troubles not the Sight by glistening or glaring," he declares, "which hinders us from looking directly upon Pictures in Oyle, if they be set directly opposite to the Light."[7] Rather than celebrating the oily sheen of domesticated, painted commodities as we might expect from this colonial entrepreneur-*cum*-experimentalist, Povey codes lustrous, pictorial gleam as noxious optical interference that can be advantageously neutralized by experimentally improved tempera technique.[8] Further, where the invention of oil painting suggests an exemplary symbiosis of early modern science and art, Povey turns the situation around completely. The collector explained to the Royal Society how he had experimented with Sergeant-Painter Robert Streeter, keen alchemist Sir Robert Moray, and Dutch emigré artist Hendrik Danckerts in mixing eggs and fig juice to form a chemical solution that changed within seconds: "The Juyce or Milk thereof so prevailes upon the Egge which in its own disposition is Viscous and ropy, that it becomes instantly thin and fluid, as water."[9] It is as if standard histories of science and art are being written backward here. Whereas founding Netherlandish art historian Karel van Mander had traced the invention of oil as a painting medium and its beautiful, glossy sheen (*een schoone blinkende glans*) to Jan van Eyck's alchemical pursuits, Povey narrates collective experimentation on figs' "powerful Salt" as the basis for new interest in this old, literally dull artistic technique.[10]

As optical flatness replaces oil's treasured shine—as an old-fashioned, seemingly outmoded paint medium inspires new experimental inquiry—so Povey concludes his discourse by imagining a surprising relationship between experimental science and visual art. The collector encourages experimentalists to collaborate with "some of the best masters of that art living in London, as Mr. [Peter] Lely, Mr. [Samuel] Cooper, and Mr. Streeter," to fashion a comprehensive history of the painter's art complete with "the Names and Characters of the most eminent Masters; their Manner and several Ways of Working; the Degrees of Improvement of this Art; the Variety of their Colours, their Natures and Mixtures."[11] Although these propositions would seem to manifest the agency of experimental philosophy in advancing late seventeenth-century England's "discovery of painting," Povey actually envisions the balance of power running in the opposite direction.[12] By advancing "this almost Divine Art, which not only imitates but approacheth very deceivably, even to the giving of life itself," he enthuses, the Royal Society could silence its growing chorus of critics who complain "that you discourse and make Flourishes, and subsist chiefly upon what is delivered to you, by them that lived before you."[13] Instead of boot-strapping England's infantile school of painting into social respectability on the back of experimental philosophy, that is, science is to profit through positive, practical association with painters.

We know that Thomas Povey's lecture actually did catalyze collaboration between institutionalized experimenters and Restoration London's leading artists. At that very meeting in late December 1667, the Royal Society founded a committee on painting, reuniting Peter Lely with his former apprentice Robert Hooke.[14] This connection is tantalizing; it promises a single, bureaucratic body forging strong, direct interactions between London's communities of scientists and painters—interactions of a type that have been highly valued by recent interpretation.[15] Here at last, the reader may think, this book will finally leverage a trafficker's intervention to begin telling a cooperative, properly collaborative tale of relations between London's experimental philosophers and visual artists. Unlike Hooke's noisome breaking and ruthless remaking of graphic techniques learned from his apprenticeship to Lely, as examined in chapter 1, or the veritable demonizing of the art object against which we saw Hooke modeling experimental intelligence in chapter 2, painting and science will now emerge as the symbiotic, mutually benefiting partners recent historiography has taught us to expect.

Yet the reverses and surprising recourses of Povey's 1667 discourse are more strongly indicative of the story I will tell. After its promising, motivated beginning in December of that year, the Royal Society's committee on painting sank almost immediately from the historical record. Although Hooke continued to share ingenious technical improvements to painting with his philosophical colleagues through the 1670s—all the while keeping up his friendship with Lely—picture-making was decoupling from and decentralizing within his experimental activities. So we will see in coming chapters, the core concerns of Hooke, Christopher Wren, and other leading London experimentalists were increasingly focusing on the management of burgeoning constellations of far-flung agents and their frequently nonpictorial visual production, which ranged from printed texts to museum artifacts to physical architecture. By the mid-1680s, as I propose in the third section of this chapter, both Hooke and Wren had effectively ceased the pictorial production that had won them such early fame within the experimental community. Therefore, the question needs to be asked: if Restoration experimentalists were so doggedly intent on defacing, denouncing, and ultimately abandoning the pictorial heritage in which they were abundantly learned and skilled, should we not follow their interpretive cue? In other words, why shouldn't we take the course proposed by theorists such as James Elkins and simply let the fetishized relationship between painting and science go?[16]

The approach taken in this chapter is a tactical one. I argue that just as Peter Lely and his period commentators can disclose fundamental concerns shared among Robert Hooke's circles, so the experimental archive provides valuable resources for rethinking Lely, Restoration London's leading painter and a crucial, understudied artist. Therefore, the chapter begins by using Lely's historical "subject pictures" to surreptitiously illuminate the contours of the Restoration-era philosophical beholder, a key figure absent from analysis thus far. The second section turns the interpretive direction around, deploying the Royal Society's experimental research on the visible, bodily effects produced by exotic intoxicants to reexamine a defining stylistic feature of Lely's Restoration pictorial production: the "sleepy-eyed look." Then, taking my interpretive brief from Thomas Povey, I conclude by considering Lely's pictures, neither for the beholder they imagine nor for the rapturous preoccupations they share with experimental culture, but as made things whose complex, collaborative production presages problems and strategies central to the experimental community, especially in the years around 1680. Throughout this chapter, my aim is to begin placing Hooke's visual activities on progressively larger stages—first amid Cavalier culture of the early interregnum, then within Charles II's restored court in the later 1660s, and finally opening onto the hurly-burly of London life in the tumultuous last decades of the seventeenth century.

Seeing through Lely

Lit by a pale, amber light, five female figures collapse before a fountain in a dusky, wooded glen. With their sullied feet and hastily arranged bedding, these protagonists of Peter Lely's *Nymphs by a Fountain* (circa 1650–55) must have tramped through the scraggly, surrounding foliage to arrive at this clearing (plate 1). Such travels have evidently left the maidens exhausted. Lost in slumber, the heads of the three figures at the compositional center align like the stars of Orion's belt, a form echoed and reversed by the stony, male cupids in the fountain behind them. Discovering such artifice buried deep in the heart of nature does nothing to lessen its allure; indeed, the artificiality and contrived modesties of Lely's scene are positively provocative. A sky-blue stole just conceals the central nymph's crotch and a nipple peeks out from a shawl falling from the shoulder of the figure in profile at right, while the powerful arm of the recumbent maiden below reaches back to conceal her buttocks. Through the triangular window of this crooked arm, we can see a protruding wedge of illuminated breast as the figure turns coyly away from the picture plane. More frankly exposed is the supine nymph sprawling backward in the rocky twilight at left, whose upturned chin and unseeing eyes open her splayed body to view. As her sleep is unperturbed by the murmur of water spouting from the mouth of a putto into the cistern nearby, so the fantastic contortion of her body leaves no registration on the wan mask of her features. Cornered and conveniently insensate, these maidens are well within optical range from the vantage point of the inviting, conspicuously vacant sandy wedge in the pictorial foreground.

Painted in precisely the era when young apprentice Hooke would have been mixing pigments in his studio, *Nymphs by a Fountain* is the type of "subject picture" through which Lely established his English reputation after arriving from his native Low Countries around 1641.[17] We know little of the patronage and provenance of this painting, now held by the Dulwich Picture Gallery in London, but it surely hews close to the sensual treatment of mythological subjects widely praised by Lely's period commentators.[18] Endorsing "the *correctness* of his *Drawing*, and the *beauty* of his *Colouring*," admirers such as politician and diplomat Richard Graham reserved special praise for "the *graceful Airs* of his *Figures*, the *pleasing Variety* of his *Postures*, and his *gentile negligence* and *loose manner* of *Draperies*."[19] Poet Richard Flecknoe similarly imagined Lely's dreamy figures as occupying "some other World . . . , / They do not live such lives as we do here: / But rather such as *deathless shadows* do."[20] If seemingly far from the attributes typically assigned to the visual production of Hooke and his circles, what I argue in this section is that Lely's negligent, loose, and dreamy subject pictures can nonetheless shed significant light on a key agent in Restoration experimental intelligence: the philosophical observer.[21]

Let me stress that my argument here is both less and more strange than it sounds. On the one hand, recent work by social historians of science has demonstrated that many of the actual historical beholders of Lely's subject paintings were, in fact, closely connected to courtly branches of Restoration experimental philosophy.[22] By elaborating observational concerns shared between Lely's milieu and the Royal Society, my story implicitly joins with others that have complicated the putative opposition between the new science of experiment and the "decadent" Stuart court.[23] Yet beyond these identifiable historical contacts, the aim of this section is to draw forth what Peter Lely's paintings can reveal about the structuring concerns of experimental beholding at a more fundamental level. Joseph Leo Koerner has powerfully evoked this target in describing the artistic beholder as "something . . . internal to the picture, even if he seems to be standing in our place. He is a 'figure

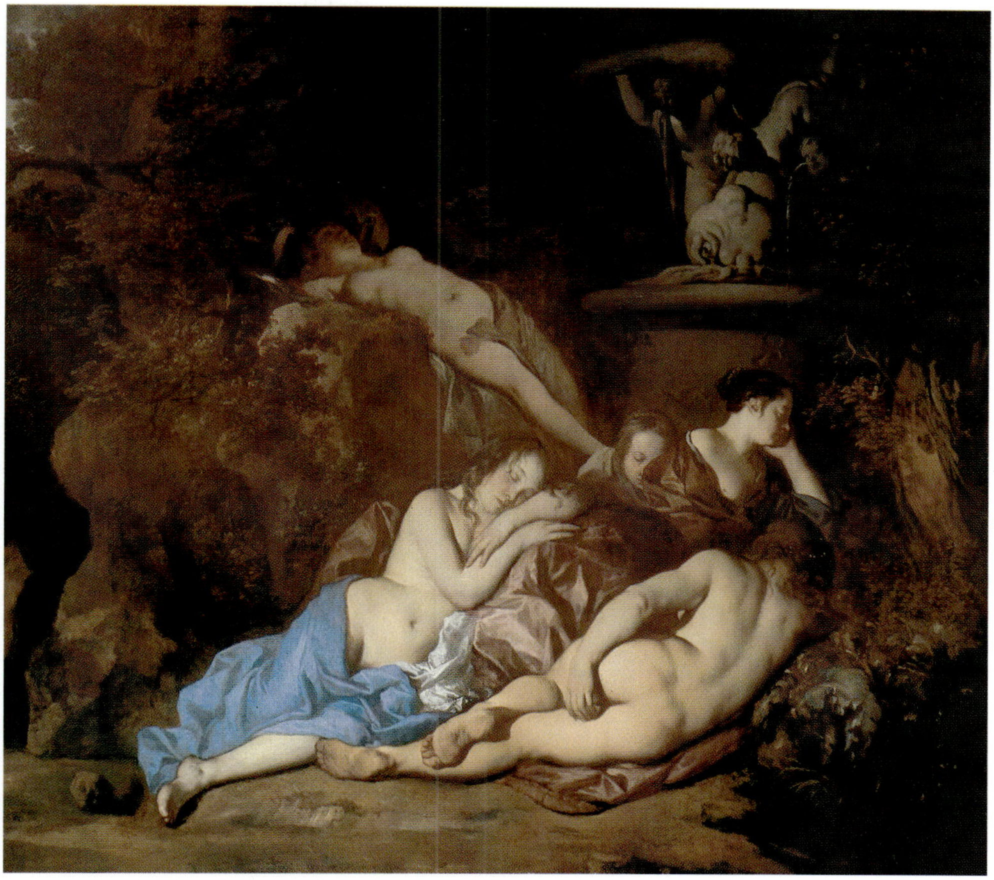

PLATE 1 * Peter Lely, *Nymphs by a Fountain* (oil on canvas, circa 1650–55; Dulwich Picture Gallery, London).

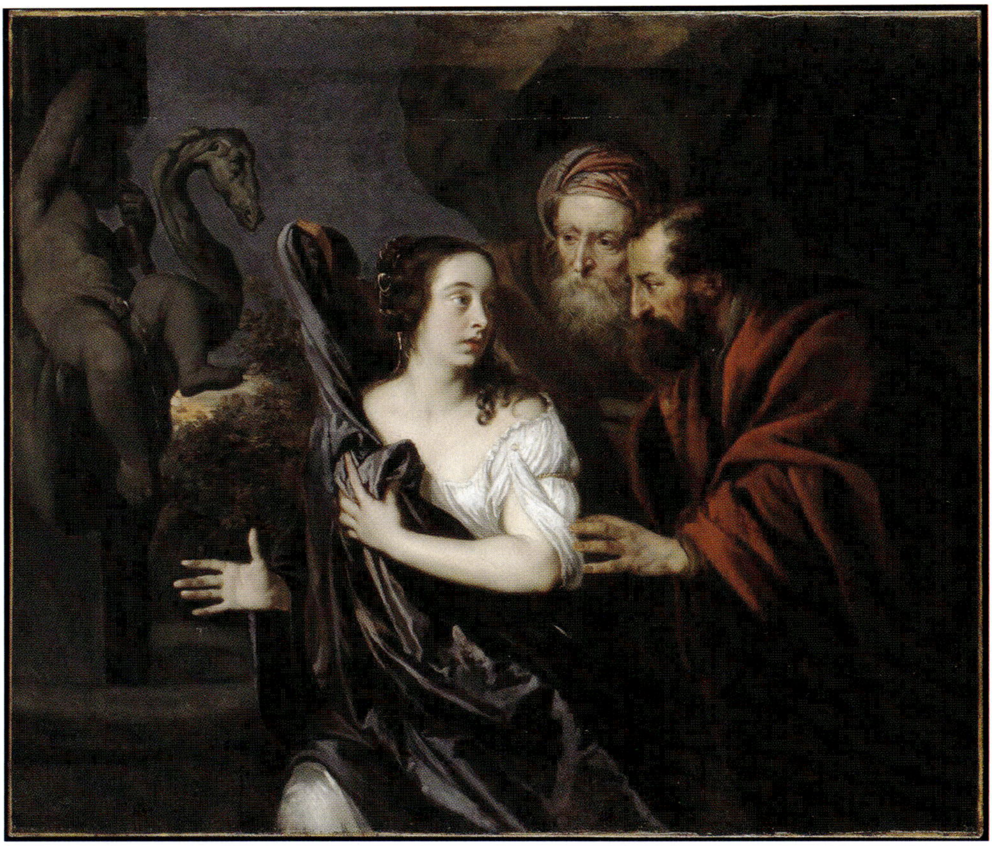

PLATE 2 * Peter Lely, *Susanna and the Elders* (oil on canvas, circa 1650–1655; Tate Britain, London).

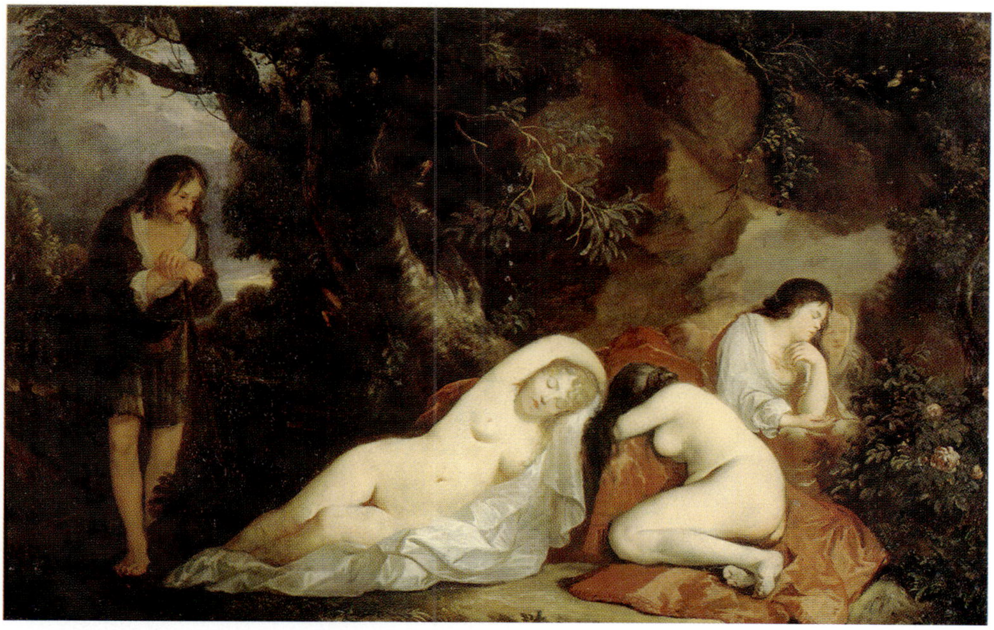

PLATE 3 * Peter Lely, *Cymon and Iphigenia* (oil on canvas, circa 1650; Knole House, Kent). Reproduced by kind permission of the Trustees of the Knole Estate; © National Trust Images/John Hammond.

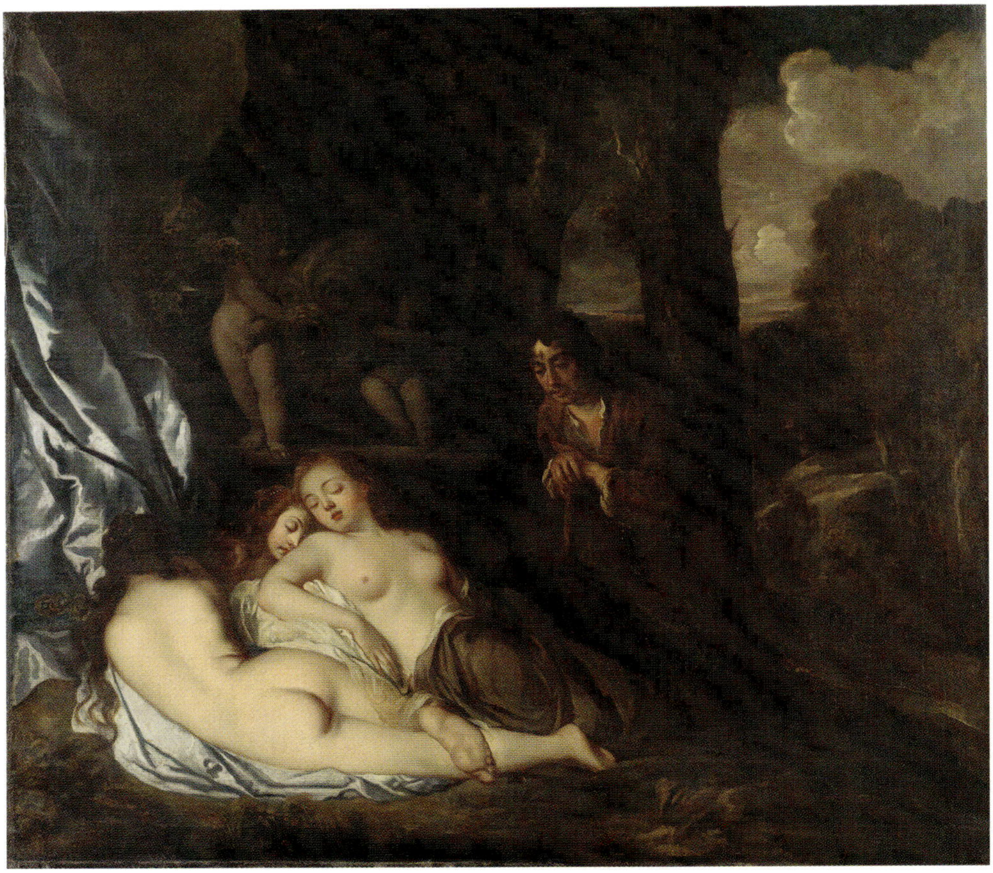

PLATE 4 * Peter Lely, *Cymon and Iphigenia* (oil on canvas, circa 1650; Doddington Hall, Lincolnshire). Reproduced by permission.

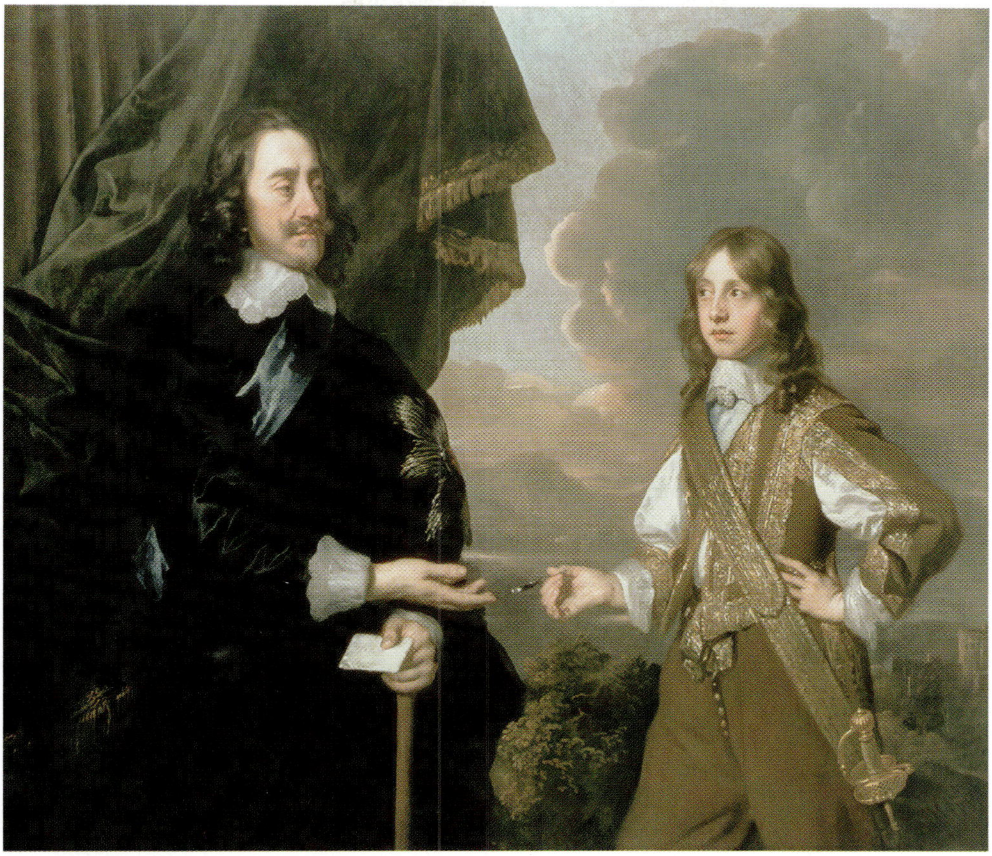

PLATE 5 * Peter Lely, *Charles I (1600–1649) and James, Duke of York (1633–1701)* (oil on canvas, circa 1647). Syon House, Middlesex, UK / The Bridgeman Art Library, BAL 2050. Reproduced by permission of The Bridgeman Art Library.

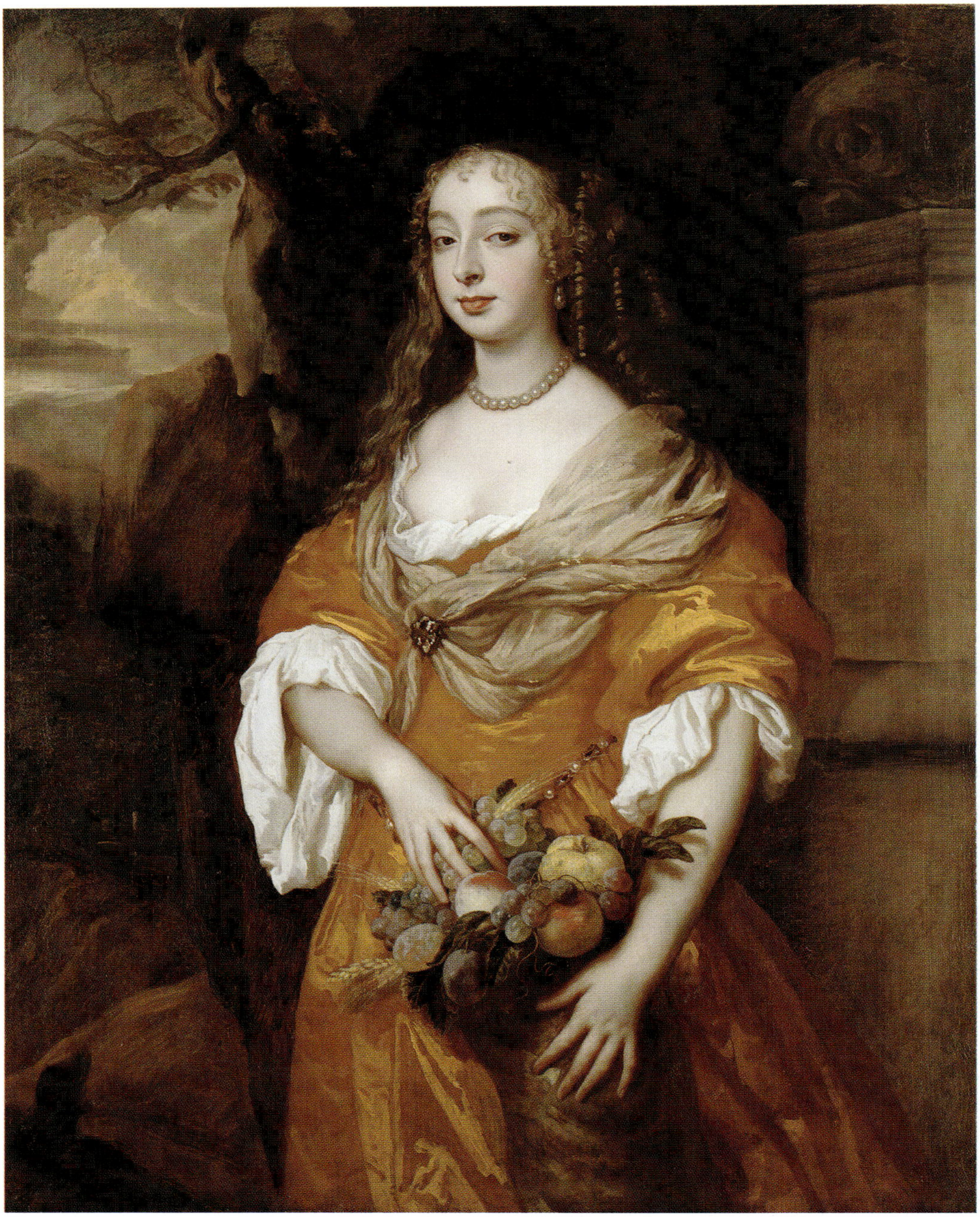

PLATE 6 * Peter Lely, *Jane Needham, Mrs. Middleton* (oil on canvas, circa 1663–65). The Royal Collection © 2011 Her Majesty Queen Elizabeth II / The Bridgeman Art Library, ROC 399051. Reproduced by permission of The Bridgeman Art Library.

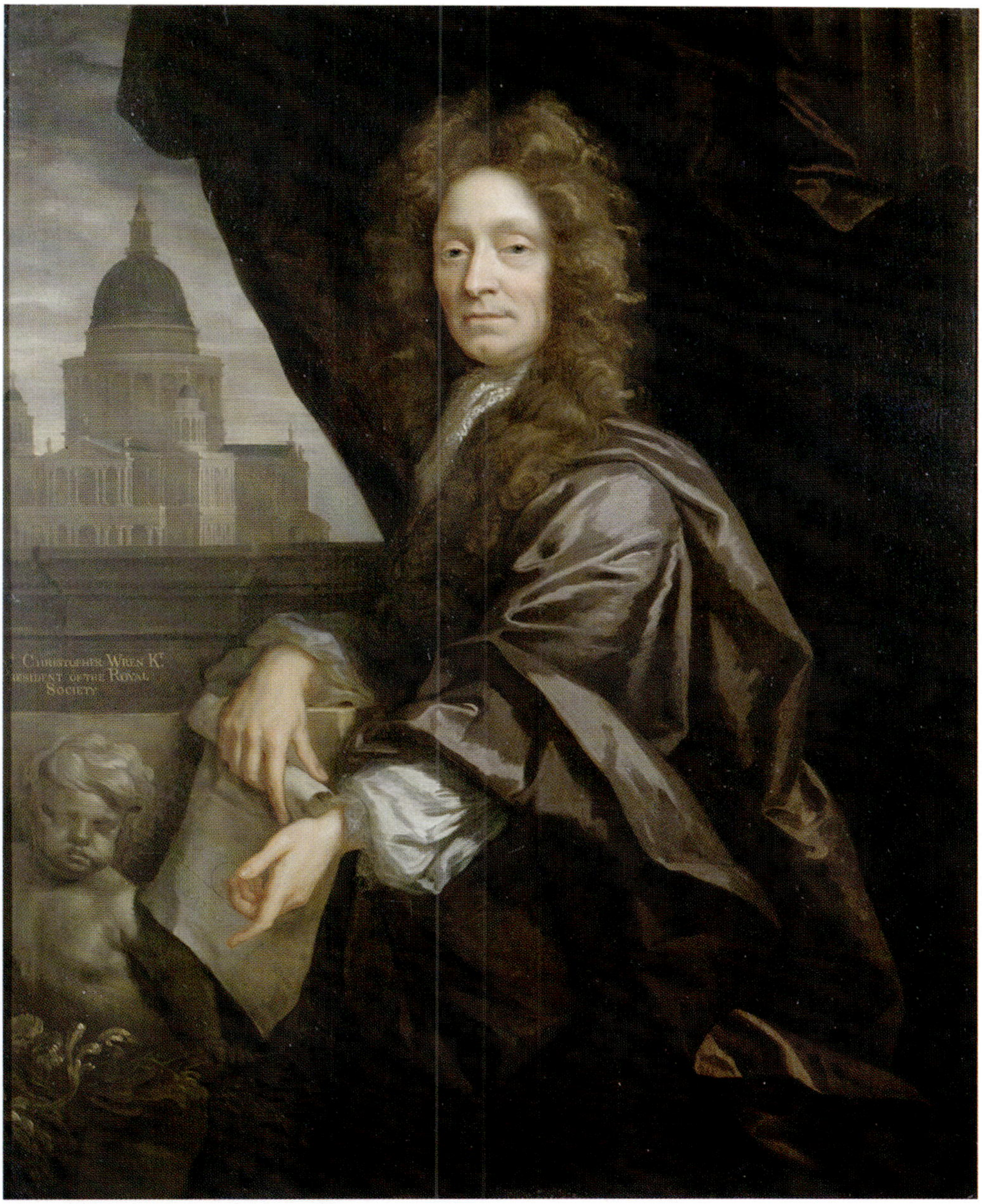

PLATE 7 * John Closterman, *Christopher Wren* (oil on canvas, circa 1695). © Royal Society of London. Reproduced by permission.

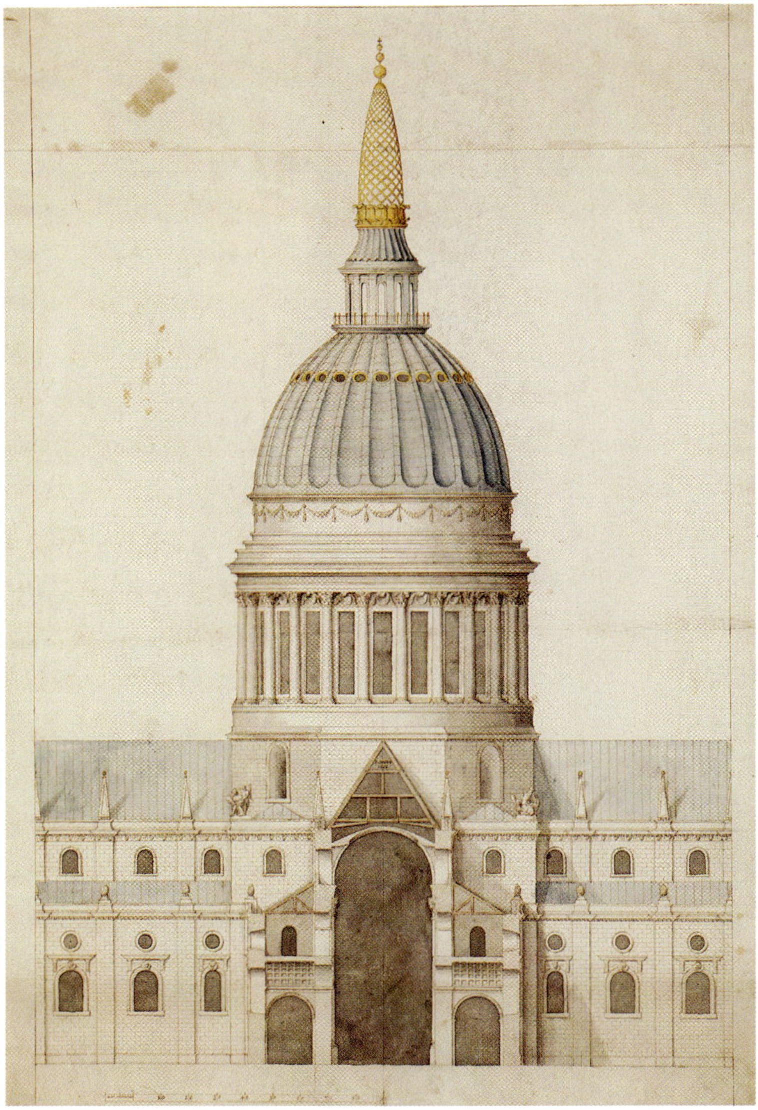

PLATE 8 * Christopher Wren's pre-fire, ink-and-watercolor design for St. Paul's Cathedral, spring 1666. The Warden and Fellows of All Souls College, Oxford, II.6. Reproduced by permission.

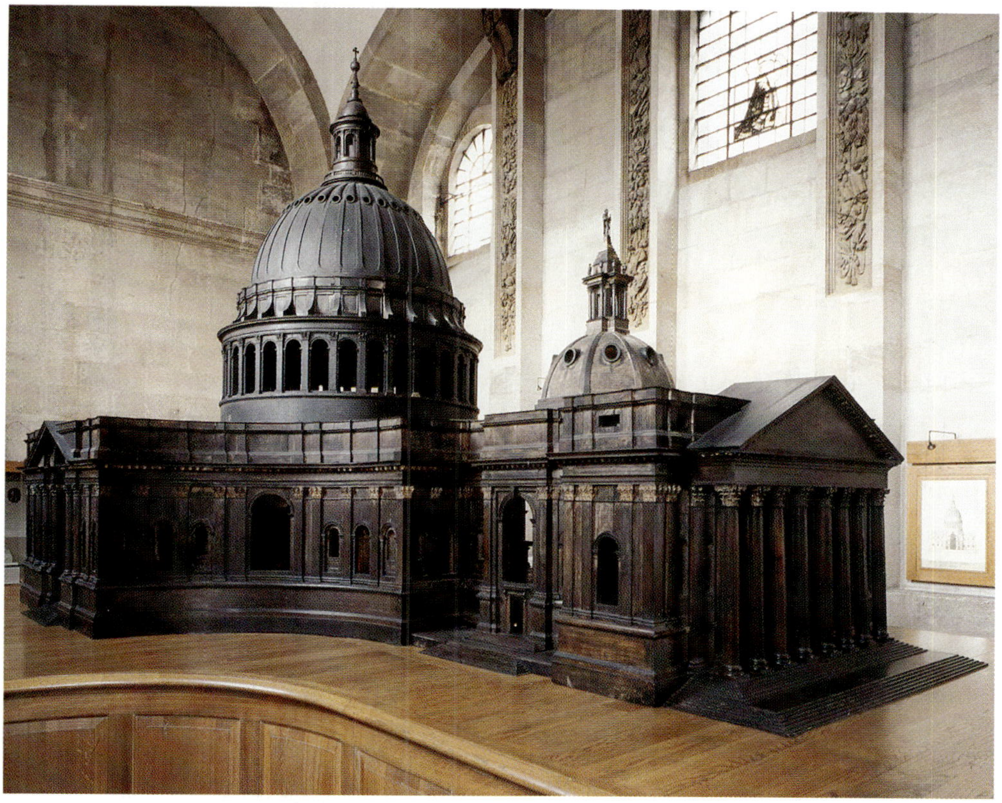

PLATE 9 * "Great Model" of St. Paul's Cathedral (1673–74), fabricated by William and Richard Clere, in collaboration with Christopher Wren, Edward Woodroffe, and others.

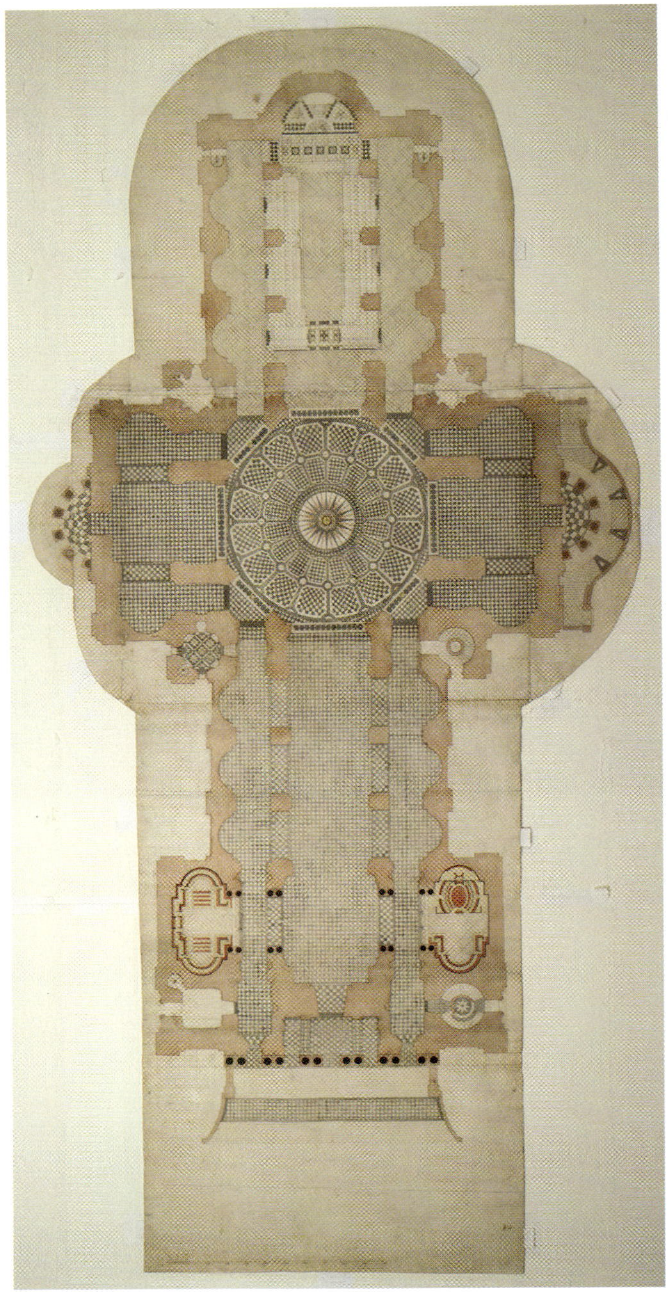

PLATE 10 * Watercolor floor plan by Wren's Office of Works (circa 1709) showing the executed paving of St. Paul's.

of the viewer' implied by the picture's plot: difficult to theorize, but as much present as a fire is when there is smoke."[24] My proposal is that Lely's subject paintings and their critical reception offer resources through which to apprehend anew the contours and constraints of the scientific observer in Restoration London.

Between circa 1650 and 1655, Lely painted a series of pictures on the apocryphal theme of Susanna and the Elders. According to that legendary tale, the beautiful maiden Susanna is trapped at her bath by two lecherous Elders who threaten to charge her with adultery, punishable by death, unless she submits to their sexual demands. Susanna refuses; her virtue is ultimately vindicated, the wicked Elders executed. In the version of Lely's scene now at Tate Britain in London, we see Susanna at half length extending her right hand toward a fantastical fountain (plate 2). Her swiveling shoulders open toward the picture plane as the nefarious Elders slide into the shallow, claustrophobic foreground. Misdirection is their ruse; while the bald, stooping admirer at right caresses the maiden's undergarment and fondles his own beard, his coconspirator forcibly pries her cloak from her body. Positioned on the vertical axis of the canvas, Susanna's body directs the narrative as it unfolds from left to right. Prepared for her private toilette before the fountain at left, she has loosened her hair so that it streams down her shoulder; at the same time, her left arm gathers in her slackening cloak, which falls away from her body as she turns to confront the sensed, but not fully apprehended, assailants at right. Instinctive response is etched into Susanna's hardening visage, widening eyes, and parting mouth.

Among art historians, the Susanna and the Elders theme has often been read as a thinly veiled pretext for voyeuristic lasciviousness.[25] The theme certainly could cater to what Svetlana Alpers has described in a telling gloss of her own influential work as "the ability of the voyeuristic eye, unseen, to capture something or someone unawares which proved so fascinating to the technological and pictorial culture" of seventeenth-century northern Europe.[26] So, is the beholder figured in Lely's subject pictures a voyeur *tout court*? The pictorial sources and consumer purchases that informed Lely's *Susanna* series in the early 1650s are instructive here. Within months of the execution of Charles I in late January 1649, the Commonwealth government initiated a massive auction—the "sale of the century"—dispensing with the deposed monarch's magnificent art collection, which included several depictions of the Susanna theme by prominent Venetian masters. A viewer and a shopper at that sale, Lely then gained access to paintings such as Paolo Veronese's *Susannah and Elders* (circa 1580), now in the Museo del Prado, Madrid.[27] The comparison with Lely's subsequent composition is telling: where Veronese's Susanna firmly claps her hand over her naked breast and upbraids the Elders for their ignominious invasion, Lely's maiden appears still forming her response as her eyes swim imploringly backward.

A still stronger contrast is with the work of Jacopo Tintoretto, author of "A peece of Susanna" that Lely actually purchased from Charles's collection on April 24, 1650.[28] As in his celebrated canvas now in Vienna's Kunsthistorisches Museum, Tintoretto's numerous renderings of the theme consistently harness the diagonals of the canvas to strafe Susanna's full-length figure with the Elders' lurid gazes. Again, Lely's approach is different; he compresses his scene into a tightly cropped, self-enclosed study of fomenting reaction and recognition. Instead of distributing the beholder's visual attention across the incidents of landscape, costume, and flesh to which Susanna's story was prone, Lely materializes a network of visual exchanges.

When narrated by critics in Restoration London, Peter Lely's career told a tale of success through pragmatic compromise. Having come to England to paint according to "the natural bent of his *Genius* in *Landtschapes* with *small Figures*, and *Historical Compositions*," Lely quickly reconsidered that high-minded, low-paid strategy: "Finding the practice of *Painting after the Life* generally more encourag'd, he apply'd himself to *Portraits* with such success, as in a little time to surpass all his *Contemporaries* in *Europe*."[29] Lely's *Susanna* series effectively anticipates that compromise; it sacrifices the vigorous expression and even the salacious spectacle privileged by his Venetian sources in favor of visualizing response to seeing and being seen. Against Susanna's flitting, focusing gaze, the heavy-lidded, pit-like eyes of Lely's distanced Elder look with dull stupefaction at the spectacle of her unfurling radiance. Rapt in entranced wonderment, his expression hews close to what French painter and founding academician Charles Le Brun would soon call the passion of "Astonishment." When perceiving an especially rare or extraordinary sight, a beholder's perceptual apparatus can be physiologically frozen. "Being so much imployed in considering this impression," Le Brun explained, ". . . there remains no spirits to pass through the Muscles, the Body thereby becomes immoveable as a Statue. This excess of Admiration causes Astonishment."[30] Eyebrows arched, pupils dilated, and hidden mouth at least partially agape, Lely's distanced Elder embodies this tellingly stiffened posture. The feeling hand, intent gaze, and beard-grasping gesture of the Elder at right, by contrast, suggest the comportment of a lustful beholder who is nonetheless slightly more the master of his desire. We might name his passion as curiosity or what one recent historian has defined as "a highly selective obsession for some very peculiar objects viewed in a very peculiar way; an obsession that is best understood . . . as a form of insatiable, if esoteric consumerism."[31] By juxtaposing distinguishable passions commanded by the sight of Susanna's beauty, Lely's canvas implicitly encourages the pictorial beholder to reflect on his or her own response to that same target.

In Stuart England, Susanna's story was a useful tool for testing moral and juridical boundaries along these very lines. As imagined in epic form by Jacobean poet Robert Aylett, the Elders are elected legislators corrupted by humanity's all-too-typical experience of beauty: "The sight whereof so rauishesth his brest / A reasonable man turnes senseless beast."[32] Rebuffed by Susanna's religiosity and moral scruples, Aylett's Elders compound their lust with heresy, declaiming her religion as a duplicitous ruse. "Art thou so unwise," one Elder asks Susanna, "Thou knowst not Politicians did deuise / Religion, onely to repressse the base, / And hold the Noble in the peoples grace?"[33] Liquidating religious principle in the solvent of debased political power, the wicked Elders swear that their superior authority can legitimate the proposed tryst in human eyes while verily blinding God to the act. So they query Susanna:

> Dost fear God should vs in this action see?
> This Lawyers gowne shall couer thee and me;
> Vnder which oft to Heau'n hath past vnseene,
> Farr greater Treispasses then this, I weene.[34]

Made beasts by seeing Susanna's beauty, the Elders cast by a conservative ecclesiastical lawyer like Aylett have dangerously conflated the powers of church and state.[35] Perniciously undermining religious institutions, they impose bogus, human legislation on the purview of God's vision. Their ultimate comeuppance utterly crushes such specious, hubristic innovations. Debased, deluded beholders, the Elders are executed for claiming rights of inspection that menace and unmoor the very pillars of social authority.

Allowing his pictorial beholder to similarly sound the stakes of looking, Peter Lely also mapped out an ingenious solution. Through the 1650s, he had explored historical subjects that placed an enraptured *and* self-conscious beholder at the very center of pictorial representation. Most important here is his *Cymon and Iphigenia* series, a theme taken from Boccaccio's *Decameron*. According to Boccaccio's tale, the bestial Cymon is banished by his aristocratic father to a life among the peasantry until he is spectacularly transformed. In the composition now at Knole House in Kent (plate 3), Lely stages this climactic moment: a hairy, rustic Cymon has entered a shaded glen from the sunlit fields at left. Hands crossed and leaning on his shepherd's crook, he gazes upon the slumbering, noble maiden Iphigenia as she reclines in a company of attendants. As told in the 1620 English translation of the *Decameron*, Cymon's vision unfolds this way:

He came into a faire Meadow, on everie side engirt with Trees, and in one corner thereof stoode a goodly Fountaine, whose current was both coole and cleane. Harde by it, uppon the greene grasse, he espied a very beautifull yong Damosell, seeming to bee fast asleep, attired in such fine loose garments, as she hidde verie little of her white body: onely from the girdle downward, shee wore a kirtle made close unto her, of interwoven delicate silke, and at her feete lay two other Damosells sleeping, and a servant in the same manner.[36]

No less than Lely's astonished Elder in the *Susanna* series, Cymon is left stiffened by this sight, transfixed and "glued fast to the ground."[37] Yet where Cymon can steal only a sidelong glance, the frontally aligned, chiaroscuro flash of Iphigenia's somnolent, ivory skin is surely positioned more for the blinding visual pleasure of the beholder imagined standing before the canvas.[38] Lely's marginal Cymon is what Victor Stoichita describes as an index of the image's self-awareness, a proxy visualization of the lusty beast-man that the image figures us to be.[39]

However, in a contemporaneous painting now at Doddington Hall in Lincolnshire, Lely explores a crucial, additional facet of the *Cymon and Iphigenia* theme (plate 4). Falling to his knees in gazing upon the fleshy pile occupying Lely's pictorial foreground, Cymon experiences himself and the bodies on which he looks in dynamic transformation. First, he begins to see in the ways in which an early modern artist was trained to look: "He began to distinguish her parts."[40] Marveling at what he can now perceive as Iphigenia's golden hair, his clarifying sight distinguishes a catalogue of nameable features: "forehead, nose, mouth, neck, armes, but (above all) her breasts, appearing (as yet) but onely to showe themselves, like two little mountainets."[41] If Lely's depiction allows us to follow these incidents apace, the gleam of light reflecting off Cymon's forehead suggests the broader transformation unfolding within the observer as he works beyond his initial, petrifying astonishment. As he continues to gaze upon Iphigenia, Cymon is "strangely . . . metamorphosed from folly to a sensible apprehension, more then common. And so far did this sodaine knowledge in him extend; that he could conceive of divine and celestiall things."[42]

A proxy figure for the beholder of Lely's canvas, Cymon thus models social and moral recuperation through deep, endless looking. According to Boccaccio's narrative, the sight of Iphigenia's beauty awakens self-consciousness in Cymon. Leaving his bestial ways, he returns to his father's house, acquires human knowledge, and ultimately marries Iphigenia. To wit, "Love had made him a man, whereas (before) he was no better than a beast."[43] Thus, the difference between Cymon's sight and that imagined in the Susanna narrative is profound; in fact, they move in completely opposite directions. Whereas excessive seeing in the Susanna story constitutes a dehumanizing menace to patriarchal order punishable by death, Cymon's unbridled visual indulgence enfolds him back into selfhood and society, elevating human debasement in a way that is positively *constitutive of* that order. Rather than the voyeur who guiltily peeps at a taboo object or the depraved animal made heretical innovator by visual desire, the mandate for Lely's Cymonian beholder is clear: human intelligence can be reclaimed by relentless observation of a noble target even though that beholder may look through the eyes of a beast.

Important support and refinement for this conception of beholding can be found in the writings of Richard Lovelace, Lely's most important period critic. Poet and Royalist agitator, Lovelace (1618–1657) collaborated with Lely in several ventures; but his most telling insights are articulated in an ekphrastic poem on Lely's double portrait of Charles I and the Duke of York (the future James II), now at Syon House in Richmond[44] (plate 5). Painted during the king's imprisonment in the fall of 1647, Lely's Charles is about to receive a penknife from his young son—a tool with which he will open the letter from Queen Henrietta Maria clutched in his left hand.[45] A fluted column and sheer, shrouding curtain accentuate Charles's high, somber forehead and visibly differentiate him from the impasto-laden, flashy equipage of the Duke of York, who stands at right, hand on hip just above a low-slung sword. By summoning on canvas this air of moody anticipation, Lovelace observes, Lely shows the sham play of symbols and stock gestures cobbled together by his artistic predecessors.[46] The poet casts a jaundiced eye on those painters and their times of old:

> When only a black beard cried Villaine, and
> By *Hieroglyphicks* we could understand;
> When Chrystall typified in a white spot,
> And the bright Ruby was but one red blot.[47]

Against these coarse spots and blots, Lely's virtuosic artistry manifests in its transfiguration of painting's materials and conventions. As Lovelace notes approvingly: "Thou sorrow canst designe without a teare."[48]

The most important mark of Peter Lely's pictorial intelligence is an act of insight that transforms not only the means of artistic representation but the painter's very eyes. Lovelace takes up this theme in the poem's second stanza as he suggests how Lely had been able to visualize the "proudest, richest looke" shared between the king and his son. Gazing into his father's eyes, the young James takes courage and "by his *Sun's* enlightens his own eyes."[49] Eliding stipulation of the lustrous attributes actually possessed by this paternal "Sun," the narrator makes one the luminous vision shared by father and son. Staged in language that stutters with compressive force, this unity of sight threatens to collapse Lely's visually bifurcated portrait in on itself: "Whilst either eithers passion doth borrow / . . . both doe grieve the same victorious sorrow."[50] The key to the picture's brilliance is then articulated in the third stanza. Lely has seen the Stuarts so perceptively, Lovelace claims, that his painting captures the unbreakable line of vision shared between king and prince:

> These my best *Lilly* with so bold a spirit
> And soft a grace, as if thou didst inherit
> For that time all their greatnesse, and didst draw
> With those brave eyes your *Royal Sitters* saw.[51]

Instead of rehearsing Seneca's well-known modeling of representation on the resemblance between a father and son, Lovelace turns inheritance inside out.[52] Invoking Lely's portrayal of father and son, he imagines an act of depiction so perceptive that it transfigures viscous paint, pictorial marking, and painter alike. The poet envisions Lely rendering on canvas what his sitters saw (namely, the Stuarts themselves)—and doing so by showing sight through the very "brave eyes" that he himself has momentarily inherited. No longer does Lely see with the eyes of a soldier's son; instead, he has been written into the royal line. And as ennobled by such transformative vision, painting is not the "lifeless thing" (*res mortua*) Seneca had opposed to proper family resemblance but a vital disclosure of human intelligence. In Lovelace's memorable phrase, "None but my Lilly ever drew a Minde."[53]

What I want to suggest is that the fantasies of perception eulogized by Lovelace and visualized in Lely's subject pictures can disclose a crucial facet of the Restoration scientific observer. It has often been noted that experimentalists possessed a conspicuous fascination for looking through the eyes of other species.[54] No doubt inspired by the famous optical experimentation of René Descartes, Dutch microscopist Antoni van Leeuwenhoek sent the Royal Society a marvelous drawing of the cross section of a cow's optical nerve in 1674 (fig. 3.1), following his earlier reports on observations made by looking through insect eyes.[55] "As to the *Eye* of the Bee," Leeuwenhoek noted, "which I have taken out of the Head, exposing its innermost part to the Microscope, I find, that the Bee receives her light Just with the same shadow as we see the Hony-combs."[56] As if literalizing Lovelace, Leeuwenhoek wants to understand the relations between vision and manufacture, between seer and maker. Where Lovelace looks at the work of Lely's hand and imagines the painter's temporary inheritance of royal eyes, Leeuwenhoek uses microscopic observation

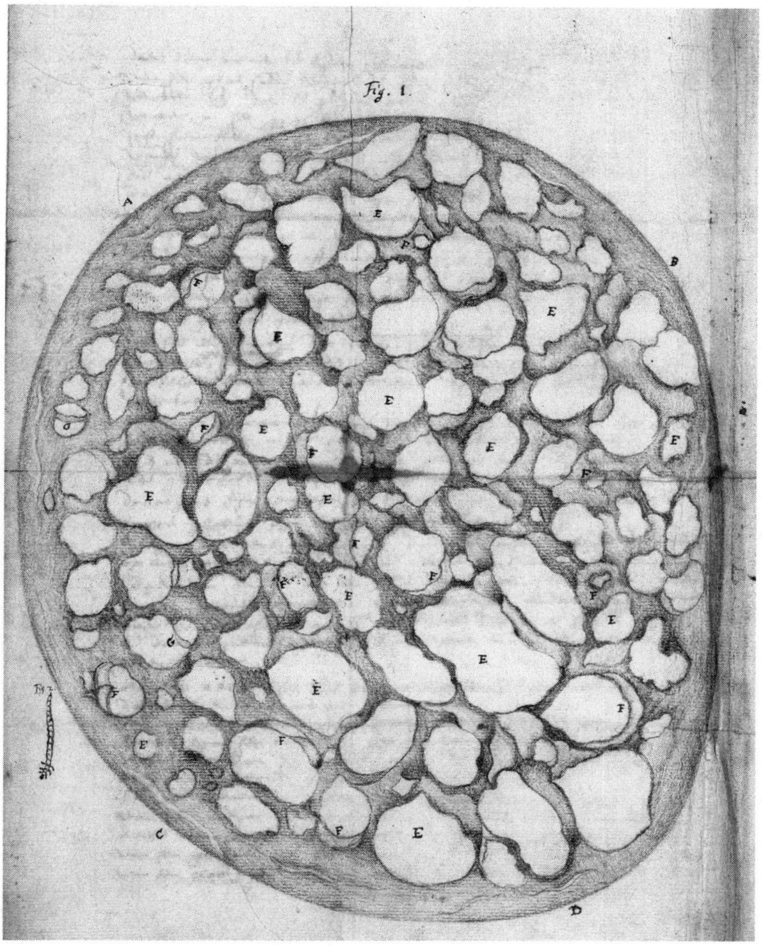

FIGURE 3.1 * Cross section of an optical nerve of a cow, drawn in pencil and ink by Antoni van Leeuwenhoek's draftsmen and sent to the Royal Society on December 4, 1674; Royal Society Early Letters, vol. L1, fol. 9. © The Royal Society. Reproduced by permission.

through insect eyes to argue that honeycombs replicate the physiological structures of bee vision.

A century later, Immanuel Kant would take bee-making as a shibboleth for defining art. As he acknowledged in the *Critique of Judgment*, "We like to call the product that bees make (the regularly constructed honeycombs) a work of art."[57] That, for Kant, is just loose talk. Because bees make by instinct rather than rational deliberation, their products have to be excluded from the domain of art in its broadest sense; they are certainly banned from the precincts of what Kant would call *schöne Kunst*.[58]

In Restoration London, however, meditation on the eyes and other perceptual inlets through which human or animal makers engage the world offered experimentalists a powerful diagnosis of the material constraints underpinning cognition itself. Robert Hooke's speculations on "insect music" are telling in this respect.[59] After a chance conversation on the London streets in 1666, an incredulous Samuel Pepys dismissed Hooke's claim that "he is able to tell how many strokes a fly makes with her wings (those flies that hum in their flying) by the note that it answers to in musique during their flying."[60] Yet despite Pepys's doubts, those marvelous inferences from nonhuman "musique" offered Hooke insight into the artifactual constraints inherently inflecting human intellection. "When I hear a Fly moving his wings to and fro so many times, with such a swiftness as to make a sound," so he observed in a lecture from the early 1680s, "I cannot but imagine, that that Fly must be sensible of and distinguish at least 3 Moments in the time that it makes one of those Strokes with his Wings, for that it is able to regulate and guide itself by the Motions of them."[61]

If it echoes period philosophical discussions of human embodiment in a seemingly outlandish way, Hooke's claim—and, in turn, my point here—is an eminently serious one. Scaled to the physiological demands of locomotion, perception of the world and thus thought about it are inextricably bound, "adapted to the peculiar structure of that animal body in which the sensation is made."[62] The radically relativistic conclusion Hooke drew from this insight effectively subtended the entire campaign of telescopic, microscopic, and more generally mechanized perception he relentlessly promoted in the Royal Society. As he put it in *A General Scheme* (circa 1666–68), "Those Imaginations we have of things, are not according to the Nature of the things themselves; but only appropriated to the peculiar Organs, by which they are made sensible to the Understanding; so that had we other kinds of Organs, we should have other kinds of Conceptions of those Effects."[63] To the experimental philosopher cognizant of his necessary constraint within a material body, then, the fantasies of occupying different organs of sight explored in and around Peter Lely's paintings of the early 1650s need to be seen as a major desideratum. Where his *Susanna* series enabled the pictorial beholder to compare and contrast between identifiable passions of observation, Lely's *Cymon* paintings thematized a transit between animalistic vision and human insight driven by the act of seeing itself. At least according to Lovelace's reckoning, Lely had achieved the acme of his art by transfiguring his medium so completely as to paint through the eyes of the Stuart kings—elevating himself by seeing *as* his ennobled sitters.[64] Thus, if scientific observation is always governed by a moral economy, then we can see Lely's subject pictures as tools for disclosing the desires, contours, and constraints of the Restoration observer no less precisely than Hooke's interface with his micrometrical devices could do.[65] That sandy beach in the foreground of *Nymphs by a Fountain* is not a blank space at all. It is, as it were, the faint, residual imprint of the thoroughly embodied, quasi-animalistic but eagerly ennobled beholder, whose attention the picture was made to hold. But from this bedrock of Cavalier culture, let us now turn analysis in the opposite direction.

"Amorous Transports and Ecstatic Languors"

Around 1656, Christopher Wren showed leading experimentalists Robert Boyle and John Wilkins an innovative mode of delivering opium. Wren famously prepared a primitive syringe, bound a dog to an operating table, and injected it with a "warm Solution of *Opium* in sack."[66] As across early modern Europe more generally, Wren's England was positively awash in a rising tide of psychotropic chemicals. Home to the nation's powerful mercantile companies, London acted as pharmacy for those far-flung novelties: nicotine in the tobacco from North American colonies; sucrose in the sugar from cane plantations in the West Indies; and caffeine in the tea and coffee trafficked from India, China, and the Middle East.[67] Opiates brought from Asia delivered the morphine alkaloids. And by the 1680s, English commander Thomas Bowery, for one, was navigating his ship, *Adventure*, up the river Hugli in the Ganges delta, trading opium to the island of Borneo as London-based merchants affiliated with the East India Company sourced the drug along the coast of Bengal.[68]

Like many exotic commodities, opium consumption patterns varied between peripheral source and metropolitan centers of distribution. Before the influx of cheap, British-backed import in the later eighteenth century, east Asian opium-taking had grown symbiotically with aristocratic "smoking cultures" and the tobacco with which it was mixed.[69] In Restoration London, opium was drunk, thereby following the dominant local mode of intoxicating consumption. "Laudanum," Thomas Sydenham's notorious opiate-laced syrup, was only the most popular of these Restoration formulas.[70] Thus, the ingenuity of Wren's intravenous technique surely lay as much in its mingling of alcohol with opium as in its fusion of tradition and novelty—all to completely debilitating effect. Narrated by Boyle, Wren's canine experiment demonstrated unequivocally what an intravenously injected opium mixture could do:

> We had scarce untied the Dog, (whose four Feet it had been requisite to fasten very strongly to the four Corners of the Table) before the *Opium* began to disclose its *Narcotick* Quality, and almost as soon as he was on his Feet, he began to nod with his Head, and faulter and reel in his Pace, and presently

after appeared so stupified, that there were Wagers offered his Life could not be saved.[71]

Given this incredible stupefying power, some experimentalists took a dim view of opium's medical usefulness. Leading physician and founding father of our "neurocentric age," Thomas Willis (1621–1675), was among the skeptics.[72] Articulating the dominant period view of opium's psychotropic action, Willis explained that the drug inhibited the vital flow of animal spirits, those "very subtle bodies, composed of a spirit and a volatile salt, both joined together and highly exalted."[73] Mimicking the salubrious effects of sleep, opiates chemically overpower what Willis called the "fortress" of the brain, causing the animal spirits to "first retire, and laying down their arms, lie idle."[74] By breaching the brain's physiological defenses and storming the very citadel of the soul, opium decouples the mind-body nexus that vexed so much seventeenth-century thought. Causing a "contraction or dejection of the whole soul," as Willis put it morbidly, this powerful poison could produce "not only the image of death, but sometimes also the same thing as death itself."[75]

If brushed to the margins of their own medical practice in England, Restoration experimentalists were fascinated with tales of opium's use and abuse at the far-flung borders of global trade. In an unpublished paper transmitted to the Royal Society in the later Restoration, Dr. Edward Smith gave a variation on Wren's canine trial, narrating the systematic administration of opium to a human subject.[76] While in Smyrna (Izmir in modern-day Turkey), Smith had endeavored to test "how farr the Turks are gone into the Use of Opium, & what are the common effects of it." Guided by local informants, he was directed to "the most famous Opium eater in the country about Smyrna": a coffee merchant named Mustapha Shatoor. "Resolving there to be an eye-witness of what he could do," Smith devised an experiment. He provided "the best Opium I could get and weighed it nicely into drams," and observed as his trial subject took the carefully measured pills over the course of the following day.[77]

What Smith's report narrates in ghoulish detail are the manifold effects produced by opium on the substrate of Shatoor's body, both inside and out. "The visible effects it had on him," Smith writes, "were to make his eyes sparkle, and to give a new air of life and brightness to his face. He told me that he was extreamly refreshed and made very cheerfull by my entertainment and that it gave him his keph, as the Turks express it." The doctor follows Shatoor through his daily routine, noting him "half an hour labouring heartily at cleaving wood to burn," before administering a second dose. Facilitating this hearty labor of cleaving wood, opium also helped Shatoor to cleave in a different sense: "He says it always has the same effects, giving him vigour and spirit. . . . That it makes him fitter for Procreation, for he has many wives and children." Nonetheless, Shatoor's quarter-century-old opium habit had left indelible marks on his physical frame: "His legs being small, his gums eaten away, so that the teeth stand bare to the roots, his complexion very yellow and appearing older by twenty years then he really is."

These mingled interests in exotic bodies rendered both sexualized and morbid—both visibly languid and positively depraved—by potent chemicals can be found more broadly among Robert Hooke's circles. According to the *English Atlas* (1680–82) that printer Moses Pitt published with extensive assistance from Hooke, inhabitants of India used opium "to produce rest and also to render them more acceptable to Women."[78] When taken too routinely, the drug produced dangerous effects, altering "the tempers of those that frequently Eat of it, making them Lazy and Stupid, and not fit for any Business or Conversation; and debilitating the Spirits to such a Degree, that they seem always rather asleep, than waking."[79] In some of his own earliest known writings, Hooke himself had explored this intersection between exotic chemicals and sultry, stupefied bodies.[80] In mid-May 1659, a twenty-three-year-old Hooke made careful notes from Jan Huygen van Linschoten's travels through the Portuguese East Indies in the 1580s, tales that he read through John Wolfe's 1598 English translation.[81] While most of these notations reduce Linschoten's adventures to the terse, staccato jottings that would later characterize Hooke's diaries, they grow particularly expansive on an inebriating substance called "Deutroa," which was used as an anesthetic in Goa and as an aphrodisiac in Restoration London.[82] Upon reading Linschoten's account of the sexual habits of Portuguese women in colonial Goa, Hooke wrote the following:

> Women all unchaste and keep 2 or 3 soldiers besides their husbands, with whom they satisfy their lust as oft as occasion serves, which to effect they usually give the herb deutroa which beareth a seed where of brusing out the sap they put it into a cup or other vessel and give it to their husbands either in meat or drink. Presently the man is as if he were half out of his witts and with out feeling or else drunk doing nothing but laugh and sometimes it take him sleeping whereby he lyeth like a dead man. So that in his presence, they may doe what they will, take their pleasure with their friend and their husband never know it. In which sort he continueth 24 hours long but if they wash his feet with cold water he presently reviveth and knoweth nothing thereof but thinketh he had slept.[83]

Beyond detailing this lurid, chemical technique for effecting seduction in colonial Goa, Hooke also recorded the list of names added to Linschoten's text by Dutch physician Bernardus Paludanus, which designated the drug in various places and tongues: "Deutroa ved Tacula alias Datura in Spanish barla dura, in Dutch Igell Holben, in Malaba Vumata Caya, in Canara Datura, in Arabia Marana, in Persia and Turky Datula. Of the description of this herb you may read in the name of stramusconium."[84]

Now, it is hardly surprising that leading experimentalists in Robert Hooke's circles would have been interested in exotic pharmaceuticals. After all, chemical medicine constituted a veritable cutting edge of seventeenth-century science. Where university-educated physicians schooled in Galenic and Hippocratic theory relied on a traditional range of humor-balancing techniques to preserve health, practitioners newly skilled in emerging variations of alchemically inflected medicine not only conceived of their aim as actually curing diseases but readily embraced far-flung commodities in their therapies.[85] As we will see in chapter 5, Hooke's colleagues such as Robert Boyle, Christopher Wren, Willis, and philosopher William Petty were at the very forefront of those advances in England. But however we read the evidence of Hooke's own avid drug-taking, what is most important to note when parsing his meticulous notes on the technical nomenclature of Portuguese pharmaceutical seduction is how stupefying chemical transports in colonial Southeast Asia could offer a material boon to experimental picture-making practices in Restoration London.[86] Consider the vexing problems that beset Hooke when drawing ants through his microscope in the early 1660s (fig. 3.2). "I could not, for a good while," he reported, "think of a way to make it suffer its body to ly quiet in a natural posture."[87] Neither the twisting, relentlessly mobile body of the living ant contorting in Hooke's improvised shackles of glue nor its lifeless corpse were amenable to optically aided draftsmanship. So fragile did a dead, husk-like insect exoskeleton become beneath the searing heat of magnified candlelight used to guide Restoration microscopic ventures that, as Hooke put it, "your object shall be quite another thing, before you can half delineate it."[88]

Stimulated by the puzzle of how to pose his microscopic insect subjects for depiction, Hooke found an ingenious solution in chemical intoxication. "Having insnar'd several of these into a small Box," he explained his ant-drawing technique:

> I made choice of the tallest grown among them, and separating it from the rest, I gave it a Gill of Brandy, or Spirit of Wine, which after a while e'en knock'd him down dead drunk, so that he became moveless . . . and after I had taken it out [of the alcohol], and put its body and legs in a natural posture, remained moveless about an hour.[89]

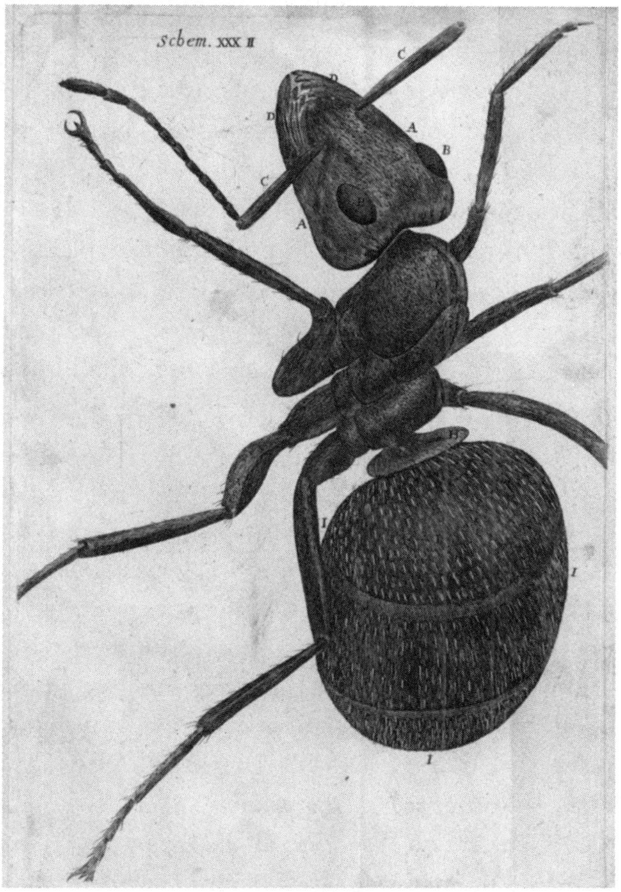

FIGURE 3.2 * Print of a magnified ant; from Robert Hooke's *Micrographia; or, Some Physiological Descriptions of Minute Bodies made by Magnifying Glass* (London: John Martyn and James Allestry, 1665), scheme 32. © The Royal Society. Reproduced by permission.

Picture-making is here staged as a battle of wits waged between the draftsman, who wants to render the pliant body of his target in as "natural" a condition as possible, and that subject of depiction, who wants to escape. Where bodily restraints and brute, physical force fail, intoxicating chemicals succeed, keeping the target plump, immobile, and therefore eminently agreeable to depiction. Like the Portuguese seductresses of Goa, the experimentalist achieves his aims by stupefying obstructions, by cunningly immobilizing targets.

When read against broader experimental fascinations with the spectacle of the intoxicated body, I think that Hooke's approach to posing the subject of depiction can offer surprising insight into a crucial feature of Peter Lely's Restoration-era portraiture. Since the late seventeenth century, critics have observed a peculiar stylization in the painted faces of Lely's sitters in the years after 1660, when he turned firmly away from subject pictures toward portraits. Exemplary of this manner are the *Windsor Beauties* now at Hampton Court Palace in Richmond, a series Lely painted in the 1660s depicting beautiful women from the court of Charles II and Catherine of Braganza.[90] As in his portrait of Jane Middleton (1645–1692; plate 6), each maiden appears alone in a rustic setting at three-quarter length, a pictorial format advantageous to Lely's display of his mastery in painting exposed flesh and shimmering fabrics after the manner of Anthony van Dyck.[91] With bodies turned slightly askew to the picture plane, Lely's sitters gaze directly out at the beholder through exaggeratedly sensual and conspicuously soporific eyes. Alexander Pope's pithy evocation of this mannered effect is justly famous: "Lely on animated Canvas stole / The sleepy Eye, that spoke the melting soul."[92]

Noted since the Restoration, attempts to elucidate this "sleepy-eyed look" are hardly convincing. As if borrowing from the discussions of fetishism then emanating from Portuguese traffickers along the coasts of Africa, one period critic located the roots of this style in Lely's fixation on the sensuous eyes of Barbara Villiers, Duchess of Cleveland, an influential force in the court of Charles II.[93] So this anonymous observer quipped, "Sir Peter Lilly when he had painted the Dutchess of Clevelands Picture, he put something of Clevelands face as her Languishing Eye into every one Picture, so that all his pictures had an Air one of another, all the Eyes were Sleepy alike."[94] Among art historians, the sleepy-eyed look has remained a standard, if no more compellingly explained, feature of Lely's later pictorial production.[95] My proposal is that the evidence of experimental fascination with the spectacle of the languorous body can keenly assist our understanding of this stylized approach to posing the sitter and its concomitant effects on the beholder. A key document facilitating that interpretation is the racy account of the Restoration court penned around 1705 by Anthony Hamilton and published a decade later as *Memoirs of the Comte de Gramont*.[96] As narrated by Hamilton, affectation of an exotic daze was a much-vaunted art at the later Stuart court. Describing Theodosia Hyde, sister-in-law to the patron of Lely's *Windsor Beauties*, Hamilton writes, "Long practice had made her glances so languishingly tender, that she scarcely opened her eyes except *à la Chinoise* . . . when she ogled you, you would have said that she was doing something else at the same time."[97] Feigned and artful, the appeal of this sleepy look rests on the asymmetry it appears to create between the performer and audience, between the lassitude of the surveyed body and the stimulated beholder. What becomes clear from Hamilton's account is that, although intended to excite their audiences, such theatrical displays of languid indolence could simply reverse the effect—boring the *observer* to sleep if handled with insufficient skill.

Such was the case with Jane Middleton herself. Conceding her beauty, Hamilton's Gramont acidly complains, "The airs of indolent languor, which she had assumed, were not to everybody's taste; the sentiments of delicacy and refinement, which she tried to express without understanding what they mean, put her hearers to sleep."[98] Tellingly, Hamilton commends most women who "play dumb" literally and figuratively, mutely stimulating their male observers without straying into an excessively theatricalized languor that causes actual exhaustion.[99] As if recalling Lely's double portrait at Hampton Court, moreover, Hamilton's paradigmatic foil to these alluringly reticent, seductively retiring ladies is none other than the Duke of York, "the most reckless ogler of his day."[100] Where Lovelace had imagined Lely looking upon the young prince and inheriting his "royal eyes," Hamilton narrates the ever-roving eyes of the duke as enlivened by Lely's sleepy-eyed depictions. Shown the *Windsor Beauties* series that his wife had commissioned, "the Duke of York enjoyed looking at it, and once again began ogling the original."[101]

What this economy of gazes enables us to infer is that, as with Christopher Wren, Dr. Edward Smith, and Robert Hooke, the Restoration male courtly elite was fascinated by the spectacle of bodies visibly transported into states of exotic lassitude. Although the means for achieving these languorous displays differed between the experimental laboratory and the courtly salon, art was required for each. Learning from far-flung precedents such as the seductresses of colonial Goa, Hooke had to craft a sufficiently stupefying chemical web to ensnare his ant while keeping it pliant and amenable for picture-making. Equally, were it to properly excite and hold attention at court, the "fiction of the pose" that was the sleepy-eyed look required deft skill to pull off—as much by the female self-fashioner as by Lely himself. Where Lovelace's Lely transfigured spots and blots of paint on canvas into real bodies, so the sitter had to transform affected, theatrical conventions of courtly life into a delicate, stylized insensibility that flattered an observer's intelligence by comparison, but without collapsing into mere sleep-inducing tedium. Thus, while achieved by experimentalists through the cruder, more direct means of dosing their subjects with potent pharmaceuticals, Lely's sleepy-eyed look emanated from a shared concern for posing the pliant, languorous body and, concomitantly, expanding a fantasized remit of the observer's power.

If all of this, I would stress, is necessarily prior to the problematizing of the beholder in which Michael Fried has located the ontological origins of modern art, Peter Lely's pictorial strategies still deserve to be seen as modern in a certain sense.[102] Where his *Nymphs by a Fountain* or the *Cymon and Iphigenia* series had placed the spectacle of the somnolent female body in the dark recesses of myth, Lely's Restoration portraiture transferred that sleepy look to prominent, even popular, women drawn from across the class spectrum of later seventeenth-century England.[103] In that profoundly misogynistic culture, these women were thought to exert fearsome political influence over the state through their sexual allure to the infamously promiscuous monarch Charles II.[104] But, as historian Kevin Sharpe has proposed, Charles's own power was not so very different from that wielded by his courtly ladies. Legitimated more by a "politics of pleasure" than dynastic claims to divine-right authority, the restored monarch had to appeal to the general populace—he had to court popular interests—just as the king was, in turn, wooed by the ladies of his court.[105] If made to look as docile as Hooke's ant actually was, then, Lely's sleepy-eyed sitters might better be seen as wielding the very instruments of persuasion by which the restored regime established its rule and was itself, so charged the critics, all too frequently being ruled through the 1660s. Yet as in statecraft, so experimental philosophy: during the crucial years around 1680, the stakes and fantasies of power would assume new media and still-greater proportions.

Forsaking Drawing

On Saturday, November 15, 1679, Robert Hooke bought a porpoise.[106] Paying for his slimy prize what an agricultural worker in Restoration England could expect to earn in a month, Hooke took the porpoise to Garraway's, one of the many coffeehouses he haunted around London's Royal Exchange.[107] First at Garraway's and then back home at Gresham College, he systematically dissected and drew his "sea hog" over the next three days.[108] As they would appear in the prints promptly made from them and published in Dr. Edward Tyson's *Phocæna; or, The Anatomy of a Porpess* (1680; fig. 3.3), Hooke's depictions offer a fantasy of observing his anatomy in chronological sequence. From our view of the animated, spouting beast seen in profile at upper left, we move counterclockwise to a bird's-eye view of the porpoise's now-flayed carcass at center left and on to magnified renderings of discrete parts: lungs at lower left and, at the base of the page, a magnificent depiction of the creature's triple stomach and attendant matrix of blood vessels, culminating with the phallic larynx at upper right.[109]

These porpoise portraits accord well with what we have seen of Hooke's contemporaneous drawing practice. In their exploded form and sequential character, these figures recall his serialized notations of smudgy comets examined in chapter 1. Yet by the mid-1680s, a curious problem becomes perceptible: Hooke's drawing activity sinks into oblivion and largely evaporates.[110] The latest drawing we have by him dates

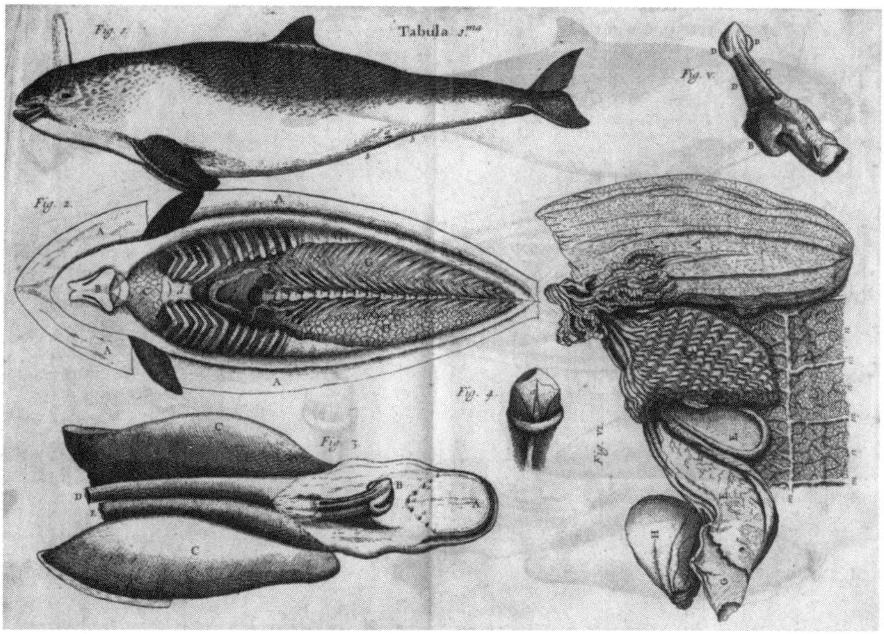

FIGURE 3.3 * Etching and engraving after Robert Hooke's drawings of his dissected porpoise; from Edward Tyson, *Phocæna; or, The Anatomy of a Porpess* (London: Benj. Tooke, 1680), n.p. © The Royal Society. Reproduced by permission.

to the summer of 1686.[111] Even if later specimens can be adduced, the general eclipsing of Hooke's draftsmanship in the 1680s raises uncomfortable interpretive questions. By abandoning drawing in middle of his career (at approximately the age of fifty, some twenty years before his death), he would appear to have severed the crucial thread of pictorial practice that Thomas Povey and numerous others have promoted as a veritable umbilical cord connecting early modern science and art. More to the point, if drawing in the experimental community was the vital, cognitively generative activity I have claimed it to be, why would Hooke have stopped doing it? And worse, isn't Hooke's abandonment of drawing—compounded by his near-contemporaneous renunciation of the formal-epistemological graphic techniques taught by Lely—but the last in a long sequence of turns against the very painterly heritage for which I have been arguing in this chapter?

Far from undermining the interpretive necessity of thinking Restoration science and painting together, Hooke's abandonment of drawing actually makes the point. It discloses where we might locate painting's dynamic historical force and conceptual possibility for experimentalists' wicked intelligence. Yet apprehending those points requires us to turn the proverbial glove inside out, to fundamentally rethink the terms in which painting has been considered thus far.[112] Framing the agenda for the book's subsequent chapters, this section will sketch how Hooke's renunciation of drawing might begin to reveal the contours of a systemic conceptualization of visual practice that would come to be theorized by leading experimentalists and embedded in their architectural interventions, especially in the later 1670s and early 1680s. Forsaking drawing in later Restoration London's experimental circles, then, did not mean foregoing visual activity as it did among contemporary Netherlandish artists such as Ferdinand Bol or Meyndert Hobbema, who abandoned painting for, respectively, a wealthy marriage and plum prospects as a wine-gauger.[113] Instead, the turn away from draftsmanship visible in the careers of both Hooke and Christopher Wren in the 1680s might be regarded as closer in spirit to that of a contemporary artist such as Diego Rodríguez de Silva y Velázquez (1599–1660), who significantly scaled down his pictorial production as his responsibilities managing and administering royal collections increased.[114] And, as I will argue in conclusion, the structuring implications of that calculated, predictable decision to abandon picture-making can be seen in and as a key legacy of Peter Lely's pictorial intelligence.

Mythically teaching himself to draw by making the tools of the painter's trade and then using them to copy prints, Robert Hooke had worked extensively as a draftsman early on in his career.[115] First in the private employment of Robert Boyle and then as the Royal Society's Curator of Experiments, he had received a commission in August 1661 to draw "the Figures of small Insects by the Help of the Microscope" for the entertainment of Charles II, a project that would ultimately become his celebrated *Micrographia*.[116] But that graphic enterprise had fallen to him only after it had been sloughed off by the more decorated, higher-status experimentalist to whom it had originally been assigned: Christopher Wren.[117] And as microscopic draftsmanship had been delegated to Hooke in the early 1660s, so he was increasingly distributing the task of drawing among his own coterie of assistants by the later 1670s. Who were these understudies? Son of painter Mary More, Richard Waller (circa 1660–1715) began working for Hooke around 1678. With the Curator's guidance, he quickly advanced to Fellowship in the Royal Society, where he took on many of the private drawing assignments Hooke himself had previously executed.[118] Now in the Royal Society is an exquisite collection of botanical drawings that Waller prepared and presented to the institution in 1689, as part of a collaboration with the Cambridge naturalist John Ray.[119] If not always in line with Hooke's objectives, he also prepared many of the drawings and engravings for his *Posthumous Works of Robert Hooke* of 1705.[120] (See figure 1.16.)

Maker of "some pieces [of] extraordinary work," assistant Henry Hunt (died 1713) was another of Hooke's protégés in the 1670s.[121] Unlike Waller, Hunt followed the traditional pattern of apprenticeship by actually living with the master, before taking over Hooke's position as Keeper of the Royal Society's museum in 1696. If Waller's intellectual ambitions and accomplishments are suggested by his precocious illustrated translation of Virgil's *Aeneid*, Hunt seems to have truly been Hooke's right-hand man. (See figure 4.5.) Assuming the Curator's earlier role as general draftsman charged with depicting the flood of curiosities received by the Royal Society from all corners of the globe, Hunt worked intensively through the 1670s in preparing prints for the broadening expanse of Hooke's publications, including his scientific journal, *Philosophical Collections* (1679–82). Yet even taking into account what future research might disclose of other assistants Hooke deployed in surveying the devastated City of London or in his burgeoning architectural practice, his team of understudies would still be dwarfed by the army of draftsmen, foremen, lawyers, financiers and skilled laborers commanded by his friend Christopher Wren. Appointed Surveyor-General of the King's Works in 1669, Wren had been placed at the head of a highly complex administrative system and charged with a massive rebuilding campaign that stretched from London's sewers to the blackened, burnt wreck of St. Paul's Cathedral. Crucially, by the mid-1680s, both Hooke and Wren had effectively ceased practicing the kinds of draftsmanship that had won them such early acclaim in the experimental community, distributing that pictorial work to their expanding teams of assistants instead.[122]

Much has been made of the ways in which childless bachelor Robert Hooke conducted relations with those servants and apprentices. Historian of science Steven Shapin has influentially proposed that unlike the "invisible technicians" overseen by Robert Boyle, Hooke maintained a paternal attitude toward his understudies, regarding them almost as a surrogate family.[123] In that capacity, Hooke might be seen as taking after Peter Lely, a genial and generous patriarch of Restoration London's nascent art world.[124] But whatever they might have learned from this painterly paternalism, crucially and abidingly consequential to Restoration experimentalists were the advanced models of artistic organization Lely imported and perfected. Struggling to meet the mushrooming demand for his work after the Restoration in 1660, Lely had evolved a sophisticated studio practice rooted in Continental baroque workshop traditions but then little known in England. As with many of his Netherlandish contemporaries, he recruited an able body of assistants and divided pictorial labor among them.[125] After Lely had painted a portrait-sitter's face and hands, he would distribute the canvas to a team of specialists to add drapery, landscape, flowers, and other pictorial flourishes. Not only did he cultivate local artistic talent to perform this specialized labor by introducing life-drawing and other pedagogical techniques new to England, but he relied on a crack squad of expatriate masters.[126] As one scholar recently put it, "Without precedent in England . . . Lely's production practice was built on a series of collaborations with nine highly skilled artists from the Netherlands, many of whom were already master painters with fine reputations in the homeland."[127] If he never abandoned picture-making to the extent that Hooke did, the diminishing presence of Lely's lone, masterly hand from his post-1660 canvases has been noted and broadly lamented by connoisseurs since the early eighteenth century.[128]

In transitioning to this full-scale atelier production, Lely also introduced key innovations to his pictorial facture. Recent technical analysis has demonstrated that after the Restoration, he sacrificed his time-consuming methods for creating subtle tonal variations through layers of dead coloring in favor of a new repertoire of stylized, quasi-calligraphic brushstrokes that could be readily imitated by studio assistants and copyists.[129] This complex, distributed labor and simplified facture had to be brought together—unified—under an overarching, artistic aegis. "It was in the interest of the successful studio," so Catharine MacLeod has observed, "to suppress individuality in its employees in order to create a harmonious and seamless whole, a recognisable 'Lely' product."[130] The studio also had to amass and archive a stock of visual models. While Lely had long built his practice around his magnificent collection of Old Master drawings, the collecting and preservation of the painter's own designs became an increasingly integral practice.[131] Finally, as the studio reached its apogee in the 1670s, Lely secured the circulation and market for his work by assembling a stable of reproductive engravers. Employing specialists skilled in the new technique of mezzotint engraving who could admirably evoke his oily, tonal palette, he even standardized his portrait poses so that printmakers could reuse a stock body and change only the facial features.[132] If, as one historian has put it, the baroque indeed was "as much a culture of artisans as of great artists, for baroque art was always cooperative art," then nowhere in early Restoration London was that culture more clearly articulated than in the studio and products of Peter Lely.[133]

Yet a mode of picture-making favorable to that "near-assembly-line production" had already been present in Lely's subject pictures from the early 1650s.[134] (See plate 1.) Recent technical analysis indicates that he worked and reworked canvases such as *Nymphs by a Fountain* extensively over time.[135] Unlike a brash innovator such as Caravaggio or even the youthful Godfrey Kneller, who famously upstaged Lely in a painterly duel before Charles II, Lely built his images not by *alla prima* improvisation but through incremental, methodical pictorial assembly from preparatory drawings and layered campaigns of underpainting.[136] Less a splash of liquid inspiration before the model, *Nymphs by a Fountain* might be better seen as an aggregated repository of pictorial motifs drawn from and reduplicated into this broader oeuvre. Consider the repetitions within the contemporaneous subject pictures alone. (Compare plates 1, 3, and 4.) If the general composition of *Nymphs by a Fountain* reads strikingly as a horizontal reversal of the *Cymon and Iphigenia* at Doddington Hall (most conspicuous in the side-to-side swapping of the backward-turned, supine nude who drapes a bent right leg over her left calf beneath a concealed fountain), then the motif of the bare-breasted courtesan in profile at center right in the Dulwich picture finds a close reiteration in the *Cymon and Iphigenia* at Knole House. Like a pattern book, then, *Nymphs by a Fountain* stands as an incrementally revised summa of postures, poses, and pictorial devices, all of which could be subtly reconfigured, inverted, and recombined by master or studio hands to engender new compositions. Pledging a conception of art that makes meaning through repetitions, borrowings, and outright thefts from the earlier works Lely collected so assiduously, *Nymphs by a Fountain* speaks directly to the painter's emergent apparatus of studio production, which relied on integrating visual forms and diversely skilled labor.[137]

Mannered and self-referential, Lely's repetitive, echo-chamber-like output appeared to some Restoration critics as little better than narcissism. According to poet John Dryden, he "drew many gracefull pictures, but few of them were like . . . and this happen'd to him, because he always studied him self more than those who sat to him."[138] I think that Dryden's assessment is trenchant, provided that we take the "self" materialized in Lely's pictures as a massively socially distributed one—one extending from the painter's subjectivity to include the collective network of the baroque studio system he had imported into Restoration London: its propagation and division of labor, innovative pictorial facture, unified production, archiving of source materials, and control over reproduction. By conceptualizing Lely's artistic selfhood and, more specifically, painting in these terms, we can see his loose, negligent pictures as brilliantly instantiating the hierarchical labor distributed between the variously skilled, self-interested agents through whom they were made. That was certainly the model of art embraced by his most prominent successor, Godfrey Kneller, whose own painting studio was, if anything, even more "industrialized" than Lely's had been.[139]

While later chapters will chart the generativity of these strategies in philosophical circles, it is crucial to note that a key index of how experimentalists would apply Lely's lessons had already registered in the very volume wherein prints of Robert Hooke's flayed porpoise were published in 1680. Laying the groundwork for a crucial conceptual model, Dr. Edward Tyson (later chief physician at Hooke's own

Bethlem Hospital) was positively rhapsodic as he looked up from the ruins of the sea hog to gaze upon the emerging edifice of experimental knowledge. "In so great a work," Tyson opined,

> many hands are daily imployed, some in battering down and removing the Rubbish, others in endeavoring to lay down a good foundation; some dig for new stone, others labour in polishing it; others in inventing new and more convenient Instruments and tools to work with; some give their pains; others give their purses; all their desires and good wishes to so noble a design.[140]

Although Tyson is talking about experimental philosophy and not the built environment, the semantic confusion is telling. Setting the foundation stones for experimental learning is like rebuilding the great City of London, he seems to suggest, insofar as both require massive funding and labor, and a complex distribution of skills. As with Peter Lely's advanced painting practice of the 1670s, architecture and experimental philosophy each demanded deft management and administration—what Tyson calls "the modelling and contriving . . . of the skilfullest Artists."[141] If Lely enables us to see where painting had become a collective practice that verged on architecture in its organizational complexity, we now need to consider the evidence of collecting, to see how experimental intelligence of the 1670s was moving beyond painting and consolidating around the figure of the architect.

CHAPTER FOUR

Cascade, Copper, Collection

Constellations of Images in 1670s Experimental Philosophy

WHEN JOHN COVEL BOARDED a ship bound for the eastern Mediterranean in 1670, the England he left was a place changing rapidly. Chaplain to the English Levant Company in Constantinople, Covel (1638–1722) could have seen these changes afoot while an undergraduate at Cambridge: the fen drainages, the enclosures, and the industrial manufactures growing more broadly.[1] Where nearly eighty percent of the English population had subsisted through agriculture at the beginning of the seventeenth century, overseas trade and proliferating industries such as finished textiles, coal-mining, and shipbuilding meant that but sixty percent were employed in agriculture by the century's end. Only the economy of the Dutch Republic, the wealthiest nation in Europe, was comparable.[2]

Despite two failed wars with those United Provinces, an outbreak of the bubonic plague in 1665, and the devastating Great Fire of London in 1666, industrializing England and its capital were rebounding by 1670. The hub of finance, an expanding colonial empire, and a modernizing navy able to protect it, London was as much mushrooming as rebuilding under the guidance of Royal Society Fellows Christopher Wren and Robert Hooke. With an exploding population exceeding half a million inhabitants by 1700, the capital was ever more connected to its provinces by new physical infrastructures and bonds of commerce. As one historian has recently observed, "Better roads, easier water transport, an extensive post office, insurance and improved financial services not only made trade easier for wholesalers, but transformed the nature of retail in later seventeenth century England. England was quickly becoming a nation of shoppers and shopkeepers."[3] By Covel's return in 1679, these information networks were eagerly trafficking talk of international plots, "interests," and intrigues surrounding the succession to the throne of the heirless Charles II—debates in which England's modern party politics would be forged.[4]

While equally capacious and vigorous, Covel's interests were of a different nature from these. Traveling through Turkey and the Middle East in the 1670s, he had been able to indulge his curiosity for local flora, fauna, and antiquities. A bound notebook that he took out from and brought back to England testifies to his encyclopedic passions. Opening the volume, we may read a philosophical treatise on logic in Latin and follow church history from the ancient Holy Land to eccentric devotional observances in Protestant Lincolnshire. As we admire Covel's impressive drawings of Levantine flora, we can feel the delicate fibers of friable leaves that he pasted onto paper, cut out, and then glued into the volume. Once our broader sensorium has been thus engaged, the ancient Mediterranean world is brought back to us tactilely as we inspect and caress the array of wax impressions of Roman coins Covel affixed into the book. And while marveling at the ink-wash renderings of ancient numismata (labeled the "Seals & Medals of Dr. Covels") pasted throughout the volume, we can connect the natural life of southern Europe to that of Covel's native England through depictions of insects signed and dated to the early 1660s by a team of draftsmen, apparently including the youthful Robert Hooke.[5]

If miscellaneous collections such as these were literal commonplaces among early modern European intellectuals, much about John Covel's notebook is strange.[6] We have no other evidence by which to connect him to the normally paper-trail-happy Hooke. We know nothing of exactly how it came to pass that preparatory drawings of magnified insects for Hooke's *Micrographia* (1665) would materialize in Covel's notebook, dated to 1660–61. Even the depiction of these entomological specimens is weird (fig. 4.1). Carbuncled and tusked, the elongated body of the "Tipula" (or "water measurer," in modern parlance) breaks the spine of the visual plane in the notebook's most elaborate page of insect drawings.[7] Although accompanying captions denote the names of specimens represented, dates of observation, and initials of draftsmen and witnesses—even providing minute demarcations of the targets'

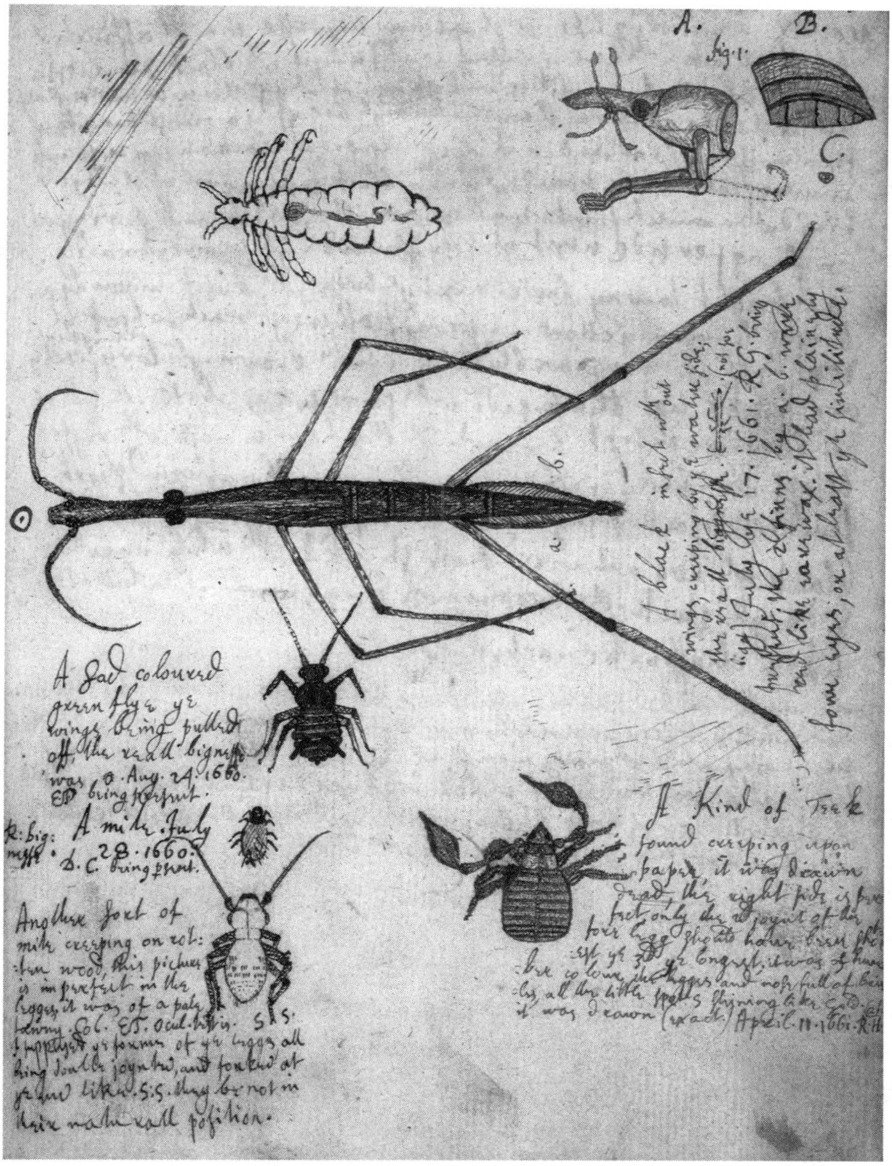

FIGURE 4.1 * Pen-and-ink drawings of insects, including contributions by Robert Hooke; in John Covel's *Natural History and Commonplace Notebook* (circa 1660). © The British Library Board (BL Add. MS 57495, fol. 13v). Reproduced by permission.

actual size or "reall bignesse"—each depicted creature effectively occupies its own world. This constellation of figures evidently grew over a yearlong period. Beginning with the tiny mite penultimate at lower left (dated July 28, 1660) and propagating outward to the pseudoscorpion at right, the image then evolved upward to the wingless "green fly" from August 1660 at center left. By the time the central Tipula was drawn in July 1661, the operating draftsman had altered the orientation of the image to the page, a directionality then maintained through the uppermost figures.

Odd enough an approach to drawing as this is, Covel's notebook also bears the imprint of the readers' hands through which it subsequently passed. In 1716, London-based collector James Petiver asked Covel to lend him a "Manuscript of yours in Quarto wherein you had very accurately delineated severall Plants."[8] Now legible in this manuscript are numerous locations where Petiver made connections to the items he saw depicted and took the liberty of scribbling initialed notes into the borrowed volume.[9] It is as if the sensorial uranium contained within Covel's slim quarto volume had not only collapsed historical time and geographical distance but made beholders (such as Hooke) image-makers and readers (such as Petiver) writers as their diverse traces intermingle, assembled collectively together.

Since early 1665, Royal Society Secretary Henry Oldenburg's *Philosophical Transactions of the Royal Society of London* had made such modes of miscellaneous information-gathering a central, serial feature of the experimental collective. Compiling information from his prodigious correspondence network on topics broadly relevant to natural philosophy, Oldenburg published his journal with the institution's printers, John Martyn and James Allestry, roughly monthly. Often described as the world's first scientific journal, the *Philosophical Transactions* helped to resolve crucial, practical concerns of the international experimental community.[10] Dates of publication in the widely read journal now became a conventional standard against which to settle to the priority disputes endemic to early experimentation.[11] At the same time, Oldenburg and the *Philosophical Transactions* functioned as conduits of communication between philosophers whose native lands were at war, as was all too common in later seventeenth-century Europe.[12]

The graphic form of the publication vitally enhanced these functions. First, while typically compressing all of an issue's visual images into a single copperplate—economically combining, for instance, Sir Robert Moray's views of French mining tools and techniques with Hooke's depiction of a hirsute, cyclopean monster (fig. 4.2)—the *Philosophical Transactions* made savvy use of the printed images so advantageous to early modern Europe's natural philosophers.[13] In delivering "the cheap rate and speady advantage of Typography," as Society Fellow John Beale expressed it, copperplate engraving "hath already gotten so much credit, that to all ingenious

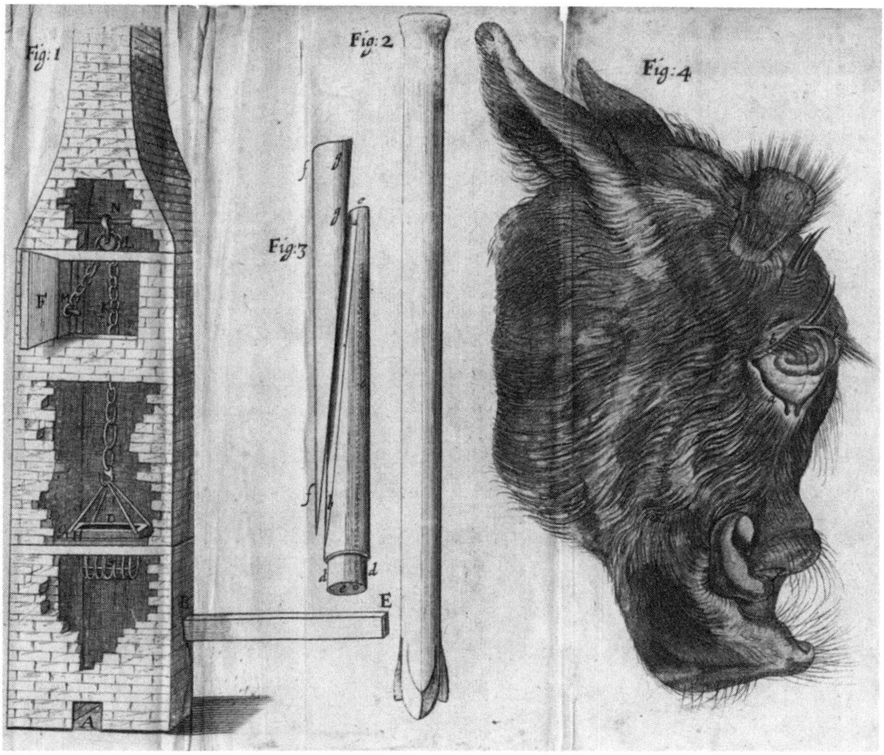

FIGURE 4.2 * The first illustrative plate included in Henry Oldenburg's *Philosophical Transactions*, juxtaposing engravings from Robert Moray's research on mining (*left* and *center*) with an etching after Robert Hooke's drawing of Robert Boyle's chemically preserved head of a monstrous colt; *Philosophical Transactions* 5 (July 3, 1665): n.p. © The Royal Society. Reproduced by permission.

persons an accurat impression is in far higher esteem, than the most gaudy paintings of vulgar note."[14] At the same time, a reader of these "philosophical commonplace books" (so Oldenburg described his journal) could readily become an experimental contributor.[15] Through a favorable arrangement with England's State Paper Office, a prospective contributor—whether a Fellow of the Royal Society or not—could transmit text and images free of charge to the anagrammatic "Monsieur Grubendol, London."[16] If accepted by Oldenburg, the submitted information would be licensed under the Royal Society's independent imprimatur and printed, often quickly, alongside other contributions and offered for sale to all comers. Read across Europe and beyond, the *Philosophical Transactions* became a key instrument for stimulating and displaying the collaborative, collective ethos trumpeted by the Royal Society.[17] Likewise, Oldenburg's journal (and soon its numerous imitators) allowed London's urban adepts and provincial enthusiasts to keep up with the latest exploits of the international experimental virtuosi.[18] It was through the *Philosophical Transactions* that schoolmaster Peter Nelson in 1660s Durham could "hear what the press is daily delivered of, and . . . make a further enquiry after such things as may either profit or delight [him]."[19]

But by the later 1670s, both Henry Oldenburg and his influential modes of information-collecting had become objects of intense scrutiny as the Royal Society's established, well-regarded forms of communication were increasingly challenged from within and without.[20] Catalyzed by Oldenburg's death in the fall of 1677, Robert Hooke took action. Through the mid-1670s, Hooke had devoted substantial energy to theorizing how institutionalized experimental philosophy and its periodicals could be improved. Replacing his nemesis Oldenburg as elected Secretary of the Royal Society in late 1677, he debuted his own periodical soon thereafter, calling the project *Philosophical Collections*.[21] By most accounts, Hooke's *Collections* was a dismal failure. Only seven numbers of the journal were published between 1679 and 1682 before he was summarily dismissed from his duties as Secretary, *Philosophical Collections* was terminated, and publication of the *Philosophical Transactions* was resumed. Evidenced by *Philosophical Collections*, as one recent biographer has put it, "Hooke's performance as secretary of the Royal Society fell far short of the high standard set by Oldenburg."[22]

This chapter takes a different approach. I read *Philosophical Collections* as a portal through which we can hear the crosscut voices of the craftsmen, chancers, chiselers, and far-flung adventurers on whom the urban luminaries of English experimental philosophy had to rely. I also see the journal as a staging ground, as wicked intelligence preparing to go public. Hooke's ill-fated journal, I argue, embodies gestating, competing conceptions of the unity that the collective enterprise of experimental philosophy should possess and the agency of visual forms for enacting collaboration. For if, as nearly all advocates of the "new sciences" agreed, experimental philosophy had to be a collaborative project, much was up for negotiation in the later 1670s.[23] Exactly how was such a society to be organized and governed? By what means were its variously skilled, geographically distributed investigators to communicate and be updated on each others' experimental work? Was the genial collaboration between discrete agents modeled in John Covel's commonplace book and normalized by the *Philosophical Transactions* sufficient? Or were more drastic, rigorous interventions required to form truly philosophical collectivity from agents operating across time and space? Indeed, might the ostensible collectivism materialized in journals like *Philosophical Collections* be read better as a "cascade"—that characteristic assemblage of visual traces by which, according to theorists such as Bruno Latour, modern science consolidates its power and wields its peculiar social authority?[24]

To narrate the social life of experimental visualization and its diverse forms, my story begins in the port city of Bristol in the waning days of the seventeenth century where a customs officer named William Cole nursed a grudge against art and dreamed of natural-philosophical greatness. Following his path from the docks and tidal pools of southwestern England inward to London, the chapter shows how Cole (a reader of and contributor to the *Philosophical Transactions* and Hooke's publications alike) deployed visual materials in philosophical negotiations, leading him through courtly micropolitics, international political revolutions, and, ultimately, to personal disaster. Demonstrating in all-too-human terms exactly why and how images were so important to experimental philosophers, Cole's heady brew of gamesmanship and bald self-interest also provides an apt point of entry into the jarring juxtapositions of visual information that would confront the beholder of Hooke's first volumes of *Philosophical Collections*. Guided by Cole and the late historiographical writings of John Evelyn, we can thus glimpse a conception of the experimental community as a loosely affiliated collective of discretely motivated agents—bound by symbolic ties but speaking in different visual languages and driven toward individualistic, often conflicting, goals.

Hooke had very different ideas. In theoretical writings of the later 1670s, he had specifically targeted the modes of social organization needed to improve experimental philosophy's efficacy. By using his own contemporaneous experiments in fusing metals as an interpretive model, I propose that we can identify compelling connections between the motivated, vertically integrated, quasi-militaristic scientific society he envisioned in theory and that he implemented in graphic form as his *Philosophical Collections*. The chapter concludes by examining how Hooke actually used his journal as an instrument to be dispatched from London "to the field" for training and integrating the dispersed locations and competencies of the individuals comprising the philosophical collective. Beginning, then, at the geographical periphery and concluding by looking outward to an emerging vision of institutional science seen from London, this chapter uses the *Philosophical Collections* to introduce historical tensions within experimental philosophy's distributed network structure, and struggles for authority that would increasingly be thematized through projects of collecting.

"Things Uselesse"

By the early 1690s, William Cole had arrived at a critical passage in his natural-historical pursuits. He was recently retired ("att my owne request," so he said) from a post as customs inspector that he had held for two decades in Bristol, then England's third-largest city and an important trading center.[25] Supported by a pension, Cole (circa 1622–1701) could finally begin a full venture into scholarship. Born into poverty in Dorsetshire, he had spent his youth along England's south coast, forming a lively network of business and intellectual contacts that included a son of the Isle of Wight, Robert Hooke. "In Southampton, Isle of Wight, and Bristoll," Cole claimed, ". . . I have had the acquaintance and friendship of many Merchants, Commanders, and Masters of Shipps with their ship Chirugeons and others navigateing to all the principall places of Europe, Affrica America (espetially the last) and of some into Asia."[26] Commerce and natural knowledge were inseparable for Cole. Through his access to both long-haul traders and local sources, he had amassed a celebrated private museum of specimens—a collection from which he hoped to write a comprehensive history of nature.[27]

In his aspiring natural-historical authorship, Cole was guided by a cold certainty. If he could claim to "have with my own hands drawn more than 200 figures" of specimens from his collection, he had no interest in art.[28] In a commonplace book from the early 1690s, he noted several "Memorandums of things to doe and procure."[29] Beyond reminding himself to refute the opinions of "Epicurean Atheists and Cartesian Theists" and to study "the long worms like a Silk thread found in the Leggs of Negroes brought in from Guinea," Cole vowed to combat those who privileged works of human art over natural creation.[30] By combing the Earth to obtain "things uselesse," such collectors only "satisfie their lusts of pride and ostentatious vanity [and] make use of and purchase att anie rates innumerable unnecessary and uselesse things, which they keep for ornament beauty and for the exquisite art of man in their curious contrivance."[31] For Cole, these pernicious artistic niceties not only included "the great varieties of sculpture, paint, innameling, carving" but extended to "all needless excess in apparel, rich furniture in their houses, trappings for horses, guilded, painted and carved coaches, etc."[32] In his collecting at least, Cole practiced what he preached. Against prevailing period convention, he neglected the oil paintings and sculptures so prized among early modern collectors in favor of natural specimens alone.[33]

What motivated this antagonism against the broad field of "uselesse things" Cole called art? An advocate of natural theology, he possessed a reading of art with which his friend and interlocutor Hooke might well have sympathized.[34] Captivation by art, Cole contended, was symptomatic of a frivolous and ultimately atheistic turn of mind, one that neglects the craftsmanship of God in favor of the inferior works of man. By pursuing art, collectors

> at the same time despise the most admirable and beautiful works of God's creation wherein his infinite wisdome and power is manifest, and only because

they are not of some use for their pride and fitt for pleasure; not allowing the Allmighty maker of all things the same liberty in furnishing his lower house of this world unlesse such only as serve their lusts.[35]

Cole's art collector was one who preferred to gather luxurious artifacts most conducive to human creature comforts rather than studying the creatures with which God had providentially populated this, his lower world. But to study God's creation successfully in later seventeenth-century England—"an age of masquerade and double-dealing, [when] opportunism and self-advancement flourished"—the provincial experimentalist needed to possess both deep pockets and prodigious arts of his own.[36]

As Cole readily found, struggle and brinksmanship were effectively built into the project of publishing natural history from the English provinces. Because printing had been severely restricted by the Press Act of 1662 in London and the universities at Oxford and Cambridge, rural virtuosi could obtain the impressed images so useful to natural philosophy only with great difficulty. This was a problem Cole repeatedly bemoaned from Bristol.[37] In the summer of 1693, he appealed to Sir Robert Southwell (former President of the Royal Society who held an estate nearby in King's Weston) for a recommendation of "some curious artist in drawing and graving."[38] A few months later, he informed Southwell, "I have now a French man at work graveing the whole scheme of my figures of snow, so much desired by severall Learned Gentlemen to be made publique, with my observations on them."[39] To a correspondent in London, he would praise "a graver here, one who made me a very good rolling presse."[40] Nonetheless, when contemporaneously negotiating the sale of his collection to Oxford, Cole stipulated that the talents of Michael Burghers (engraver to the university and occasional printmaker for the *Philosophical Transactions*) be secured to produce printed images for his natural history; for by the mid-1690s his hopes of obtaining high-quality images in the provinces had dwindled entirely.[41] Relating the progress of his general history of nature, Cole updated Southwell on his plans of summering "in Oxford (and partly in London) to see the draughts, sculptures [that is, prints], and printing of it finished."[42]

This passion for natural-historical imagery was hardly peculiar to William Cole. In Restoration London, meticulous paintings of lizards on vellum or other pictorial curiosities could easily command higher prices than fashionable portraits in oil, even those by rising stars in the artistic firmament like Godfrey Kneller.[43] Nor was Cole some dogmatic "completist" who insisted that each and every artifact in his prodigious collection be memorialized by the copperplate engraver's laborious and expensive art. A hearty advocate of the pragmatic category of depiction that

Peter Parshall has labeled *imago contrafacta*, Cole was willing "to lessen the charge of drawing and ingraveing of such things as can be sufficiently described without figures by references and comparisons with things commonly known, from all which there may be competent ideas formed in the Readers imagination."[44]

Instead, the motives that drove this retired septuagenarian to chance England's arduous roads and to expend substantial amounts of ready cash for natural-historical imagery were more elemental. Without reliable access to images, he was effectively barred from philosophical participation. Declining a solicitation to publish in the *Philosophical Transactions* in the early 1690s, he protested:

> I could not contrive any way to do it without figures, so as to render them [that is, his natural rarities] fit to be published in the Transactions, they being (I mean the most considerable of them) of such a nature that by bare descriptions without figures itt is not possible to make them intelligible to the reader so as to form true ideas of them.[45]

Cole's dilemma can be restated this way: although he had claimed to despise those works of human art designed or used for the gratification of sensual pleasure, he recognized that aspects of his philosophical discussion would be literally incomprehensible without images. More than this, he seems to have been almost eager to dispense with the artifacts themselves, once he had obtained prints properly conducive to philosophical cognition. "Haveing sufficiently filled my memory and imagination with the idea" of his collected artifacts, Cole vowed that he would willingly let them go "and please myself with the contemplation of that Infinite wisdom by which they were produced, and fit myself for Death and a more blessed life."[46] Yet therein lay the rub. How could Cole complete his edifying publication and free himself from his strangely insidious natural artifacts, objects that tended to turn even the most pious naturalist's attention away from their divine craftsman and idolatrously back onto themselves? To do so, he would need not only the prints that so consistently eluded him but protection and preferment from those able to ensure a safe future for his treasured goods. In short, Cole needed a patron.

William Cole, Courtier

It is useful to add a corrective here. That is, for all his proclaimed antipathy to showy human art in the 1690s, Cole had spent a good deal of time plying and promoting it during the previous decade. At the same time he was prodding Oxford's newly formed Philosophical Society to invest in a monstrous bull (a rarity "you may sell . . .

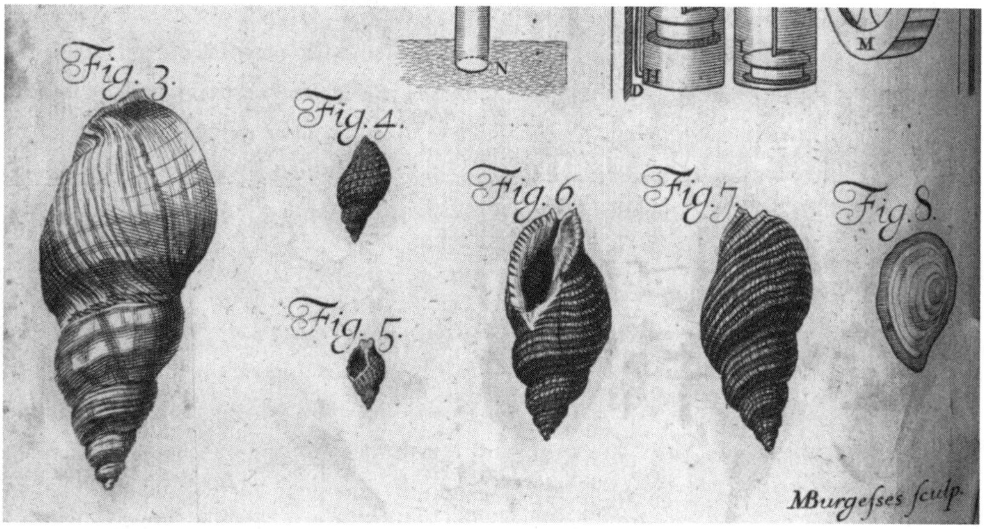

FIGURE 4.3 * Michael Burghers's engraving of William Cole's shells as printed with "A Letter from Mr. William Cole of Bristol, to the Philosophical Society of Oxford, containing his Observations on the Purple Fish," *Philosophical Transactions* 178 (December 1685): n.p. © The Royal Society. Reproduced by permission.

againe for more then double the cost to some person who may drive itt to London and other parts and gaine well by exposeing itt as the usuall manner is of thinges strang"), Cole had been trafficking talk of a beautiful purple dye made from shellfish native to the beaches around Bristol.[47] Guided by female informants in the mid-1680s, the customs inspector had tracked this species (*Murex lapillus*) and, with substantial effort, discovered its secrets.[48] As he claimed in an article published in the *Philosophical Transactions* in 1685, the dye yielded by that humble English shellfish could easily rival the purple hues once worn by the ruling classes of Mediterranean antiquity. Alongside prints of his shells prepared by Michael Burghers (fig. 4.3), Cole traced the genealogy of his hue back to the reign of Vespasian, detailing how Roman dyers then "strove to excel each other in new fashioned Purples for their own gain, and to gratify the luxuriant excess of the Great ones of those times."[49]

Now, developing an English imperial purple, charting its lineage to the days of Vespasian, and labeling it "a fair bright *Crimson*, or neer to the Prince's colour" were potentially shrewd moves in mid-1680s England.[50] In the wake of the struggle to prevent the succession of the Catholic Duke of York known to historians as the Exclusion Crisis of 1678–81, court apologists had grown more brazen in styling Charles II as a Roman emperor. Likeness to Vespasian figured particularly in the increasingly authoritarian king's propaganda.[51] Cole was certainly keen to position his purple shellfish dye in service to that Stuart imperial myth-making. And after receiving word in the mid-1680s that Charles II was interested in the dye, he moved rapidly to consolidate his whiff of royal favor.[52] He would be delighted to provide "his Majesty the great patron and Founder of their Society," so Cole informed the experimentalists at Gresham College, ". . . the pleasing diversion of seeing the experiment made in his Royall presence."[53]

Cole's desired audience with Charles II never materialized. Delayed by difficulties in keeping his crustaceans alive in seawater-filled barrels during the two weeks of overland transit from Bristol to London, his plan foundered more seriously with the king's death in 1685—an event Cole called "a Calamity too great and public to be mention'd."[54] Yet once James II had decisively suppressed the Duke of Monmouth's invasion and triumphantly assumed the throne in 1685, Cole again began advertising the virtues of his princely dye through an expanded visual repertoire. In December 1686, he sent the Royal Society a swatch tinted purple by his preparation, accompanied by the following flourish of promises for his dye:

> I propose to send for more of them about beginning of the summer and so to contrive itt that they may be brought living to me; and then send up to your museum (on finer paper, finer linnen than these, lamb's parchment and white satin, which the last of all will make the colour appear more bright and vivid) such names, letters, sentences, etc. which you shall desire and direct.[55]

Cole's attention to visual effects is telling here; he recognizes the need for material supports capable of possessing and highlighting—that is, exemplifying—his dye's colorful powers.[56] Accompanying Burghers's black-and-white, mass-produced prints of evacuated shells that could soberly indicate which species of shellfish he had harvested, he imagines complementary, sumptuous materials such as "lamb's parchment and white satin" showing off the dye's purpleness, brightness, and vividness to his elite, urban audiences.

No less than this expanding array of visual practices, William Cole was also crafting a refined sense of political channels and strategies by which to advance his interests. As he vowed "to contribute what I am capable of in promoting your noble designs," the customs inspector simultaneously alerted the Royal Society that he was trading his shellfish more widely.[57] "I sent some of them lately," he noted coolly in a letter from 1686, "unto the Right Honorable the Earle of Peterborow, who (as I remember told me) is a member of your Society and who no doubt will part from two or three of them."[58] This esteemed contact was Henry Mordaunt (1623–1697), second Earl of Peterborough, a central political player and personal confidante of James II—a king with no small imperial ambition.[59] Thus, if thwarted in his approach to Charles II in the mid-1680s, Cole was making savvy use of the appeal of Roman imperial trappings to the new Catholic king's inner circle.

Unfortunately for Cole, this valuable connection to the Earl of Peterborough quickly became a liability. As the armies of William of Orange (soon to be William III) landed in the autumn of 1688, Peterborough was seized while attempting to flee England, impeached by Parliament, and imprisoned in the Tower of London. With Peterborough made a villain of the Glorious Revolution, Cole moved quickly to renegotiate his patronage by following the period's conventional procedures. "Recurrent features of clients' lives in early modern Europe," Mario Biagioli observes, "were the discontinuities and disruptions produced by the termination of patronage relationships, usually as a result of the patron's death."[60] Cole dutifully obeyed the protocol of an early modern client seeking to obtain a new protector. He prepared a gift and made ready to depart for the capital.[61]

What was this gift? Although some speculation is required here, it is curious to note that in 1689, Cole had his 1685 article from the *Philosophical Transactions* on shellfish dye reprinted as a booklet called *Purpura Anglicana; being a Discovery of a Small Fish found on the Shores of the Severn*. So its title page claims, the piece had been reprinted "at the Desire, and for the conveniency of many" who were not able to buy the *Philosophical Transactions* when it first appeared.[62] But the timing of Cole's reprint is certainly suggestive of an overture to a new patron. We also know that during his trip to London in the late summer or early fall of 1689, he received an audience with an extremely wealthy and prestigious figure who shared his interest in color: Robert Boyle. Did Cole's meeting with Boyle in the early autumn of 1689 thus represent an attempt to offer a newly published volume in exchange for some kind of patronage? Whatever was intended, the audience in Boyle's well-appointed laboratory was marred by a catastrophic gaffe. "When I think of that unfortunate accident which hapned att your lodgings through my want of circumspection," Cole wrote to Boyle in late November, "I cannot forgive myself (though I know the goodnesse of your generous nature doth)."[63] His excruciating account of his folly reads:

> At the time I went towards the light the better to discern that curiosity you shewed me and the rest of the gentlemen present, my mind was so earnestly intent in thinking what I should propose unto you (as the most considerable, among the great number and variety of experiments I have made) for improvement of that little time I was to stay with you, that without due circumspection, (forgeting the stand soo near unto me) I made too quick a turne about

towards you, and soo overturned those things, which I doe still imagine to have been prised by you and were of greater worth than out of your great civility and goodness you would then own; or so much as seeme to resent the destruction of them.[64]

Now, it is true that I have no way of proving that Cole—a man described by his peers as "infected with an immoderate desire of honour, & apt to think well, it may be too well, of his own abilities & performances & loath to be defrauded of the reputation of his discoveries"—explicitly sought the patronage of Boyle.[65] Nor can I show that his rash attempt to impress Boyle was born of a desire to make advances on the engraved illustrations that would bedevil his quixotic natural history through the 1690s. That said, charting Cole's pathway up to the threshold of Boyle's laboratory can helpfully illuminate fundamental features of a provincial philosopher's visual practice in the waning twilight of the Stuart era. To publish in a field like natural history, first of all, an author needed reliable access to images and image-makers. If drawings could be sent through the increasingly inexpensive post, a keen naturalist like Cole would have found the scarcity of provincial printmakers and presses extremely frustrating.[66] Therefore, in addition to the time and money required to collect, draw, and transport naturalia, a provincial naturalist also had to have connections to skilled engravers—connections that, as Cole found, were frequently brokered by metropolitan philosophical societies.[67]

It was in this double bind that William Cole became ensnared. A provincial naturalist needed to display to these brokers a capacious array of visual practices just to obtain the printed images requisite for publication. As Cole's case indicates, such an aspiring natural author needed not only the drawn *pictures* from which the desired *prints* of natural targets could be made and the written *texts* able to elaborate them. He or she also needed *samples* that could possess and make reference to critical properties of those targets, *specimens* of the entities on which the research was based, plus the financial and practical wherewithal to transmit them to metropolitan centers. Ideally, this naturalist would additionally be able to command a more graceful mode of experimental *performance* than that which Cole had been able to muster in Boyle's laboratory in the fall of 1689. Finally, an experimentalist could potentially facilitate those negotiations by demonstrating linkages between his research and the interests or ideological trappings of the arbiters of power—unless, of course, the balance of power were to shift as dramatically as it did in 1688–89. Seen in this way, it is almost less surprising that Cole's *General Natural History* never appeared than that Cole would have desired to brave this shark tank at all.

The Philosophical Group Portrait

In the last years of the seventeenth century, John Evelyn sent a copy of his most recent publication to Hans Sloane, naturalist and subsequent President of the Royal Society.[68] Like Cole, Evelyn had learned the hard way just how dangerous it could be to produce heavily illustrated philosophical material from the provinces. As he wagered in the note he inscribed to Sloane in the front matter of his *Numismata: A Discourse of Medals* (1697), "You will easily guesse how sensibly I was Afflicted, [not only by] how the *printer* had abus'd me (by leaving out many the most material *Corrections*) but how ill I was dealt with by those, who in my Absence all the summer (in *Surry*, many miles from London) undertook to supervise and repaire my failings." Even the material substrate of the book seemed to conspire against Evelyn. "I have endeavour'd to reforme some of the grosser Errata," he lamented, "but the paper is so bad, that I should have but multiplied the faults instead of mending them." Evelyn was disconsolate. *Numismata* had inflicted "Wounds so deepe" to his reputation that, even though he had assembled substantial revisions for the book, he remained "without any inclination of publishing them, after this miscarriage."

Evelyn's impassioned attention to printed form and philosophical reputation is made all the more striking by the ambitious program of visual interpretation he sets out for the future historian of experimental philosophy in *Numismata* itself. Anticipating an investigator who "shall one day write the *History* of this *Age of Wonders*," Evelyn directs that scholar to assemble medals, coins, and other portrayals of experimentalists—those men "of extraordinary Merit, for what they have written, published, and improved of the most useful, experimental Learning; and which (after all a-do) is really, and indeed, of all human Knowledge, the only true, and valid Learning."[69] As his friend Samuel Pepys was then putting to practice, Evelyn's historian was to group the portraits according to their individual sitter's accomplishments.[70] Many of the early Royal Society's leading Fellows would thus convene under the heading "Mathematicians." In the company of England's founding mathematical fathers such as John Dee and William Oughtred, Evelyn imagines his historian adding "those *Viri* Πολυμαθεςατοι . . . Sir *Christopher Wren*, Dr. *Wallis*, *Newton*, *Flamsted*, *Hook*, *Hally*, etc. Fellows of the *Royal Society*, whom none but the Αγεωμετρητοι and Ignorant . . . will envy the Honor of *Medal*."[71]

Were we to assemble Evelyn's experimental group portrait, we might see some curious juxtapositions: the metallic profiles of Christopher Wren and John Dee sitting next to each other. We could observe Robert Hooke's "eie full and popping, and not quick," scowling at his enemies, such as Isaac Newton or John Flamsteed.[72] Arrayed on a plush velvet tray, such a collection of atomized, self-enclosed medals would constitute less what seventeenth-century France's academic theorists of painting were then calling a composition than a paratactic collision of "juxtaposed, isolated and independent pictorial or verbal blocks."[73] To truly form Evelyn's philosophical group portrait, the historical interpreter would have to work up from each discrete, freestanding portrait of experimental worthies. Subjecting each portrait to physiognomic analysis, the interpreter would need to study the depicted metallic profile against all that that agent had "written, published, and improved."[74] A sense of the collective whole could only be formed imaginatively, once each discrete individual's work and character had been assessed and mentally integrated by the interpreter.

I think that John Evelyn's program for moving between texts and tables of serried metal portrait heads offers a highly instructive model for understanding the horizon of expectations brought by a seventeenth-century beholder to the pages of a scientific periodical. By this I mean that in our absence of direct historical testimony, Evelyn's program can provide compelling insight into how a period beholder of the kind suggested in chapter 3 might have engaged the totality of a journal's differentiated visual field, moving between discrete individual images and their accompanying texts. Certainly, when Hooke began publishing *Philosophical Collections* in late fall 1679, he assumed that his readership would possess a visual aptitude of that very kind. In his journal's first illustrative plate, we see the ovular nub of a canine testicle gleaming like a burnished chestnut as it teeters along the left border (fig. 4.4). Printed in a suave, inky chiaroscuro that suggestively conveys the fleshy textures denoted in the original drawing sent to London by Antoni van Leeuwenhoek, this protruding testicle breaks in all scale and tone from the mechanized bird-man at the top of the page and the sail-driven airship floating below him. Like the views of partially eclipsed Jupiter sent by Polish astronomer Johannes Hevelius at right, these latter aeronautic figures were taken from existing printed sources.[75] Building on conventions that Henry Oldenburg had deployed for over a decade in his *Philosophical Transactions*, Hooke trusted that beholders would know to move backward and forward through the accompanying text, decoding the strange visual assemblage resulting from the economical strategy of printing all the journal's illustrative matter on and as a single plate.[76]

Graphically consolidating the variegated research of far-flung philosophical contributors, this single-plate, illustrative convention also required a local integration of skilled labor. Most likely the plate for *Philosophical Collections* 1 came together this way. On a Sunday in early October 1679, Hooke was at home in Gresham College on Bishopsgate Street, assembling the materials for his new journal.[77] A visually savvy editor, he might well have developed a cut-and-pasted mock-up of the journal's

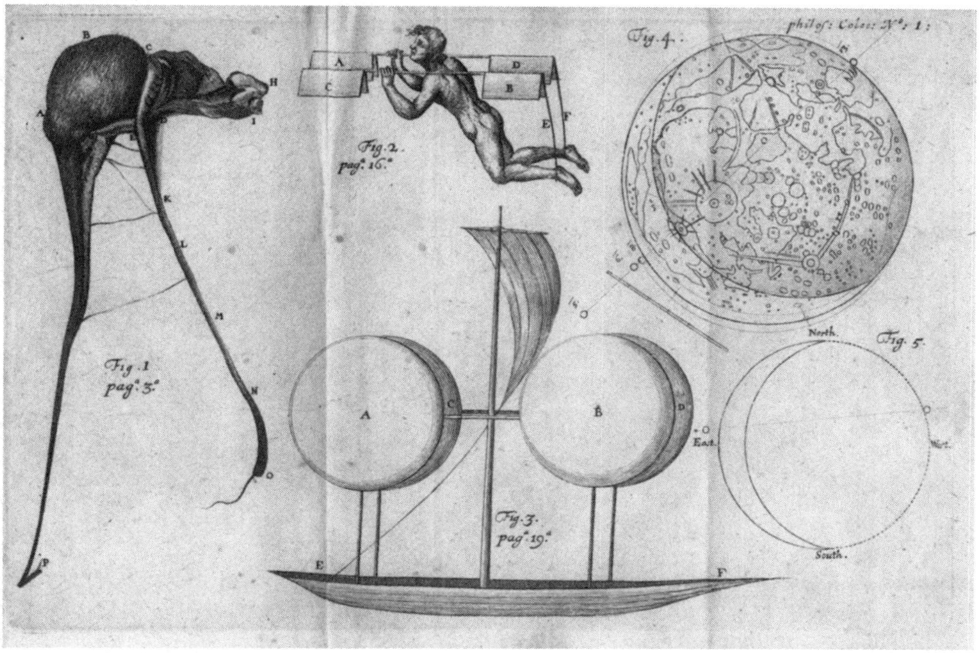

FIGURE 4.4 * Illustrative plate (possibly engraved/etched by Henry Hunt) from Robert Hooke's *Philosophical Collections* 1 (London: John Martyn, 1679), n.p. © The Royal Society. Reproduced by permission.

illustrative plate in collaboration with the printmaker, possibly his apprentice Henry Hunt.[78] Once it was engraved and etched by Hunt, Hooke would have delivered the copperplate to a local "letter man" to add the figure numbers and page references.[79] With these discrete, specialist skills of picturing and lettering applied to copper and with impressions pulled from that plate, Hooke then transmitted the prints to the stationer (in this case, his close friend John Martyn, printer to the Royal Society) who had contracted for the rights to sell the journal.[80] As publisher of the *Philosophical Transactions* for over a decade, Martyn integrated the illustrative print with the text, possibly binding the leaves together (always at the buyer's direction). Hooke's journal was thus available for sale from Martyn's shop at The Bell in St. Paul's churchyard by early December 1679.[81]

Forging these complex networks of intellectual and manual labor into a single plate, Robert Hooke was activating a well-established form of experimental-philosophical communication. Yet like the pairings of inimical experimentalists contrived in John Evelyn's philosophical group portraits, these arrays of disconnected sources and visual materials could also lead to very strange collisions. Consider the cascade of

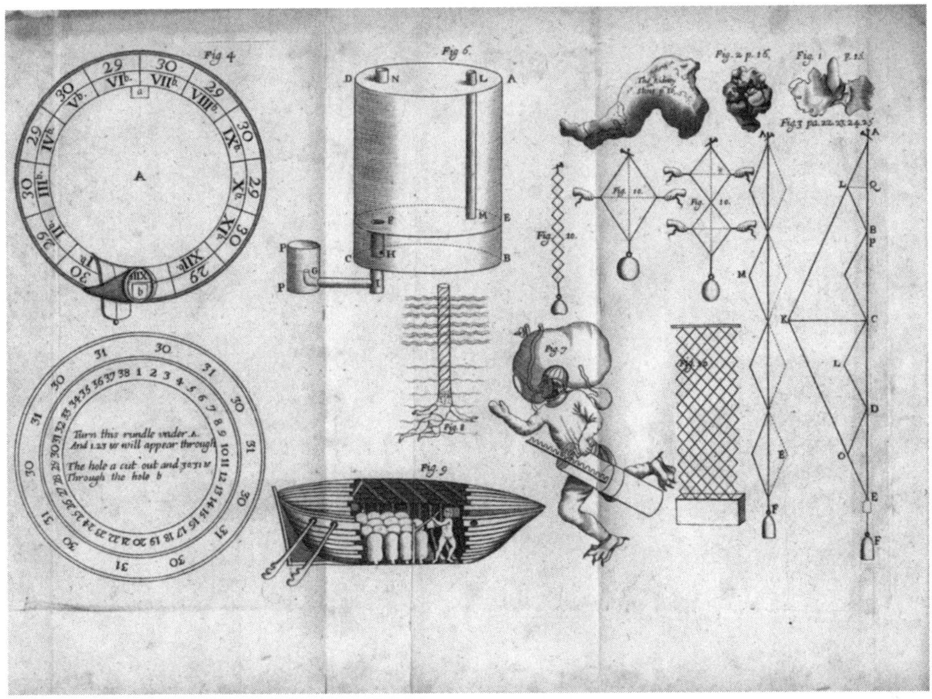

FIGURE 4.5 * Engraved illustrative plate for Robert Hooke's *Philosophical Collections* 2 (London: Moses Pitt, to be sold at the Angel in St. Paul's Church yard, 1681), n.p. © The Royal Society. Reproduced by permission.

competing strategies at work in *Philosophical Collections* 2 from early 1681 (fig. 4.5). At upper right, depictions of three crumbling clusters of calcified matter extracted from human bodies (supposedly drawn "ad vivum") sprout above diagrammatic renderings of weighted, tensed springs.[82] A schematic plan of Robert Boyle's oil lamp jostles for graphic space in the center of the print with competing, cutaway views of submarine exploration, while a literal cutaway flanks the plate's left-hand border. That is, as with Hooke's earlier paper model of Richard Towneley's micrometer, the beholder was to slice these circular "rundles" from the page and punch out the tiny apertures marked *a* and *b* in the upper portion (fig. 4.6). By attaching hemisphere *A* onto *B*, the user could form a paper instrument of Dr. Robert Wood's "new Al-mon-ac for ever." On this singular plate, then, we are shown *depictions* of bulbous, calcified curiosities; schematic *diagrams* of muscular mechanics; a range of *cutaway views*; and a *paper instrument* exemplifying the function of a mathematical tool.[83]

The most interesting image in the entire plate, though, is that which promises the least: the springy isosceles triangle *AKE* at extreme right. In his accompanying

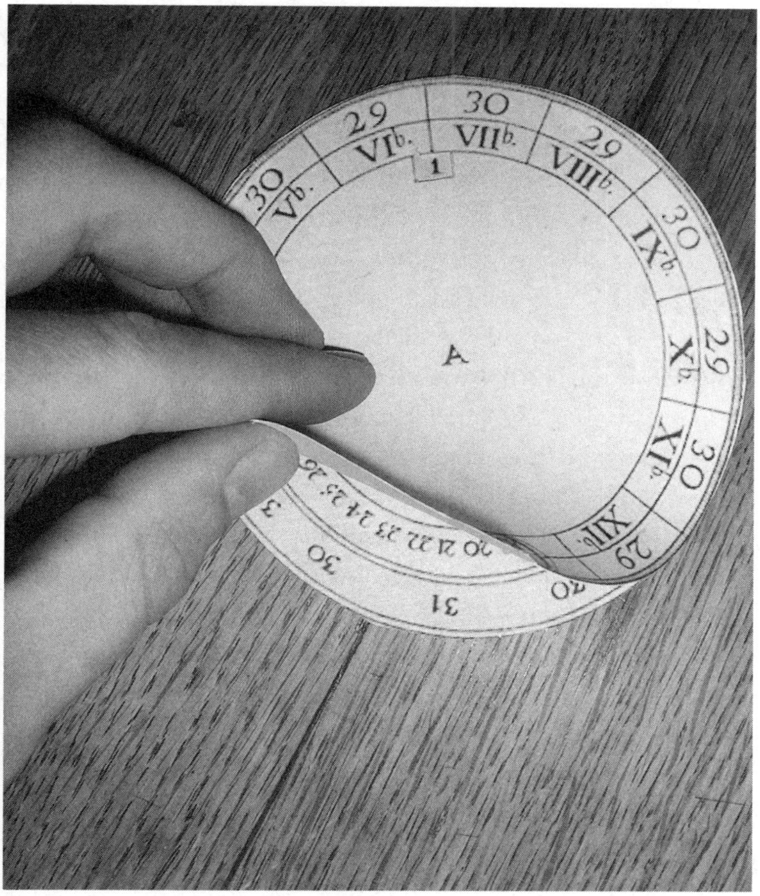

FIGURE 4.6 * Author's assemblage of Robert Wood's "Garter" as a paper instrument.

article, Dr. William Croone explains that vertical line *AE* flanking the image's border and extending down to the bell-like weight *F* denotes one of the numerous "carnous Fibres" from which human muscles are composed. In turn, each of these fibers is itself composed of "an infinite number of very small *Globules*, or little bladders."[84] Therefore, subdividing carnous fiber *AKE* into four smaller isosceles triangles (*ALB*, *DOE*, et cetera), which stand for the "infinite" globules contained within it, Croone's seemingly modest diagram takes on truly comprehensive dimensions. "Since the force of the whole *Muscle* is but an Aggregate of the Contractions of each particular *Fibre*," Croone reasons, it becomes possible to infer from the mechanics of one carnous fiber to the behavior of the whole muscle—indeed, of all muscles.[85] In this geometrical diagram, he visualizes the principles requisite for elucidating virtually any problem relating to muscular motion.[86]

Croone had been moved to this ambitious pitch of abstraction by the twin spurs of confirmation and competition. First, upon seeing the meticulous, microscopically aided images of muscle fibers recently sent to London by Antoni van Leeuwenhoek (fig. 4.7*a*, *b*), he claimed new observational support for his own, earlier speculations on muscular mechanics.[87] More important, Croone had identified a rival, an interloper intruding on his graphic vocabulary. "Because a sheet or two, and two or three schemes, of that long expected Work of *Borelli*, de *Motu Animalium*, having been sent to the Royal Society," he explained, "I find some Schemes for explicating Muscular Motion, the very same with those I make use of."[88] Knowing that Hooke was to print an excerpt from the Italian's work (see "Fig. 10" in figure 4.5), Croone rushed his own texts and diagram into publication directly alongside Borelli's. As he put it pithily, "I would not be thought to have made use of what was another's for my own."[89] Encouraged by the visual evidence of Leeuwenhoek's research, he thus produced a generalized, geometrical image denoting the sum total of muscular movement, using the medium of Hooke's periodical to claim public rights of priority and interpretive authority over his newfound philosophical rival.

What conception of collective identity is materialized in a tense, testy plate such as this one? In his classic study of group portraiture, pioneering art historian Alois Riegl emphasized the peculiar agency of the beholder in constituting the earliest, or "symbolic," phase of the corporate portrait genre in sixteenth-century Holland. In gazing upon the formal assemblages of disconnected, individual figures by which painters then signified the unity of religious confraternities and other corporate bodies, "it is the viewing subject who establishes the important relationships, not only between the figures and himself or herself, but also between the figures within the painting as a whole," Riegl claimed.[90] John Evelyn had required even greater intervention from a historically minded beholder who would harness the cognitive virtues of coins, medals, and other impressed artifacts. Evelyn's prospective historian of Restoration experimental philosophy had to move between physiognomic analysis of each experimentalist's portrait and the evidence of his writings; only by working across a mass of data materialized in medals and texts could that interpreter make inferences to the general character of individuals and, from them, the group as a whole. No less integrative intent was required for a beholder to divine unity from the variegated research and contesting visualization strategies placed in paratactic collision on the illustrative plates of periodicals such as Henry Oldenburg's *Philosophical Transactions* or the early numbers of Hooke's *Philosophical Collections*. Reciprocally, in the hurly-burly of experimental practice, asserting priority and individual authorship through a journal's texts and images certainly mattered almost infinitely more to contributors such as Richard Towneley, William Cole, or

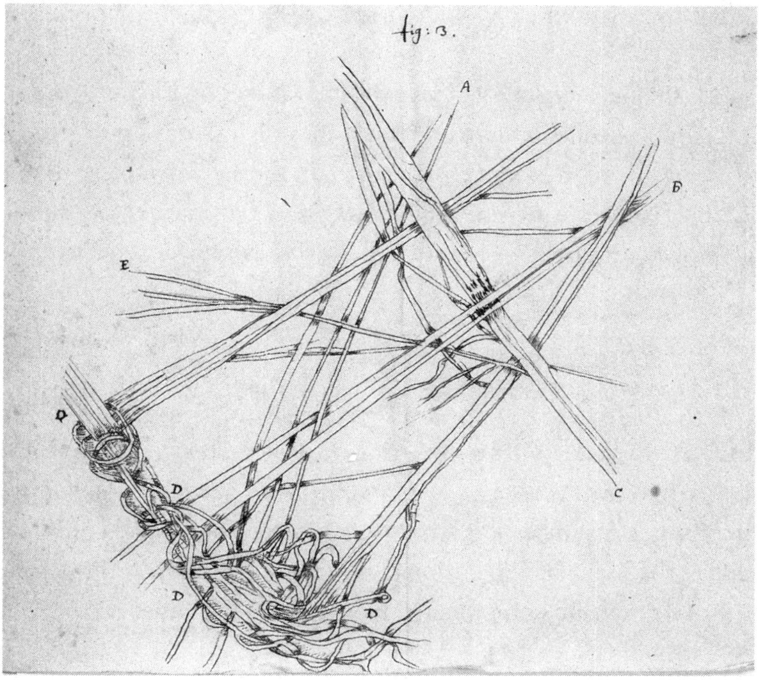

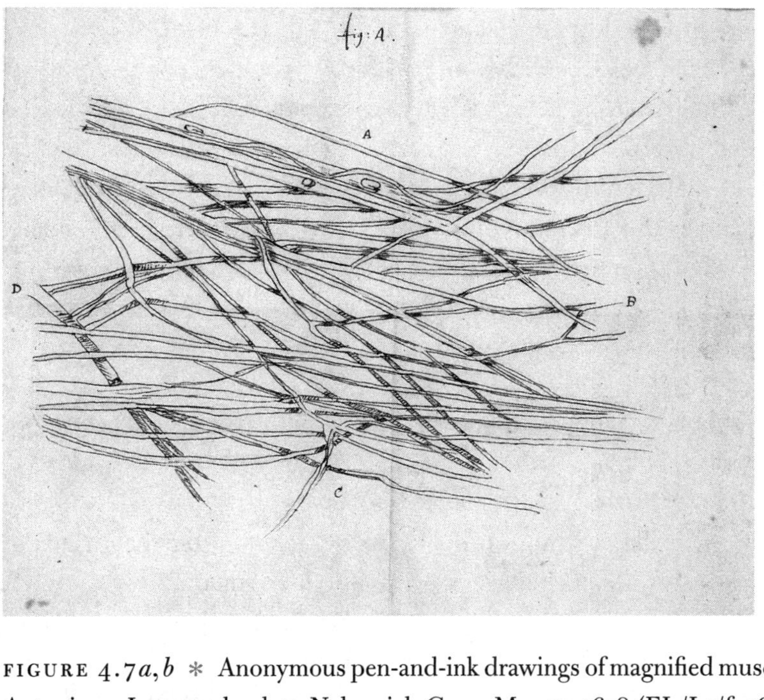

FIGURE 4.7*a*, *b* * Anonymous pen-and-ink drawings of magnified muscle fibers sent by Antoni van Leeuwenhoek to Nehemiah Grew, May 31, 1678 (EL/L1/f. 36). © The Royal Society. Reproduced by permission.

William Croone than any relations a beholder might form between the periodical's component parts.

Yet Robert Hooke took a very different view of this situation. The integrity of experimental-philosophical collaboration and the agency of periodical publication in forming it were issues he took extremely seriously. In the sections that follow, I examine why and how Hooke came to see his *Philosophical Collections* as a graphic instrument for shaping the thought and actions of the philosophical collective beholding it.

Compounding

In December of 1679, two years into his five-year stint as Secretary to the Royal Society, Hooke began performing a series of experiments on the chemical mixture of metals. By combining substances, his experiments tested how it was possible to derive a composite that departed in curious ways from the physical properties possessed by each of its metallic components. Here is how his initial public trial proceeded:

> The experiment proposed at the last meeting by Mr. Hooke, to shew, that copper and tin being melted together into one mass, would make a composition extremely different from them both, was tried and examined; and it was found, that equal parts of copper and tin melted together made a metal, which was exceedingly hard and brittle, though the ingredients are both very soft and very malleable: and whereas copper is of a very brown red colour, this was extremely white, and, which was the principle property that it had newly acquired, the specific gravity of it was found to exceed those of both copper and tin.[91]

Moving in classic alchemical terrain, these metal-mixing experiments need to be seen as sympathetic to Hooke's theoretical work of the later 1670s.[92] In *Lectures de Potentia Restitutiva* (1678), he had articulated his famous "law of spring," whereby the power of a spring is directly proportional to its extension. From that principle, Hooke then suggested more speculatively how conjunctions made between atomic particles vibrating at different frequencies could form unions with new and surprising physical characteristics. When combined, such incongruous particles "become of another nature, and receptive of another degree of motion and vibration, and make a compound particle differing in nature from each of the other particles."[93]

For nearly four months over the winter of 1679–80, Hooke explored this hypothesis in near-weekly experiments, forging compounds of lead and tin, silver and tin, silver and lead, gold and silver, silver and copper, and numerous other configurations.[94] Contrary to the dominant scholarly view of his later experimental work as intermittent and half-hearted, this program was not only sustained but an object of clear interest to Royal Society elites. Seven of these experiments were deemed sufficiently significant to be entered into the Society's Register Book, thereby constituting the full sum of Hooke's work registered during his five-year tenure as Secretary—and amounting to more than half of all twelve items so registered during that time.[95] Although dedicated to the chemical fusion of metals, Hooke saw these experiments as offering broader implications. Beyond likening his trials to the method that had led Archimedes "to discover the cheat of the goldsmith in making of Hiero's golden crown," he stressed that his experiments gave stern corrective to philosophers, physicians, or anyone else guilty of practicing common sense.[96] "The most part [of] all the world," Hooke observed, "have hitherto proceeded upon this false principle that every particular substance or body that goes to make the compositum does when mixed keep its property or power entire."[97] Instead, as the bonding of metals repeatedly proved under experiment trial, entities produced through chemical combination could depart radically from their component parts.

Important to Hooke and his contemporaries as they evidently were, these metal-mixing trials appear nowhere in the *Philosophical Collections*. Not a single mention of them can be found in Hooke's periodical, which had just begun circulating by the end of 1679. If explicitly absent, I think that the metallic experiments can be detected in the *Philosophical Collections* at a more fundamental level. Against the identity-preserving vision of the philosophical group portrait articulated by John Evelyn, that is, Hooke's experimental compounding of copper and other metals can provide a model for elucidating how the Secretary came to see his periodical and its copperplate images as instruments for advancing the collective, collaborative project of experimental philosophy.[98] I make this argument in the present section by turning to a key series of theoretical texts that Hooke drafted in the later 1670s on the improvement of institutionalized philosophy.[99] The vision of the ideal scientific society he articulates in those voluminous texts has appeared to some leading historians as not only far from the collegial, gentlemanly sociability often attributed to the early Royal Society but positively "dictatorial," "almost draconian."[100] What I propose is that we can use Hooke's theoretical writings as a way to read the graphic form of the *Philosophical Collections*—to see how he came to reckon with the possibility that, when put into integrated combination on his journal's lead-type composing tray or illustrative copperplates, discretely authored, privately motivated texts and experimental images could dissolve into novel, hybrid composites.

As a good Baconian, Robert Hooke took as axiomatic that science had to be a collaborative endeavor. The philosophy of experiment had grown so expensive—its demand for technical skills, comprehensive information-gathering, and sheer hard work so capacious—that shared labor was mandatory. "Tis most probable to be done," so Hooke observed of experimentation in the later 1670s, "by the joint unanimous and regulated labor of a multitude. This shows the necessity of a Society."[101] Since its founding and incorporation in the early 1660s, the Royal Society of London had served as a leading model across Europe for how such a community could be constituted. An avowedly nonpartisan, nonsectarian public institution, the organization claimed to choose its Fellows based on merit and to fund its operations by membership dues. Electing its own officials to hierarchical positions of governance and licensing its own publications, the institution also made its activities available to nonmembers who could visit its museum, read the *Philosophical Transactions*, and potentially attend its weekly meetings. By the mid-1670s, however, acute problems had begun to appear within the Royal Society. Plagued by financial woes due to a bloated membership of absent, inactive nobles and aristocrats who had been made honorary Fellows but simply disregarded paying annual fees, critics both inside and out intensified their scrutiny of the organization's structure and aims.[102]

Hooke was a leading voice of this criticism, even going so far as to form his own splinter "philosophical club" in the mid-1670s.[103] During the years when he was editing *Philosophical Collections*, however, he took a more conciliatory tack and prepared prodigious drafts of what he called "Proposalls for the Good of the Society."[104] Who should be admitted to an experimental-philosophical society? How should that community be organized? What kinds of access to the society's activities should nonmembers enjoy? And what could properly motivate and coordinate its efforts? These were the kinds of questions that Hooke and others were asking in the later 1670s, and the answers outlined in Hooke's "Proposalls" are decisive. Neither birth nor title would leverage admission to the experimental society he envisioned. "It must be made an inviolable order or law," he stipulated, "that noe person whatsoever upon any account or respect be admitted into the number thereof that shall not by ten persons of the Society or publique action or publication, be known to be naturally and zealously inclined to the prosecution of the designe of the Society."[105] Once such an extraordinary candidate had been found, he (no women are ever mentioned in Hooke's schemes) was required to pay both a five-pound entry fee and the annual dues of five pounds two shillings before swearing an oath on the Bible. Committing to "perfect obedience and submission to the laws and constitutions of the Society," the experimentalist's load hardly lightened.[106] Bound to secrecy concerning nearly all aspects of the group's activities, members faced fines for late payment of dues, quarterly deadlines for presentation of publishable research, and expulsion from the community if productivity flagged.[107] Even those who opted out of the society were bound by a gag rule. Departing members could not "either indirectly or directly doe or speak anything that shall lead to the detriment or defamation thereof, nor shall he at any time declare or make known the reason of his leaving thereof."[108] With these words, almost all public access to the society's activities and spaces—its meetings, experiments, museum, and library—was revoked and limited to members alone.[109]

Where many of his contemporaries linked the new sciences' ambitious conquest of nature to comprehensive social change, Hooke reveled in this culture of meritocratic exclusivity.[110] Experimental philosophy, he claimed, "tis the work of a Society of men of the most accomplish't abilities, of such as have eminency both of parts and fortune & of such as have a will as well [as] abilities."[111] Commensurately, he calibrated the institutions and incentives of his philosophical society toward motivating a wealthy, leisured, and intellectually vigorous membership. To spur innovation among the society's curators, he proposed that "there be other encouragements given according to dessert upon each new invention or discovery either in money plate medals or other gratuities but that his salary and gratuity shall not annually exceed 50 p.a."[112] Symbolic incentives and honors would likewise be offered to deserving inventors or innovators: "Any person that hath found out any new and usefull invention . . . shall receive from the Society a publique attestation thereof under their common seal, and be further gratifiyed by a medall, picture, or some mark of honor and respect suitable to his invention."[113]

Whatever agency these distinctions might have had in producing important scientific research, Hooke's calculation was that they alone could serve to entrench the philosophical society in the hearts and minds of its members. "The power of the Society," he cannily observed, "is only alluring not otherwise Coercive and therefore the more allurements[,] the more power it hath, and the greater will be the terror of the only punishment it hath, namely expulsion or ejection."[114] In a series of surviving drafts, we can see Hooke tallying the enticing benefits by which a philosophical society could attract members and marshal them into productive action. Beyond the exclusive rights to consult the library, museum, and registers of the society, the enticements he listed include

> Desirable acquaintance
> Delightfull Discourse
> Pleasant Entertainment by Experiments
> Instructive Observations, by Tracts

Considerable Intelligence by Letters
New Discoveries, by Inventors
Solutions of Doubts and Problems
An easy way to know what is already known[115]

Whereas Restoration civil society had advanced a prodigious array of disciplinary measures against its wayward subjects, the philosophical collective's only punitive recourse was expulsion and denial of its select benefits. Therefore, an experimental society had to contrive a system of modest salaries and high symbolic "allurements" to spur achievement while tightening the attachment of its fellows to their hard-won privileges.

Periodical publication was a tool for sustaining the centripetal force generated by this mixture of enticing rewards and threatened exclusion. Hooke acknowledged that the society should "compose and print such experiments or inventions as the Society shall approve of as useful to be publiquely known."[116] Unlike Henry Oldenburg's monthly, publicly trafficked *Philosophical Transactions*, though, Hooke's publication was to appear *annually*, while an entirely different periodical form would be delivered exclusively to the society's members. Brief, serial updates, Hooke explained in early 1680, would provide "some satisfaction to our absent members and may incite them to act at a distance and make good return hither, so soon as the designe is perfected."[117] This "gazette" documenting experiments and research presented at the society's meetings would be assembled from the group's official register, printed monthly, and distributed to every member of the community.[118] And because members were sworn to secrecy regarding the contents of these gazettes for a full year, experimentalists would share their work more freely and frequently. "Restraining them in most things . . . ," Hooke reasoned, "will in all probability produce more liberal imparting of secrets and more full meetings and more plentifull correspondence from abroad."[119] Not only would this monthly periodical thus advertise the achievements of each society member to one another, but the public reticence and honor-bound privacy of the society's communications would actually encourage its members' philosophical verbosity.

In practical terms, periodical publication was a critical means by which the activities of the society's highly specialized, geographically distributed membership could be coordinated. What it lost in numbers by excluding all but men of prodigious skill, activity, and wealth, Hooke's philosophical society had to recoup in rigorous organization. Since information relevant to experimental philosophy was to be found "first in bookes, secondly in men, and thirdly in the things themselves," he proclaimed, a philosophical society should accordingly subdivide—each member "according to his opportunities and abilities."[120] Specifically, a tenth of the society should collect and assess relevant literature on philosophical topics, affirming "what they find true and reall and damning all that they find false and fictitious."[121] A fifth of the society's members would turn their attentions to correspondence, to "consulting men and procuring from them either by friendship or reward whatsoever is curious usefull and pertinent to the business of the Society."[122] With their colleagues deputed to surveying existing literature and expertise, the remaining seven-tenths of the society could focus their energies "upon the examination and tryall of Naturall things and improving of artificiall inventions and productions."[123] Only by working to agreed standards and a common plan shared between their specialized endeavors in this way could experimentalists possibly hope to advance the cause of philosophy. Appealing to a suggestive model, Hooke put it like this:

> [No] substantial philosophy should be built before materials necessary for such a structure be collected. Nor can such a collection be made without a great deal of knowledge, care & judgment. It is not any kind of observations or experiments that will be proper for this designe, nor is every observator fitt to be a collector; observations not rightly and accurately made are pertinacious and destructive to others that are real and good, as they are uselesse in themselves and like a rotten beam in a large edifice with their own failing they procure the downfall of the whole.[124]

Specialized and far-flung as its component participants were, Hooke's edifice of experimental philosophy was so reliant on its interconnections that any deviation from integrated function menaced the standing of the whole. "This newfound world," he had pithily noted, "must be conquered by a Cortesian army well disciplined and regulated though their number be but small."[125]

The material imprint of this theorized experimental interconnectivity can be seen in both the textual content and the graphic form of the journals Robert Hooke actually published between 1681 and 1682. Entirely absent from the first number of *Philosophical Collections* printed by John Martyn in 1679, Hooke introduced a sequence of editorial segues connecting topics of experimental research in the journal's second number, which he published with Moses Pitt in early 1681.[126] "You had in the preceding *Philosophical Collections* [1] . . . ," so Hooke's editorial voice intones, opening the journal and referring the reader to the previous edition's article on floods in Gascony.[127] After noting that the floods had been caused by "some Extraordinary change in the Subterraneous Caverns of those Hills," he explains that the reader can find relevant elaboration in the first article of *Philosophical Collections* 2: a piece by John Beaumont on "*Ookey-hole*, and several other subterraneous Grottoes and Caverns in *Mendipp-hills* in *Somersetshire*."[128] At the end of this article, Hooke's editorial voice returns, thanking Beaumont for "communicating several other curious and instructive remarks made by him in his search into Mines" and connecting that work to another unfolding program of research within the experimental community. The editorial voice then observes that "somewhat of this kind is hoped for from the Learned Dr. Plot, who is finishing his Natural History and Survey of Stafford-shire, and in some excursions hath Surveyed the Peak in Darby-shire, and viewed from some other Caverns thereabout."[129] Later, the reader is told how a letter concerning a petrified body in Nuremberg and the sample of "stiff, red, somewhat curled, but rotten Hair" sent to the Royal Society with it had given "the occasion of producing the following discourse"—the terms by which Hooke introduces a paper by Edward Tyson on hair, teeth, and other excrescences discovered growing within human organs.[130] Not only do these editorial segues create connections between international programs of as-yet-unpublished research, but they narrate for the reader how distant, diverse findings delivered to the Royal Society's exclusive meetings had actually stimulated new philosophical work.

By the final numbers of the *Philosophical Collections* printed with Richard Chiswell in rapid succession in 1682, Hooke's interventions had become far more graphically invasive.[131] In the fourth number of the journal from January 1682, Hooke fully eschewed Oldenburg's trusted convention of the single illustrative plate, opting instead to mingle letterpress type and woodcut text with copperplate engravings of Antoni van Leeuwenhoek's magnified cross sections of hair.[132] The journal's graphic experiments with *mise-en-page* simultaneously veered into outright criticism.[133] For *Philosophical Collections* 4, Hooke had translated observations on the 1680–81 comet made by French astronomer P. J. de Fontaney and juxtaposed his own critical commentary to them (fig. 4.8). Separated from Fontaney's text by a cordon of bare page, Hooke's objections are printed in a smaller italic font. As one indented note reads, "He takes no notice of a way of Sir *Ch. Wren*, by four observations published by me *An. 1677.* by which the position of the line and parallax is determined, though neither of them can be by this of *P. Pardies*."[134] Thus, at his journal's most intensive period of activity in 1682, Hooke had come to embrace an aggressive array of graphic editorial practices by which to make manifest the connections obtaining between diverse, specialized experimental researchers and their findings. On printed page, he was narrating the thematic conjunctions and historical linkages between seemingly disparate programs of research, mingling texts and images, and embedding his own critical commentaries within what he had earlier proclaimed would be "Verbatim" delivery of contributors' work.[135] Rather than

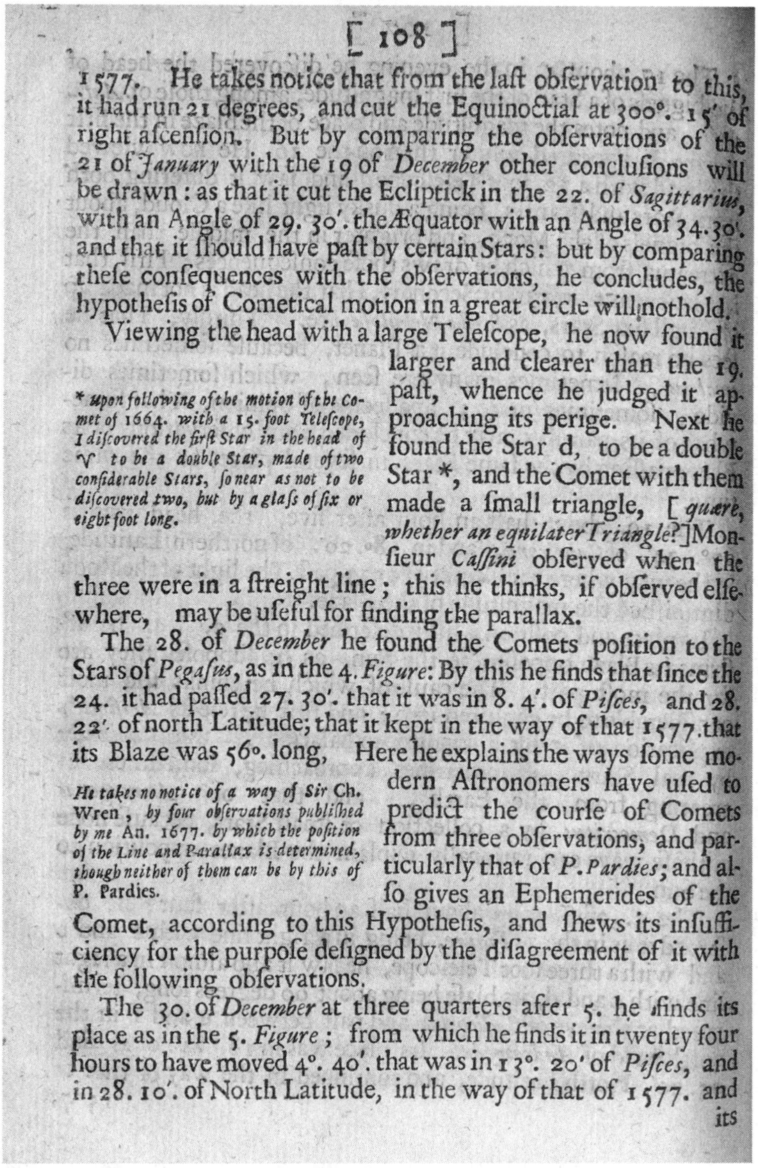

[108]

1577. He takes notice that from the laſt obſervation to this, it had run 21 degrees, and cut the Equinoctial at 300°. 15′ of right aſcenſion. But by comparing the obſervations of the 21 of *January* with the 19 of *December* other concluſions will be drawn: as that it cut the Ecliptick in the 22. of *Sagittarius*, with an Angle of 29. 30′. the Æquator with an Angle of 34. 30′. and that it ſhould have paſt by certain Stars: but by comparing theſe conſequences with the obſervations, he concludes, the hypotheſis of Cometical motion in a great circle will not hold.

Viewing the head with a large Teleſcope, he now found it larger and clearer than the 19. paſt, whence he judged it approaching its perige. Next he found the Star d, to be a double Star *, and the Comet with them made a ſmall triangle, [*quære, whether an equilater Triangle?*] Monſieur *Caſſini* obſerved when the three were in a ſtreight line; this he thinks, if obſerved elſewhere, may be uſeful for finding the parallax.

** upon following of the motion of the Comet of 1664. with a 15. foot Teleſcope, I diſcovered the firſt Star in the head of ♈ to be a double Star, made of two conſiderable Stars, ſo near as not to be diſcovered two, but by a glaſs of ſix or eight foot long.*

The 28. of *December* he found the Comets poſition to the Stars of *Pegaſus*, as in the 4. *Figure*: By this he finds that ſince the 24. it had paſſed 27. 30′. that it was in 8. 4′. of *Piſces*, and 28. 22′ of north Latitude; that it kept in the way of that 1577. that its Blaze was 56°. long, Here he explains the ways ſome modern Aſtronomers have uſed to predict the courſe of Comets from three obſervations, and particularly that of *P. Pardies*; and alſo gives an Ephemerides of the Comet, according to this Hypotheſis, and ſhews its inſufficiency for the purpoſe deſigned by the diſagreement of it with the following obſervations.

He takes no notice of a way of Sir Ch. Wren, *by four obſervations publiſhed by me* An. 1677. *by which the poſition of the Line and Parallax is determined, though neither of them can be by this of* P. Pardies.

The 30. of *December* at three quarters after 5. he finds its place as in the 5. *Figure*; from which he finds it in twenty four hours to have moved 4°. 40′. that was in 13°. 20′ of *Piſces*, and in 28. 10′. of North Latitude, in the way of that of 1577. and its

FIGURE 4.8 ✻ Robert Hooke's commentary in indented notes printed in *Philosophical Collections* 4 (January 10, 1682; London: Printed for Richard Chiswel): 108.

the identity-preserving mode of metallic juxtaposition envisioned in John Evelyn's philosophical group portrait, we might say, Hooke's literal melding of experimental research evokes the transformative chemical fusions of compounded metals he was then demonstrating before the Royal Society.

Reading the *Philosophical Collections* this way, it is important to note how profoundly the journal's strategies of "compounded" authorship resonate with the numerous other works that Hooke was publishing under his own name in the late 1670s. As with his *Lectures de Potentia Restitutiva* (1678), Hooke begins his *Lectures and Collections* (also 1678) with a substantial tract detailing his own research on a philosophical topic.[136] Both books then open onto contributions from other named experimentalists, including Robert Boyle, Christopher Wren, and Antoni van Leeuwenhoek.[137] As he would later do in the *Philosophical Collections*, moreover, Hooke links these component pieces to the substance of his own initial texts through a range of often very loose discursive segues while deploying the convention of the single illustrative plate to compress images together.[138] If he had earlier claimed lone responsibility for the work and "the faults of my *Conjectures*," Hooke's authored publications of the late 1670s had thus begun to read and look like his periodical.[139] In fact, he even combined *Lectures and Collections* and *Lectures de Potentia Restitutiva*, printing them together in 1679 as a veritable Latourian cascade called *Lectiones Cutlerianæ*.[140]

"So many are the *links*, upon which the true Philosophy depends," Hooke had written in *Micrographia*, that ". . . if any one be *loose*, or *weak*, the whole chain is in danger of being dissolved."[141] Hooke's metal-mixing trials of the late 1670s suggested ways to transmute those crucial links into *Philosophical Collections*. But how could that journal actually help to regulate and motivate what the Secretary of the Royal Society had come to see as his philosophical "army"? Offering direct counterpoint to the narrative of Bristol's William Cole with which the chapter began, Hooke's use of his periodical as a means for advancing his vision of experimental connectivity among far-flung factors (including Cole himself) enables us to see what the philosophical collective had begun to look like when gazing outward from rebuilding London, and why it needed an architect.

Printing Things Together

By the later 1670s, Robert Hooke possessed keen advantages in the scrum that was experimental-philosophical publication. Commanding the powerful post of Secretary of the Royal Society of London from 1677 to 1682, he was wealthy, urbane, and in near-daily contact with London's communities of publishers, booksellers, engravers, and etchers—those agents of print culture on whom an aspiring experimental author depended. St. Paul's churchyard, home to many of the City's printers, was less than a mile from his lodgings at Gresham College. So his diaries show us, the Secretary could not go long without indulging his curiosity by leafing through the latest works from the Continent on mathematics or terrestrial exploration offered for sale at the shops of John Martyn, Moses Pitt, and Richard Chiswell, the printers with whom he published his *Philosophical Collections* between 1679 and 1682.[142]

But Hooke's urban location also had its drawbacks. Unlike his friend William Cole, the Secretary could not exploit an official position at one of imperializing England's leading ports to skim off the cream of natural specimens and curiosities arriving from distant corners of the globe. Nor did Hooke possess a country estate like that in Surrey from which John Evelyn conducted his natural-historical interests in later life. Whereas John Covel or his friend Edmond Halley pursued their philosophical investigations from distant outposts, Hooke rarely departed from London; he never left England.[143] And as the Royal Society itself could only commission acquisitions to a very minimal degree, so Hooke was forced to rely on data collected from a network of informants.[144] In conclusion, I want to examine how Hooke utilized periodical publications as instruments for sourcing, disseminating, and enacting the models of experimental collectivity graphically built into them.

In the late fall of 1679, Hooke began corresponding with an English resident of Portugal named Henry Jacobs.[145] Although only Hooke's side of the correspondence survives, the Secretary was clearly eager to make Jacobs his Iberian agent. As he instructed in a letter from October 1681, Jacobs should try to learn "if Don Alu: de Laucha has done anything about the Map of Portugall, or if there be any mapps, observations, or voyages lately published of any part of the East or West Indies, Brazil, Ceylon or Africa."[146] With no maps to be had, Jacobs was then told to take regular readings of "variations of the magnetic needle" by which available geographical information could be corrected.[147] But once it became clear that his would-be Iberian man needed remedial instruction, Hooke provided both a variation compass and explicit guidelines.[148] So he wrote in March 1682,

> I have procured a very curious needle for observing the variation. It is wrapped up in paper in this box and soe you may put it on when you use it, the edge of the brass plate (placed under the box) is to be set true to an exact meridian line drawn upon a stone or table without any nails or iron about it. And great care is to be taken when the observation is made that no iron be within two or three yards of the needle. Then the box is to be turned and sett soe as the points of the needle may exactly respect the points what are placed in

the ends of the box. When by looking upon the divisions of the brass plate you will easily find the quantity of the variation upon the needle from the true meridian. The box is made to open [such] that you may turn the ends of the needle in it as you shall have occasion.[149]

Hooke seems to have had little confidence in Jacobs's aptitude as a philosophical observer. To ensure that Jacobs's data would not act "like a rotten beam in a large edifice," he thus supplied the compass himself while articulating simple steps for setting up the instrument, eliminating environmental interferences, and taking the quantitative reading.

When working with an informant like Jacobs, Hooke saw periodical publication as both a pedagogical instrument and a species of incentive he had earlier called an "Allurement." The enticement was simple. "So soon as they have a sufficient number of such observations," he was then pledging, the Royal Society would publish its collected results in a periodical "wherein mention shall be made of every person so making and communicating his observations."[150] In exchange for following Hooke's instructions and delivering reliable readings from his standardizing instrument, correspondents like Henry Jacobs would be noted in print as contributors to the larger collective enterprise of the prestigious Royal Society. But periodicals could also act as a guide in the field in a far more direct way. Scrawled on the back of a copy of a letter he sent to Jacobs in early 1682, Hooke noted, "With this I sent Doctor Woods paper of infinity; last correction of his almanac; his sheet almanac; and the graven garter and moveable lens."[151] Woods's "graven garter" had, of course, previously appeared in *Philosophical Collections* 2 as a paper instrument. (See figure 4.6.) Compounded from diverse strands of philosophical research, the text and images of the *Philosophical Collections* thus became tools Hooke could redeploy to teach and direct his would-be agents. If building the edifice of knowledge required the guidance of "the skilfullest Artists," so Hooke's friend Edward Tyson had asserted in 1680, "even the meanest in some things may give in their informations."[152] Alluring as its commonplace form could be to "proletariat of organized science" such as Jacobs, the *Philosophical Collections* had simultaneously become a mechanism for ensuring that information gathered by such inexperienced, unskilled investigators would still be worth collecting.[153]

If wielded with greater subtlety, periodical publication could also assist in extracting contributions from more senior members of the experimental community, including the irascible William Cole. As Cole was busy trafficking his politics of shellfish in the mid-1680s, Hooke delegated his deputy Richard Waller to visit the naturalist's home in Bristol. Beyond sizing up Cole's celebrated collection of natural rarities, Hooke was eager to obtain specimens for the Royal Society's museum. What Waller found was impressive. "I have seen some of Mr. Cole's curiositys," so he wrote to Hooke in 1687: "He has a good collection of English plants dried in books, with their names and descriptions. The figures of severall not described insects, amongst the rest some figures & observations touching spiders which he told me he had designed for the Royal Society."[154] Try as he might, though, Waller was unable to persuade Cole to publish his efforts in the organization's periodical. "He put me off," Waller reported, "saying he intends to publish all his observations in a book together."[155]

Effectively deploying the analysis of institutional power he had articulated in his theoretical writings of the late 1670s, Hooke encouraged Waller to take a different angle. He suggested, "Pray if you see Mr. Cole againe, offer my service to him; tell him he must not forget his promises to mee soe long before he went out of his way in going into the Oxford Society before he was of the R.S. of London who had soe great an esteem of his discovery."[156] Unable to punish Cole directly for noncompliance, Hooke's strategy was to mingle the benefits ("my service to him" and the Royal Society's "great esteem") Cole stood to regain by cooperation while weaving him back into the net of mutual obligation. Hooke also called on periodical publication as a means of prying loose and gathering together the observations Cole still clutched in his obstinate hands. Waller was to remind Cole that

> he can have no better way in the world to present use and future ages both his observations and curiositys than by the Transactions and the Repository of the R.S. and that will also repay him in some measures at present by returns of other observations, and for the future by a lasting monument of his name.[157]

With the privileged access to a free flow of natural rarities he commanded from his customs post, Cole knew he was in a desirable position. Further, he had used the *Philosophical Transactions* to his advantage in the past, publicizing his work on dye in the mid-1680s while jockeying for position between the experimental communities of Oxford and London. But Hooke shrewdly sized up Cole's vulnerabilities. The Secretary reckoned that his Bristol colleague would struggle to accomplish his individual publication, which could secure the "lasting monument of his name" Cole clearly desired. As a periodical regulated and rendered useful the observations of novices like Henry Jacobs, so it enacted what Hooke saw as a scientific institution's sole basis of power. A philosophical society could stimulate intellectual production neither by paying handsome salaries nor by leveling threats of bodily punishment. Desire was more effective than fear; only by increasing its attractiveness could such a society persuade Cole—a man "infected with an immoderate desire of honour"—to subsume his private, personal agenda within the collective enterprise.[158] On this score at least, Hooke was entirely correct. Ever resistant, Cole died intestate. Of his *General Natural History*, famous collections and extensive drawings, his biographer concludes, "oblivion has swallowed all."[159]

Historians of science tell us that neither Robert Hooke's *Philosophical Collections* nor his actual administration of institutionalized science adequately addressed the pressing needs of the Royal Society in the later 1670s.[160] Nonetheless, what Hooke's theoretical writings and his periodical's graphic form indicate is that in the years around 1680, orchestrating experimental philosophy's constellations of images and coercing its disjointed, distributed, self-interested agents—figures such as Cole, John Covel, Michael Burghers, William Croone, John Martyn, Moses Pitt, Richard Chiswell, Henry Jacobs, and indeed Hooke himself—into coherent action had become a contested project central to the wicked intelligence practiced in London. How the enterprise of collecting artifacts became a leading frontier for theorizing those dynamics circa 1680 is the problem to which we must now turn.

CHAPTER FIVE

"The Wonderful Elaboratory of the Animal Body"

The Royal Society's Repository at Work

IN HIS CLASSIC *Historie of Four-Footed Beasts* (1607), Anglican cleric Edward Topsell outlined a compendious program for those who would breed that central object of early modern transport, labor, and social status: the horse. The astute groom was to make a close study of probabilities, calculating how to produce the most desirable colt. Clues to fit parentage could be drawn from a range of evidence: the color of the stallion's mane and the texture of his ejaculate, the clarity of the broodmare's eyes.[1] Guided by the woodcut of an exemplary stallion that Topsell had pilfered from naturalist Conrad Gesner, though, the ultimate criteria for any prospective breeder had to be physiognomic[2] (fig. 5.1). "The beak or snout of a horse," Topsell directs, "ought not to stand out like a swine's, but to bend down a little crooked. . . . It ought also to be considered, whether his cheekbones be sharp, tender, or unequal, standing one above another, for their imparity maketh the horse's neck to be hard, and stubborn."[3]

No doubt the object of such prenatal attentions, a colt was born among crowded

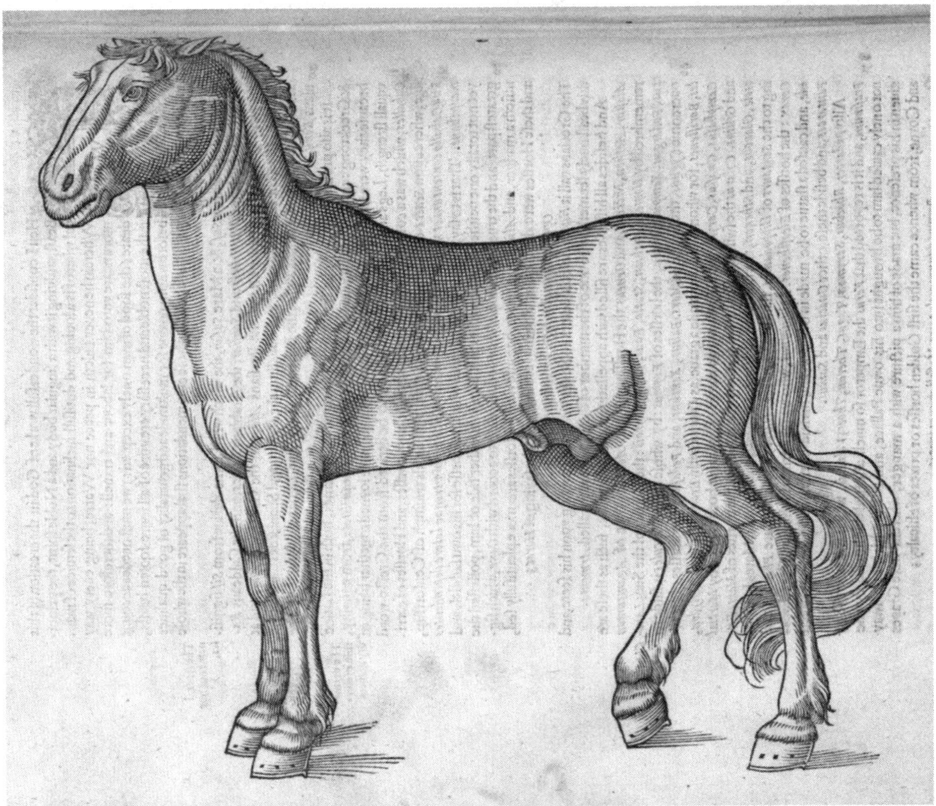

FIGURE 5.1 * Anonymous woodcut of a horse; from Edward Topsell, *The historie of foure-footed beastes* (London: William Iaggard, 1607), 282; Huntington Library, RB 17726. Reproduced by permission of The Huntington Library, San Marino, California.

stables in London's fashionable West End in the spring of 1665 that dashed the hopes of its human wards. If its body "appeared to ye Eye compleatly formed without any monstrosity to be taken notise of in it," the newborn colt's head was less lucky.[4] The beast's defining feature was described by one contemporary as "a double eye in the midst of its forehead, having double eye-lids, and double pupils."[5] This was the attribute that caught the attention of Restoration England's leading experimental philosopher, Robert Boyle. Like many early modern intellectuals, Boyle was cognizant of the luciferous properties of monsters, but he took a highly pragmatic interest in this colt. In his *Some Considerations Touching the Usefulness of Experimental Philosophy* published two years earlier, Boyle had promoted ethyl alcohol as a means of preserving and displaying anatomical specimens. "This Liquor being limpid, and not greasy," he had written, "[it] leaves a clear prospect of the Bodies immers'd in it."[6] With this new monster, he could now prove his point. Thus, amid the rising temperatures that would soon compound London's oncoming devastation by the bubonic plague of 1665, Boyle had the colt's head removed from its body, immersed in preservative solution, and donated to the Royal Society of London.[7]

At Gresham College, in the heart of the City some two miles to the east, Robert Hooke properly immortalized his former mentor's gift. (See figure 4.2.) Through the promptly published etching based on Hooke's drawing, we see the animal in three-quarter profile at right, its lone, compounded eye gazing lugubriously upward. The colt is a wreck. Rising from dewy meadows of hatching that denote its fur, the creature's topography of excesses appears all the more surprising. Here, a wooly caterpillar of flesh sprouts from a mysterious orifice on its forehead; there, superfluous lashes jut at midbrow as bristles spit from a slack jaw and dog-like nose. Raking its tongue crookedly across a battered forelip, the animal's expression of blunt stupidity is equaled only by its apparent abashment. With backward-turned ears and drooping tear ducts framing the massive, cyclopean eye, the beast appears as if apologizing for its own monstrosity.

Disfigured and nearly unrecognizable as a horse, the colt's physiognomy is still legible, literally, by way of reference characters flanking its eye and forehead tumor. In this way, Hooke's image has become both graphic memorial to the donated head and visual testimony to the efficacy of Boyle's chemical preparation. By showing differentiated features persisting within delicate monster flesh, Hooke's depiction, as much as the pickled head itself, gives what Boyle would call "good example together with a proofe yt by ye help of spt wine . . . ye parts of animals & ev'n monsters may in summer itselfe be preserved."[8] Where monstrosity as pickled specimen had become gift and example, as depicted it became visual proof of experimental process.

Once it had entered the public space of the Repository, or museum of the Royal Society at Gresham College, however, the deck was shuffled once again. Boyle's head prompted an entirely different set of questions. Had this beast really been a monster at all? Some Fellows of the Society wondered: could a postnatal calamity rather than reproductive accident be blamed for causing this "double eye in the midst of its forehead, having double eye-lids, and double pupils"?[9] Against Boyle's claim of native monstrosity, other experimentalists queried whether the colt "had received a forcible kick on the tender head, whereby the eyes were dislodged, and forced into one place."[10] The matter was to be settled by dissection. In chambers adjoining his own lodgings in Gresham College, Hooke (now in his role as Keeper of the Royal Society's Repository) was enlisted for the task. Boyle was vindicated: Hooke's dissection revealed that between and behind the compounded eye was indeed "but one optic nerve."[11] And while Boyle's gift and example had

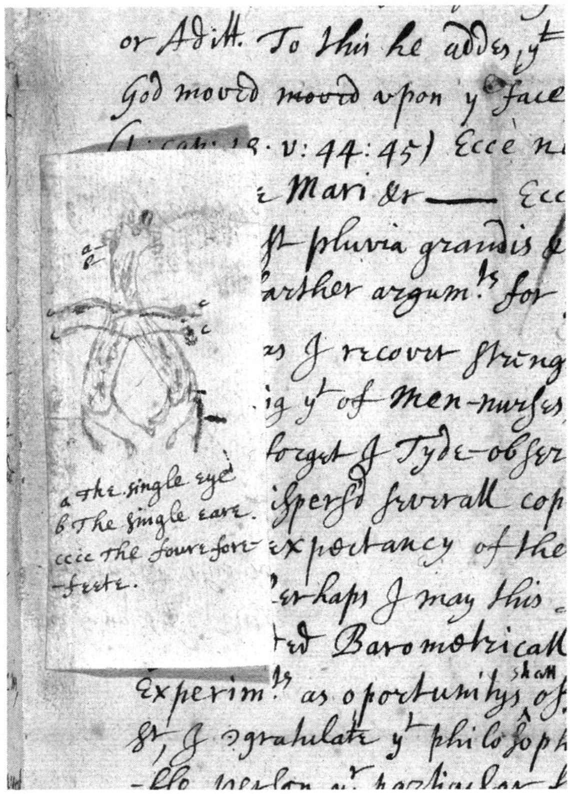

FIGURE 5.2 * Anonymous anatomical drawing of a monstrous lamb sent by Samuel Colepresse to Henry Oldenburg, April 13, 1667; Royal Society Early Letters, vol. C1, fol. 14. © The Royal Society. Reproduced by permission.

been sacrificed in the name of truth, the Repository—the space that hosted the head, Hooke's drawing session, and the dissection alike—quickly gained a new donation. Inspired by "observing in your Repository a monstrous calfe; an account of another; & a colts head mentioned in the Ph: Transactions," a correspondent named Samuel Colepresse wrote to Henry Oldenburg in April 1667 with a report of his own monstrous findings.[12] Colepresse even sent an image to the Royal Society: a picture of a monstrous lamb drafted by a local who was, as he correctly observed, "noe Limner, that did it"[13] (fig. 5.2).

What are we to make of these interweaving procedures of private collecting, destroying, and transforming that could beget such public prompting, gifting, and reimagining? Why did these events happen in or around the Royal Society's museum, a type of space ostensibly defined by the static uselessness of its collected artifacts?[14] Thanks to its ample period promotion, understanding *how* the Repository attained such prominence among agents within and beyond the early Royal Society of London can be made clear enough. Effectively initiated in the autumn of 1663, the Royal Society's collection grew rapidly through the benefaction of

City merchant Daniel Colwall, who provided one hundred pounds sterling for the wholesale purchase of an existing museum of rarities.[15] Methods of encoding and systematizing artifacts were quickly advanced to organize and expand this existing substrate. As newly created Keeper, Hooke was instructed to "always affix some note to the things in it, by which it might be known what they are, and by whom they were presented."[16] Ambitious plans for the fledgling museum were then trumpeted by Thomas Sprat in his laudatory *History of the Royal-Society of London* (1667) as the Repository was installed briefly at Arundel House on the Strand, among Restoration London's greatest collection of art.

By the early 1680s, Nehemiah Grew's illustrated catalogue, *Musaeum Regalis Societatis* (1681), had won the Repository a favorable reputation across Europe. Donations to the museum became a conventional token of a newly elected Fellow's gratitude to the Royal Society, and as elsewhere on the Continent, a guided tour of the collection figured as an obligatory component of any dignitary's visit to the seat of Restoration experimentalism.[17] For the vast majority of Londoners outside that charmed circle, meanwhile, the Repository was one of the few facets of the Royal Society's life to which a paying public had access. But despite this seemingly auspicious conjunction of commerce and knowledge, art and science, prestige and the public sphere, scholarly accounts of the Repository make for doleful reading. Lax oversight, frequent movement of the collections, and even dampness in the walls, it is said, all conspired against the preservation of the Royal Society's valuable artifacts.[18] "By the late 1720s," so the most recent and thorough historical study proposes, "much of the material recorded in Grew's catalogue was in a bad state or had perished."[19]

What I argue in this chapter is that the material destitution catalogued by period writers and modern historians alike is hardly accidental; it might better be seen as the logical conclusion of the wicked intelligence practiced in and theorized around the early Royal Society's Repository. The skilled transformations, merciless destruction, and cunning reinventions enacted on Robert Boyle's monstrous colt head constitute more than public, collective performances of the private bricolage we saw Robert Hooke practicing with his paper micrometer in chapter 2. Instead, they exemplify ways in which the Royal Society's collections were historically put to use while signaling how the space of the Repository was being imagined as an epistemic model by those most intimately engaged with it. By the early 1680s, Hooke and his contemporaries were deploying the space of the museum to articulate the contours and wherewithal of a governing intelligence needed at the center of the Royal Society's collective enterprises. Building on the project of engineering experimental collectivity exposited in chapter 4, this penultimate chapter elucidates how the need for a governing agent—an agent wielding what we will see called "Archietonical" powers—came to be envisioned at the core of the Royal Society's collections circa 1680.

To tell this story, I begin by juxtaposing the narratives of William Petty and Ned Ward (commentators who will figure significantly in chapter 6) as they elaborate why London's experimentalists needed a museum, and what they could do with it. The chapter's second section then sifts through the highly fragmentary evidence of the Repository's organization and contents to assay the actions performed with objects in the museum at Gresham College. The narrative framed thus, we can apprehend how experience with collected artifacts in the Royal Society's museum enabled the experimentalists to reconceptualize the traditional "repository" of the mind as a radically productive matrix for modeling experimental intelligence. That those collective, collecting enterprises governed by an archietonical agent can be read in and as the built environment of later Restoration London will be a claim central to the final chapter.

The Repository Imagined: Two Views

In the later 1670s, Robert Hooke had labored over the practical means for managing and motivating a "Cortesian army" of collaborating experimentalists. Newly returned from Paris some thirty years earlier, philosopher William Petty (1623–1687) could have identified an obvious, integrative model. In 1647 he had wished "that a Society of Men might be instituted as carefull to Advance Arts as the Iesuites are to Propagate their Religion for the government and mannaging of it."[20] Connected to the dynamic intellectual atmosphere of Wadham College in 1640s Oxford, Petty imagined a philosophical program comparable in infrastructure intensity and pedagogical focus to the Society of Jesus.[21] Writing in a celebrated tract dedicated to emigré reformer Samuel Hartlib, he promoted universal education as means for unleashing economic productivity.[22] By replacing elitist grammar schools with "Literary-work-houses," Petty proposed, the countryside of England's future would be populated with children who could not only read and write Baconian natural histories but draw, paint, garden, set jewels, and grind optical lenses.[23] Speaking with broader conversations about the capacity of technology to advance political economy, he saw his schools as keeping England competitive with "Countries where Manufactures and Trades flourish, as *Holland*, &c.," while simultaneously rectifying the disastrous practice of educating only the children of the nobility.[24] After all, so the humbly born Petty observed, "many now are holding the Plough, which might have beene made fit to steere the State."[25]

Strongly informed by his medical training, Petty sought to make this energizing rural industry available "to Active and Philosophicall heads."[26] Therefore, his scheme required what he called a "Gynasium Mechanicum or College of Tradesmen."[27] Taking a page from the model of Salomon's House sketched by Francis Bacon in his *New Atlantis* (1627), he envisioned this urban research center as spurring practical ingenuity and the "Interpretation of Nature, whereof there is so little, and that so bad as yet extant in the world."[28] What Petty imagined was a wondrous collection,

> a compleate *Theatrum Botanicum*, stalls and Cages for all the strange Beastes and Birds, with Ponds and Conservatories for all exotick Fishes, here all Animalls capable thereof should be made fit for some kind of labour and imployment, thaa they may as well be of use living as dead; here should be a Repositorie of all kind of Rarities Natural and Artificiall pieces of Antiquity, Modells of all great and noble Engines, with Designes and Platformes of Gardens and Buildings. The most Artificiall Fountaines and Water-works, a Library of Select Bookes, an Astronomicall Observatory for celestiall Bodies and Meteor, large pieces of Ground for severall Experiments of Agriculture, Galleries of the rarest Paintings and Satues, with the fairest Globes, and Geographical Maps of the best descriptions, and so farre as is possible, we would have the place to be the Epitome or Abstract of the whole world.[29]

It is worth stressing just how exceptional such a collection would have been in England circa 1650. If the "Ark" of rarities collected and contemporaneously exhibited by the Tradescant family in south London might have approached the scope of natural specimens envisioned by Petty, the galleries of ancient and modern art effectively did not exist.[30] With the outbreak of the English Civil Wars in the early 1640s, many prominent Royalists had fled to the Continent, hustling their collections of "rarest Paintings" along with them.[31] The sound of the auctioneer's briskly banging gavel as Parliament sold off Charles I's fabulous collection of sixteenth- and seventeenth-century paintings to overseas buyers has often been heard as tolling the death knell of a nascent artistic culture.[32] Even more unusual were the "Artificiall pieces of Antiquity" stipulated by Petty. Encountering the strange sight of "the Earl of *Arundel*'s Garden, where there were a great number of Ancient Statues of naked Men and Women," Francis Bacon himself had been gobsmacked. Reportedly, Lord Verulam "made a stand, and as astonish'd, cryed out, *The Resurrection*."[33]

From his sojourns in Paris and Leiden, William Petty would have known just how important physicians and medical men were to the contemporaneous formation of museums.[34] It is thus not surprising that this collection he imagined transforming the intellectual life of the individual and the political economy of the state was to be managed by a board of medical authorities. "A Physician, Chirurgeon and Apothecary," he observed, should be chosen, "each well versed both in the Theory and Practise of their respective Professions."[35] But the tricky work of governing such a complex hive of advanced research required what he called a Steward. As historian Ted McCormick elaborates, Petty's Steward was to be "a proto-statistician . . . what apothecaries, surgeons, and physicians did for individual bodies, he tried to do for the population at large."[36] A mathematically agile administrator, the Steward had to possess the prudence to guide salubrious research as well as the shrewdness to bring it to practice.

Salutary as the medically minded museum was for Petty, attitudes toward collections were changing rapidly in later seventeenth-century London, implicating the reputation of the Royal Society's Repository in them. Penned some fifty years after Petty's utopian visions, the collections imagined in Ned Ward's *The London-Spy* (1703) effectively come from a different world—a world, as we will see in chapter 6, where markets for art had exploded and experimentalists of the Royal Society were hustling for space within a bustling, urban consumer culture.[37] Indeed, it is amid the lurid spectacles of a national capital soon to become Europe's most populous city that Ward's Spy and his knowing guide take in the Royal Society's Repository at "Wiseacres' Hall, more commonly called Gresham College."[38] Crossing court and quadrangle, the visitors are escorted up the stairs, where an "Elaboratory keeper" is only too glad to display a "Ware-house of *Egyptian Mummies*, old musty Skeletons, and other antiquated Trumpery."[39]

In the Royal Society's Repository circa 1700, the Spy sees not just a view of ingenious sights but a show. Using a magnet, the Keeper cleverly "made a Paper of Steel-Filings Prick up themselves one upon the back of another, that they stood pointing like the Bristles of a *Hedge-Hog*."[40] After this jolly animation of inert matter, the contrasting spectacle of bodies from which life has departed is put on display:

> When we had taken a survey of these Pincushion Monsters, we turned toward the Skeletons of Men, Women and Monkeys, Birds, Beasts and Fishes; Abortives put up in Pickle, and abundance of other Memorandums of Mortality; that look'd as Ghostly as the Picture of *Michael Angelo's* Resurrection; as if they had collected their scatter'd Bones into their Original Order, and were about to March in search after the rest of their Appurtenances.[41]

Where an early Stuart courtier such as Bacon had looked at the Earl of Arundel's art collection and seen the Resurrection, Ward's Spy gazes upon a collection of natural remains and imagines the work of a supremely individualist Renaissance artist.[42] What propels this chiaroscuro flicker between nature and art, life and death, is not Petty's famous "political arithmetic" but the mundane calculations of wage labor. Ward's Repository Keeper is an entertainer who stages "a notable Harangue upon every Bauble in his Storehouse."[43] Of course, he does so for money; not until he has "crost the hand of our Raree-show Interpreter with a piece of silver" can the Spy depart Gresham College.[44] Thus, while Petty had imagined the philosophical collection as a salubrious space of making and knowing, Ward represents the Royal Society's museum as a commodified palliative to what Richard Altick has called "the dullness, the mental vacuity, the constriction of horizons, the suppression of the imagination which were too often the price of life in the enveloping city."[45] Instead of healing, learning, and leveling, we might say, the business of the Repository had become truck in visual quackery.[46]

Knowing the Repository

If Petty and Ward represent extreme views culled from chronological bookends of the Repository's creation, might we find a via media from the vibrant years of the collection's early practice? "Now methinks in *Arundel* house," so Thomas Sprat proclaimed in 1667, "there is a perfect representation of what the Real *Philosophy* ought to be."[47] This was quite a contention. After the Great Fire destroyed much of London in early September 1666, administrative committees overseeing the City's frantic rebuilding had requisitioned Gresham College, forcing the Royal Society to relocate. Through entreaties made by John Evelyn, the Duke of Norfolk invited the Fellows to hold their meetings and to display their Repository at Arundel House, seat of the Howard family on the Strand. Thus, between 1667 and 1673, the Royal Society met and discoursed on experimental philosophy in the fashionable West End amid the Arundel collection of ancient marbles that had so astounded Francis Bacon.[48] As the Society's official historian, Sprat was ebullient. What had looked like disaster to London and its experimentalists turned out to be a fortunate fall.[49] At Arundel House, he could crow, "we behold new *Inventions* to flourish amongst the *Marbles*, and *Images* of the *Dead*: so the present *Arts*, that are now rising, should not aim at the destruction of those that are past, but be content to thrive in their company."[50]

The Repository's rapidly accumulated array of ancient and modern wonders certainly sounds impressive enough. We read of an Egyptian mummy, the full skin of a Moor, and curious anamorphic paintings by Christopher Wren.[51] Yet because the vast majority of those marvelous artifacts has long been lost—few detailed records of their provenance or display remain, while the actual exhibition spaces are gone entirely—Sprat's augured marriage of art and science has failed to convince modern scholarship.[52] According to one influential historian, the Royal Society's Repository best tells a tale of internecine strife within the experimental community: "a conflict between what might be called 'scientific' and 'virtuoso' values, between the adulation of the exotic and the rare which characterized the virtuosi on the one hand, and, on the other, the aspiration to a comprehensive collection of objects, ordinary as well as extraordinary."[53] To make interpretive matters still worse, cavernous epistemic fissures rupture through the resource on which most scholars of the Repository have had to rely: Nehemiah Grew's *Musaeum Regalis Societatis* (1681).[54] Like many of the Continental sources on which it draws, Grew's illustrated volume divides the artifacts in the Repository into categories of animal, vegetable, mineral, and artificial, parsing each such section according to descending orders of artifact magnitude.[55] Grew's category "Of Animals" thus progresses from humans to quadrupeds, serpents, birds, fish, shells, and insects. Since artifacts are arrayed from complete totalities to smallest particles, the Egyptian mummy donated to the museum by the Duke of Norfolk precedes skeletons, skulls, and, finally, tiny fragments of bodies like teeth and gallstones.[56] But in his organization and discursive catalogue entries, was Grew intending to make general statements about the order of nature?[57] Or were his claims focused exclusively on artifacts on display at Gresham College?

If a hard core within the Royal Society apparently lamented Grew's departure from ambitious taxonomic classification, other readers took *Musaeum Regalis Societatis* as a splendid guide to the physical space of the museum.[58] As bibliophile Zacharias Conrad von Uffenbach reported in 1710, "Both in Germany and elsewhere, an exalted idea of this Society has been formed, both of it and the collections they have in their Museum, especially when one looks at the Transactions of this Society and the fine description of the Museum by Grew."[59] What Uffenbach discovered when he actually visited the Repository at Gresham College was just how profoundly the material reality departed from its paper representation.[60] The Repository, he complained,

> consists of what appear to be two long narrow chambers, where lie the finest instruments and other articles . . . not only in no sort of order or tidiness, but covered with dust, filth and coal-smoke, and many of them broken and utterly ruined. If one inquires after anything, the operator who shows strangers round . . . will usually say: "A rogue had it stolen away," or he will show you

pieces of it, saying: "It is corrupted or broken"; and such is the care they take of things! Hardly a thing is to be recognized, so wretched do they all look![61]

Reading his disgust in reverse, Uffenbach expects that a museum would house and preserve the individual integrity of its treasured contents so that they could be displayed on command.[62]

Most illuminating in Uffenbach's disappointment, though, is his assumption that a catalogue like Grew's functions not as a taxonomy of the macrocosm but as a finding aid to a specific collection. In fact, navigating the physical space of the collection was exactly the kind of task that Grew's catalogue was quickly asked to perform. This point can best be elaborated less by studying what the text of the *Musaeum* actually claims than by attending to what was done *to* it. In a copy of the volume now at the Huntington Library in San Marino, California, a period reader has not only enhanced the navigability of the text by adding a paginated table of contents, but he or she has reconciled (likely) firsthand knowledge of the Repository's physical space with its paper representation.[63] Among numerous marginal notes and drawings is a detailed depiction of one of the Repository's most prized possessions: the air-pump donated to the Royal Society by Robert Boyle[64] (fig. 5.3). As a schematized supplement to the *Musaeum*, pictorial augmentations such as these give graphic form to instruments described by Grew but not depicted in his catalogue.

If such inserted drawings could have helped an individual reader to recall objects seen in the Repository, a far more radical redaction survives at the Royal Society where Grew's *Musaeum* was transformed into an instrument of institutional memory (fig. 5.4). In this object made in the first half of the eighteenth century, parts of Grew's text and engraved images have been cut out of the *Musaeum* and pasted into folios entitled "A General Plan of the System of Classification."[65] Clearly still indebted to Grew's ordering and catalogue entries, this "General Plan" also registers important innovations. New categories have been inserted into the object taxonomy; distinctions are now made between "Perfect Human Bodies" and "Monsters," while "Extraneous Things" discovered in human bodies and "Anatomicall Drawings" have been separated off as discrete subsections.[66] More important, entries in this redaction stipulate where designated objects could be found within the physical space of the museum. With Grew's explanatory text pasted alongside, the Duke of Norfolk's mummy now appears as H (Human Rarity), Class I (Perfect Human Body), Chapter J (Entire Body), Class A (Adult), with the following finding aids: Number J; Cabinet I, Division J.[67] Initially planned as a general taxonomy of nature

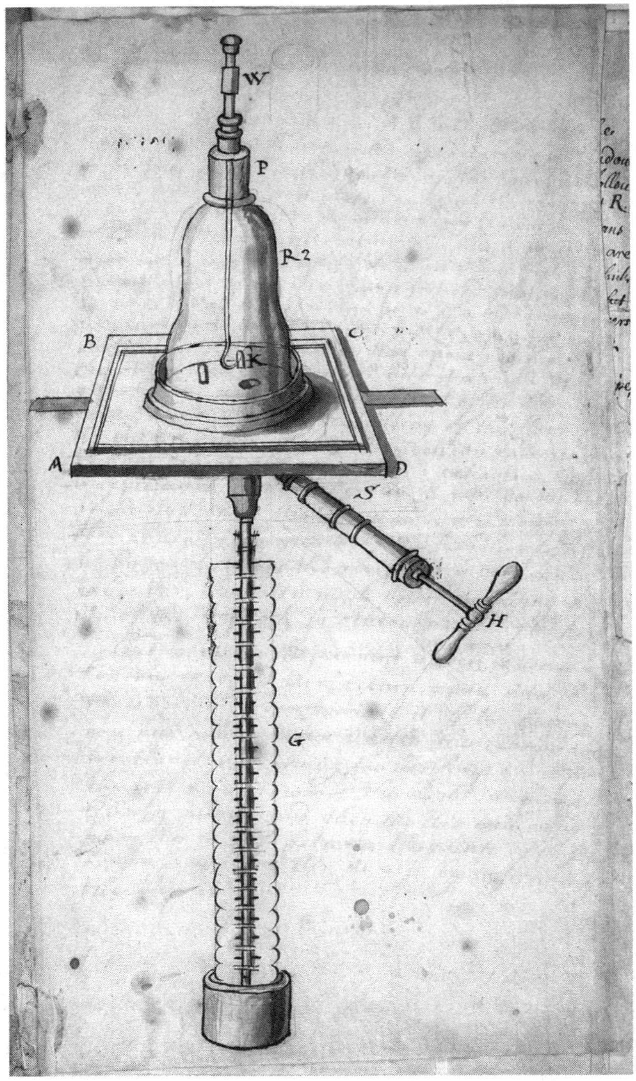

FIGURE 5.3 * Anonymous drawing of an air-pump inserted into Nehemiah Grew's *Musaeum Regalis Societatis* (London: W. Rawlins, 1681) following plate 31; Huntington Library, RB 752249. Reproduced by permission of The Huntington Library, San Marino, California.

and art, Grew's catalogue had been retooled as an instrument by which the Royal Society could manage and retrieve loot from its private warehouse.

A working collection in evolution itself, the Royal Society's Repository needed to make and rework artifacts like Nehemiah Grew's catalogue—its inventory-cum-map—simply to govern its precious objects. Following institutional critiques of the modern

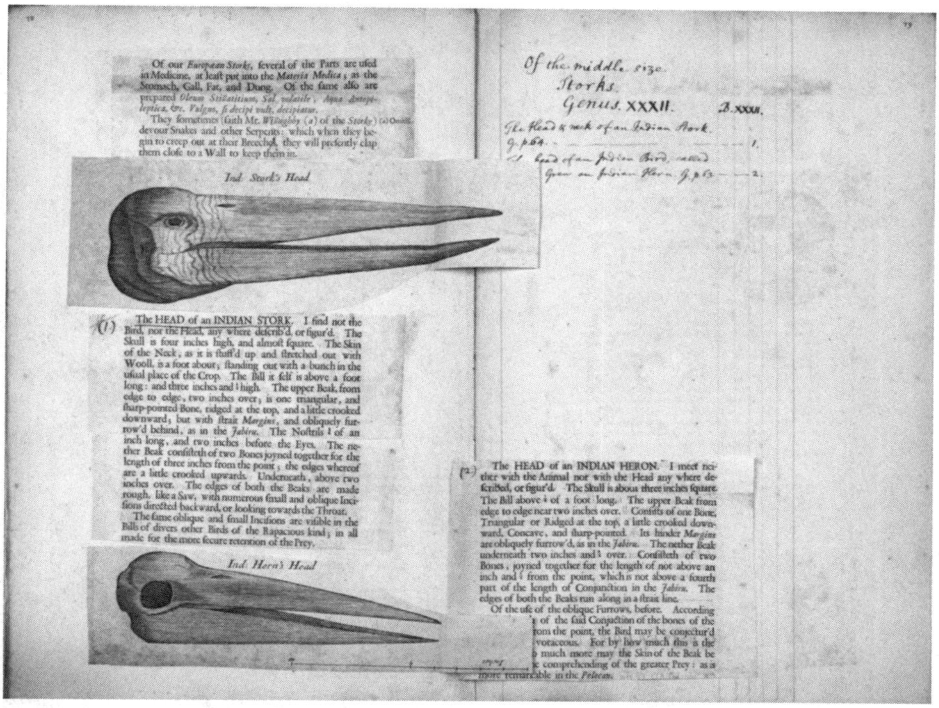

FIGURE 5.4 * Anonymous mid-eighteenth-century redaction of Nehemiah Grew's text and engraving *Of the middle size storks*; MS (General) 415/5, fols. 78–79. © The Royal Society. Reproduced by permission.

museum, many interpreters have called attention to how such strategies of rationalized preservation are themselves transformations—stripped from their original contexts and ordered anew, artifacts are *made* museum objects.[68] This "museumification" might even be seen as meaningfully coextensive with acts of iconoclasm more traditionally conceived. Along these lines, art historians Fabio Rambelli and Eric Reinders have recently argued for "a continuum ranging from complete obliteration of the object to cases where the material object is preserved but redefined so as to erase the effect of its sacrality."[69] Rather than extending the frustrating tasks of reconstructing the Repository's lost taxonomies or piecing together its shattered contents, what I stress in the remainder of this section is that attention to such a continuum or spectrum of object-retooling can best illuminate the collection's working life. Because our evidence for these practices is necessarily fragmentary, I will use case histories of four objects donated by prominent virtuosi—Denis Papin, Gottfried Wilhelm Leibniz, Cassiano dal Pozzo (via George Ent), and Christopher Wren—to suggest how variously and surprisingly labile artifacts became when subjected to

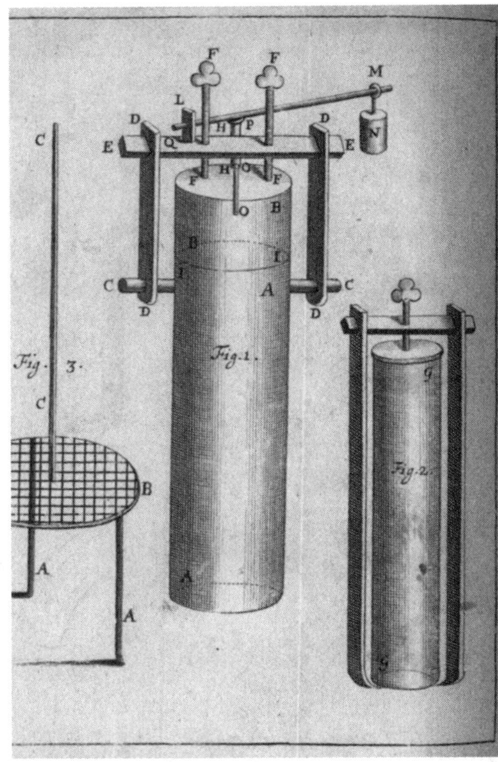

FIGURE 5.5 * Engraving of Papin's Digester; from Denis Papin, *A New Digester or Engine for Softning Bones* (London: J. M. for Henry Bonwicke, 1681), n.p. © The Royal Society. Reproduced by permission.

the experimentalists' collective intelligence. By following those objects into the museum and rapidly out of order, we can begin to see how the Repository worked in practice, and why it came to attain such suggestiveness as a cognitive model.

An occasional assistant to Robert Boyle since the mid-1670s, Huguenot refugee physician and experimentalist Denis Papin (1647–circa 1712) made a sensational presentation to the Royal Society in March 1679.[70] Papin produced "a small glass, which he had in his pocket, wherein were contained several small pieces of harthorn [that is, deer antler], which he had softened by a new way, that he had found out, of boiling them."[71] Testing Papin's samples by "cutting and biting," the Royal Society Fellows were amazed to find that the softened antler was "not much harder than a sticky and seeded carrot-root."[72] These strange effects had been accomplished by Papin's "instrument, by which he could boil any thing *in vacuo*": a machine called the Digester.[73] The Digester was essentially a pressure cooker (fig. 5.5). An operating experimentalist would place the object to be softened into the instrument's slender brass receiver (marked *AA* in Papin's fig. 1), add liquid, and fit the snugly capped device above a fire.[74] Rendered by this "new way of boiling," softened substances continued to change; a week after their initial presentation, Papin's supple antlers had "now hardened again to a very great degree."[75] Such findings raised the hope that his instrument might be used to aid craftsmen in carving and dyeing ivory, distilling alcoholic beverages, or preparing food. To the Royal Society's delight, Papin presented numerous samples of gelatinous calves' feet ("the flesh thereof was notwithstanding very well tasted"), boiled mackerel (whose bones "were as soft almost as the flesh itself"), and other substances experimentally softened to the taste of the seventeenth-century aristocratic palate.[76]

When he returned to the Continent in the early 1680s, Denis Papin left the Digester to join the collection of philosophical instruments in the Royal Society's Repository.[77] Writing from Antwerp in early 1681, he explained that he had "left at Mr. Hooke's lodgings his engine for softening of bones, &c. to be presented to the Society."[78] But just because the instrument had entered into the organization's collection did not mean that its practical life was over. During the spring of 1681, Robert Hooke continued to present experimental trials with the Digester, not only putting the instrument to various uses but disassembling and altering it.[79] Soon after Papin donated it, the Digester was "opened, and all the several parts of it explained by Mr. Hooke, together with the method of fitting it and using it for boiling, &c."[80] As in older, Continental traditions of *Kunst- und Wunderkammern*, the Repository remained a space for the playful manipulation of natural creation and human ingenuity.[81]

Modest as they were with the Digester, physical incursions into donated objects could also produce outrage. Such was the case of an artifact presented to the Royal Society by philosopher Gottfried Wilhelm Leibniz (1646–1716). Visiting London in January 1673, Leibniz made the following mechanical performance before the assembled Fellows:

> He now shewed them a new arithmetical instrument, contrived, as he said, by himself, to perform mechanically all the operations of arithmetic with certainty and expedition, and particularly, multiplication, after such a manner, that a whole series of numbers, to be multiplied by other numbers, might be multiplied, if the multiplier be one number, by only one turn of the wheels of the machine; and if there be two numbers multiplicands, the operation shall be dispatched by two turns, and the addition of the two products performed at the same time, and so on.[82]

In truth, Leibniz's machine seems not actually to have made it into the society's Repository. The inventor claimed that the device remained "imperfect," but promised to have one produced in France "for the service of the Society."[83] Leibniz was possibly put off from doing so by the excessive interest that his machine received from Hooke. "When I explained that [machine] before the Royal Society," so Leibniz complained to Henry Oldenburg, "he was certainly very well forward."[84] Not only had Hooke "absorbed every word I said," Leibniz objected, but the Curator had actually begun to disassemble the machine itself, removing "the back plate which covered it."[85]

Among theorists of modern technology, strategies of reverse engineering are often cast as a public good.[86] By limiting the exploitative hoarding of technology by private monopolies, reverse engineers and "pirates" make useful knowledge broadly available—arguments with which William Petty might well have sympathized in 1640s Oxford. In early 1670s London, however, Leibniz saw nothing benevolent in Hooke's motives, which appeared little better than common thievery.[87] Hooke's access to the calculating instrument and his peculiar cast of mind made his intrusive inspection particularly potent. "That he did not distinctly trace out all its wheels I readily admit," Leibniz conceded: "But in such cases it is enough for a man who is clever and mechanically-minded to have once perceived a rough idea of the design, indeed the external manner of operation, and then for him afterwards to add to that a little of his own."[88] This, of course, was exactly what Hooke had been planning to do. Two weeks after Leibniz's presentation to the Royal Society, Hooke received the institution's encouragement to make a calculating machine that could "perform all the operations of arithmetic, with great expedition and certainty, without making use of the rhabdology, and that much more simply than that of Mons. Leibnitz."[89] Leibniz was utterly dismayed by this dirty trickery. It was, he declared, unworthy of Hooke's "own estimate of himself, unworthy of his nation, and unworthy of the Royal Society."[90]

A still more invasive approach to Repository goods is suggested in the case of the petrified wood transmitted to the Royal Society by Dr. George Ent (1609–1689). A physician and a Royal Society Fellow, Ent had received petrified vegetal specimens from Roman antiquary Cassiano dal Pozzo.[91] In turn, dal Pozzo had acquired his collection from Federico Cesi, founder of the celebrated Accademia dei Lincei.[92] As David Freedberg has recently shown, Cesi and his Lincean colleagues stood at the vanguard of an ambitious, influential tradition of natural history that took no small interest in fossils. Indeed, following the arguments of Cesi's colleague Francesco Stelluti in his *Trattato del Legno Fossile Minerale* (1637), museum-cataloguer Nehemiah Grew claimed that the so-called fossilized woods in the Royal Society's collection had never really been vegetal tissue at all.[93] These "Pictures . . . in time petrify'd," as Grew called them, propagated when nature replayed her forms through slow, chemical evolutions in the bowels of the earth.[94]

Hooke found this explanation totally unsatisfying. Taking a small piece of dal Pozzo's petrified wood, he lit it on fire (which emitted "a resinous smoke or fume"), cut it apart, charred it in a crucible, and then studied the results with a microscope.[95] There, he discovered "an infinite company of small microscopical pores"[96] (fig. 5.6). Struck by the resemblance between these porous surfaces and living plants—their similarity to "the pith of elders, the stalks of divers sorts of reeds and canes, and the like"—Hooke hypothesized that his rough handling had jarred loose "some kind of succus nutritious or some other appropriate juice of the vegetable" from the pores of dal Pozzo's specimen.[97] How could this hypothesis be tested? If he could find a way to extract the coagulated sap from other samples of petrified wood in which such tiny apertures were not already visible, he reckoned, then he could presumably reveal pores matching those in the dal Pozzo sample, thus supplying evidence of the truly vegetal origins of these strange stones. Two experimental methods were devised; one specimen of seemingly nonporous petrified wood (likely extracted from the Repository's collections) was exposed to "a very good heat but so as that the air not being admitted and it should not burn away," while another was repeatedly soaked and dried.[98] Both procedures revealed the presence of previously invisible pores. Compared, tested, physically altered: museum artifacts not only supplied material evidence with which to elucidate their own strange nature, but the Repository provided a generative matrix for exploring broader, often radical, theories.[99]

That the Royal Society Fellows could also use a lighter touch when working with Repository collections is demonstrated by a fourth and final case object: "an anient urn of glass, taken up in Spittlefields upon digging cellars there, presented by Sir Christopher Wren for their repository."[100] In the estimate ventured by Grew in the *Musaeum*, this now-lost urn was "above fifteen Hundred years old. Almost like a bottle containing a Gallon and 1/2; but with a very short Neck, and wide Mouth, and of whiter metal. Encompassed girth-wise, with five parallel Circles."[101] When first donated to the Repository in 1678, the craftsmanship of Wren's antique urn had generated great curiosity. The object "seemed to be made after quite another manner than that used by the present workmen in that art, it having no place at the bottom thereof; nor any visible sign how it could be held, whilst the lip and handle thereof were joined to the body."[102] Five years later, Wren's urn was again enlisted into the Royal Society's discourse, both giving evidence and prompting new inquiries. Amid a discussion of the resemblance between sandstone in Dartmouth and "cuogolo" (a substance used by Venetian glassmakers), Dr. Martin Lister intervened.[103] Romans in the ancient British Isles, he explained, had used a similar substance when constructing objects that would be exposed to the elements.[104] Responding to this discussion of glass-making and Roman artistry, Hooke postulated that the

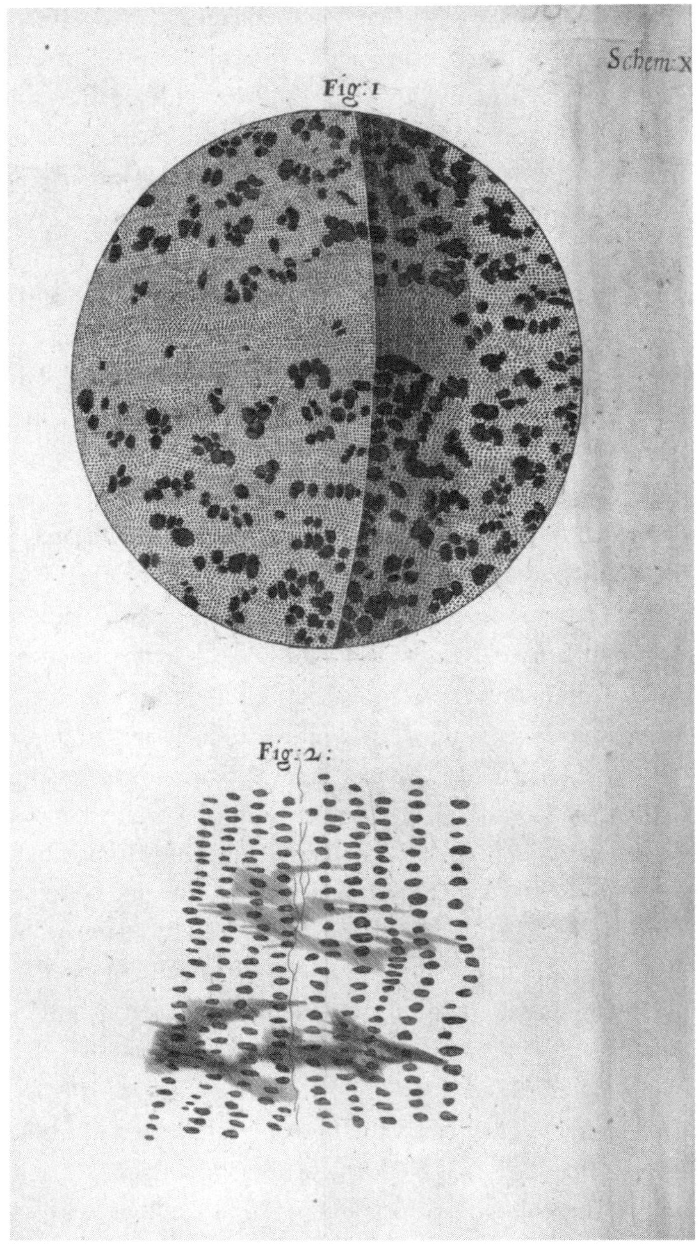

FIGURE 5.6 * Magnified petrified wood and its pores; from Robert Hooke, *Micrographia; or, Some Physiological Descriptions of Minute Bodies made by Magnifying Glass* (London: John Martyn and James Allestry, 1665), scheme 10.

perceptible differences between ancient and modern glass were simply consequences of age. As he put it, ancient glass artifacts "were altered by lying."[105] To test these claims, the Fellows turned to the Repository:

> The urn given by Sir Christopher Wren being called for and viewed, was found very smooth, but in some places had a bluish shining. Mr. Hooke observed, that this was a sign of its readiness to scale: that he had formerly seen several scales, that came out of it: and that Sir Christopher Wren thought, that the mark of the puntillion might be fallen off with the scurf.[106]

As opposed to Denis Papin's Digester, Gottfried Leibniz's calculating machine, or dal Pozzo's wood, Wren's urn was not cracked open and examined from within. Instead, it was housed for safekeeping in the Repository so that it could be compared, tested, and otherwise made available for a mixture of keen inspection and that hands-on cognition some recent theorists have called "thinkering."[107]

The crucial, historical point, though, is this: conducive as they may have been to generative thinking, the varieties of heavy-handed knowing practiced in museums like the Royal Society's Repository were not favorable to the *preservation* of artifacts. This was an insight grimly discovered by aristocrat Zacharias Conrad von Uffenbach in his travels across England in 1710. Having already found the Repository in a dismal state, Uffenbach then made his way to the Ashmolean Museum in Oxford, where care for objects appeared alarmingly wanting. "Even the women," he growled, "are allowed up here for sixpence; they run here and there, grabbing at everything and taking no rebuff from the *Sub-Custos*."[108] This lax attitude toward collections stemmed from more than mere eagerness to gather the entry fees of any and all comers. To Uffenbach, the haptic compulsion to handle and mangle almost amounted to a national trait.[109] "It is surprising," he claimed, "that things can be preserved even as well as they are, since the people impetuously handle everything in the usual English fashion."[110] In what follows, I argue that Uffenbach may well have had a point. Handling, disassembling, and remaking museum artifacts were *exactly* the kinds of procedures London's experimentalists would come to theorize as cognitive activities—indeed, as intelligence itself.

Skullduggery; or, Philosophy in the Storeroom

Uffenbach was disgusted by the state in which he found English museums in the first decades of the eighteenth century. But he did identify one feature worthy of praise: the ingenious presentation of specimens in chemical preservatives. "I was delighted," he enthused, "at the way in which all these things were fastened to small glass balls and floated in the spirit, so that all may be seen with ease. Even when the spirit is somewhat evaporated, the things sink with the balls and do not hang without moisture and perish, which they do when fastened to the glass or the stopper, as they usually are."[111] For innovators of these "wet-sampling" techniques such as Robert Boyle, chemical preservation offered more than an aesthetically pleasing mode of display.[112] As Boyle argued in 1663, "'Tis a known thing, to the Collectors of Rarities, that the external *Idea* of Fishes, Crocodiles, Birds, and even Horses, may be preserv'd for many Years, by taking out the more corruptible parts, and stuffing their prepar'd Skins with any convenient Matter."[113] With chemical assistance, an anatomist could now preserve animal specimens including Boyle's own totemic, monstrous colt head until they had "impress'd an *Idea* of it upon it upon his memory."[114]

What did Boyle mean when he talked of preserving the "external *Idea*" of natural entities? In exactly what sense was the experimentalist's memory to be "impress'd" with these preserved artifacts? And how was epistemic function to derive thereby? Questions like these were given serious exposition by Robert Hooke in a sequence of lectures from the early 1680s on cognition, reasoning, and memory—procedures key to the experimental intelligence that he would model as operations of a repository. Written by the longtime Keeper of the Royal Society's Repository, this section examines how Hooke's lectures effectively reimagine key problems in the tradition of English empirical philosophy through modes of gathering and working with collected artifacts as historically practiced in the museum at Gresham College.[115] But in attempting to theorize how cognitive function could be achieved from the riot of materials collected and delivered to a centralized repository, I want to emphasize how Hooke's lectures also offered a telling solution to socio-epistemological problems of information organization he had confronted practically when editing his *Philosophical Collections* circa 1679–82. If philosophical knowledge had there required the chemical "compounding" of collected information, Hooke's cognitive schemes—especially as articulated in a crucial text of 1682, touched on in chapter 2—now envisioned a polymathic, "Archietonical" intelligence at their heart. Explicating what he and his colleagues had come to mean by this archietonical agency in the early 1680s, chapter 6 will then explore how these ideas would materialize in and as later Restoration London's built architectural environment.

Now known as the "Philosophicall scribbles," the earliest and most rudimentary of Hooke's texts on cognition from the early 1680s follows a broadly Aristotelian conception of the mental faculties.[116] Therein, the human sensorium appears "like a peice of soft wax to receive those impressions and stamps" from external objects.[117] What differentiates humans from beasts, Hooke explains in this lecture apparently aimed at a young audience, is possession of "an *active faculty*, which collates and compares these impressions, and is thereby inabled to compound, & compose new ones, & to regulate what is defective or ammiss in the other."[118] Knowing in this scheme relies on a careful screening of perceptual impressions, a caution that the percipient learns to exercise only by experience. Because the youthful mind is positively greedy for sensation, a juvenile percipient will "(like a young and unskillful receiuer of cash) take & lay up for true and good all that comes whether they be soe or noe."[119] Once duly apprised of the deceptions consequent from unquestioning trust of the senses' impression currency, the active mental faculty begins to impose discrimination. Receiving external data, it differentiates between "a true & as it 'twere immediate impression, & that which has some thing as it were layd between the seale & the wax that does disfigure the true & genuine figure soe excellently engrauen."[120] Since the mind is a bank in this scheme, sensory impressions are currency, and the active intelligence is an accountant, successful cognition takes place only when imprinted coinage can be reconciled with its object matrix—when the active intellect can make inference to "the true forme of the seales, from various impressions that are made by them upon his sensible parts."[121]

Rough and ready though it is, Hooke's bank model of mind might be placed in an important tradition of conceptualizing subjective interiority as a cabinet or chamber.[122] Royal Society Fellow John Locke's contemporaneous use of the camera obscura to model human understanding offers a celebrated and complementary variation on that theme.[123] "Methinks the *Understanding*," Locke wrote, "is not much unlike a Closet wholly shut from light, with only some little openings left, to let in external visible Resemblances, or *Ideas* of things without."[124] As in Hooke's scheme, images introduced into that inner chamber become the units of human cognition, provided that "the Pictures coming into such a dark Room but stay there, and lie so orderly as to be found upon occasion."[125] In the summer of 1682, Hooke set out an extremely vigorous account of exactly how this fixing and ordering of external information might be achieved. Read twice to audiences packed with philosophical luminaries (and published as "Sect. VII" of the *Lectures of Light* in Richard Waller's *Posthumous Works of Robert Hooke*), Hooke's lecture thematized relations between an active intelligence and what he now unequivocally called its "Repository of Ideas."[126]

Hooke's 1682 lecture was nominally designed to explain memory and its agency in forming consciousness of time. Seated in the brain, Hooke proposes, the soul is the "*primum movens*, the self-moving principle" of the body.[127] Interrupted only by the body's sleep or disease, this attentive soul acts constantly on the organ of memory to make ideas, which are "material and bulky, that is, to be certain Bodies of determinate bigness, and impregnated with determinate Motions, and to be in themselves distinct."[128] Once formed, ideas are arrayed outward from the centralized soul in concentric circles. Like "a continued Chain of Ideas coyled up in the Repository of the Brain," the oldest ideas are "farthest removed from the Center or Seat of the Soul where the Ideas are formed; and the other End is always present at the Center, being the last Idea formed, which is always the Moment present when considered."[129] Constantly compressed and forced ever farther away from the soul, the ideas damage and age, thus accounting for faults of memory. Because the soul can recognize these disparities between old and new ideas as it radiates vibrations through the physical space of the mnemonic repository, it gains consciousness of time—a concept entirely absent from the eternal present of sensory flux. "The Notion of Time," Hooke concludes, "is the Apprehension of the Distance of Ideas from the Center or Present Moment. And so Time comes to be apprehended as a Quantity, and so falls under the Consideration of Geometry or Mensuration."[130]

Remarkable even by his pyrotechnic standards, Robert Hooke's 1682 lecture takes the trope of the mind-as-cabinet to fascinating and unusual extremes. Building on recent readings of this important lecture, I want to elaborate first the powers Hooke had come to attribute to the soul at the seat of mnemonic collection, and second, why he sought to explain the agency of that soul as "Archietonical."[131] And by rubbing his ideas about the mental repository against those of his contemporaries, I aim to show what these lectures can teach us about the intelligence understood to be instantiated in the practices of collecting and using artifacts in a museum like that at Gresham College.

If likening the mnemonic faculty to a repository was a familiar trope, Hooke's ideas about memory are unusual.[132] Against prevailing views, he asserted that memory is a robustly physiological procedure of artifact-making seated at an identifiable location within the skull.[133] Memory is, he writes, "as much an Organ, as the Eye, Ear or Nose, and to have its Situation somewhere near the Place where the Nerves from the other Senses concur and meet."[134] (See figure 0.7.) Drawing on the cerebral anatomies made by his friend Christopher Wren and his former mentor Thomas Willis, Hooke judiciously refrained from speculating on exactly where this "Wonderful Elaboratory of the Animal Body" was sited.[135] His 1682 lecture elucidates instead the mechanisms operative within that mnemonic laboratory. Where his earlier bank model had relied exclusively on passively accumulated sensory impressions, Hooke stages this repository as a dynamic space of making and remaking. He puts it this way: "I suppose . . . this Repository to be furnished with variety of Matter adapted for the Uses to which the Soul applies them, which I call the Elements out of which Ideas are made; among which Variety there are principally five sorrs fitted and adapted to receive the Impressions from the five Senses."[136]

To explain how these mnemonic media might work, Hooke drew from firsthand knowledge of experimental materials in the Repository at Gresham College. The soul's formation of visual ideas "may a little be explained by the Matter of the Phosphorous made of the *Bononian* Stone, or that found out by *Baldwinus* made of Chalk and Niter."[137] As he and his colleagues had experimented with the chalky earths from Bologna to obtain phosphorescent effects, so the soul was to be understood as possessing still more potent materials for encoding and representing visible data.[138] Likewise, to create sonic ideas, Hooke stipulates that the soul deploys materials "like those Bells or Vases which *Vitruvius* mentions to be placed in the antient Theaters, which did receive and return the Sound more vigorous and Strong."[139] Provisional as these explanatory models no doubt were, the key point is this: once delivered to the inner space of the mnemonic repository, vibrating frequencies transmitted through each sensory channel would be encoded by and stored in the soul's internal representational media, which Hooke could conceptualize through reference to materials in the Repository at Gresham College.

The difference between Hooke's conception of 1682 and comparable mnemonic schemes in Restoration experimentalist circles is substantial. Consider the argument in Willis's influential *Two Discourses Concerning the Soul of Brutes* (1672), which envisions a "Chamber of the Soul, glased with dioptric Looking-Glasses."[140] Akin to the Lockean camera obscura model that it informed, Willis urges that "Images or Pictures of all sensible things, being sent or intromitted by the Passages of the Nerves, . . . are represented upon the Callous Body, as it were upon a white Wall; and so induce a Perception, and a certain Imagination of the thing felt."[141] Passively collected and projected, these mnemonic images flow through the billowing, watery spaces of the brain. "Entring into its folds," Willis explains, ". . . they Constitute the memory or remembrance of the Thing."[142] By contrast, Hooke's model had imposed on his own earlier, impression-based conception of cognition a crucial second stage of media transfer and encryption. Through the intervention of the soul, the ephemeral, vibrating motions collected by the senses are encoded into what Hooke describes as "appropriated Matter." Only when stabilized as physical objects could sensory vibrations be stored, recalled, and put to cognitive use. The singular currency of impressions dominating cognition in Hooke's earlier writing had thus been replaced by a two-stage process whereby ephemeral sensory data are transformed into durable, physical tools of the soul's intellection called ideas.[143]

Consequently, the powers Robert Hooke attributed to the soul in his 1682 lecture had become far more robust than in his earlier speculations. In the act of "attention," he explains, the soul becomes like the sun.[144] As the soul radiates its attractive stimuli through the mnemonic repository to "apprehend, or as it were feel, . . . any Ideas that remains treasured up" and influences those ideas by the inverse-square law of gravitation, so Hooke deployed his experiments on harmonic resonance to suggest how the soul could prompt "Re-actions from several Ideas placed here and there in the repository, and its being sensible of the harmony or discord of them one with another."[145] But, as suggested in chapter 2, ascribing unexpected cognitive force to material ideas was a tricky business. Beyond threatening to reduce the workings of the immortal, immaterial soul to highly mechanical terms, it also made the soul's knowledge veritably beholden to its fallible, distributed surrogates.[146]

For his part, Thomas Willis had elaborated a more traditional system whereby humans possess twin souls: a "corporeal" soul, governing the body and its interface with the external world via the senses, and a "rational" soul nested delicately within it.[147] Willis envisioned the rational soul occupying "its Throne . . . in the Imagination, made out of an handful of Animal Spirits, most highly subtil, and seated in the Middle or Marrowie part of the Brain."[148] From this seat, the rational soul commands its corporeal lieutenants and "easily performs the Government of the whole Man."[149] If this conception has been described as a resolutely "Royalist" model of the mind—one where the rational soul is a beneficent, divinely appointed king and the corporeal armature a dutiful if fallible body politic—then it also bears the humiliating imprint of then-recent Royalist history.[150] Like the politics of pleasure at the late Stuart court, we might say, Willis's rational soul is "almost completely and sometimes woefully dependent upon the environment of information and preconditions created for it by the nervous system."[151]

Simplifying Willis's system, Hooke's lecture of 1682 also shifted the broader sociopolitical register for modeling the action of an integrative governing agent working on a vast bodily collective. If still relying on an elaborate system of data-collecting, his centralized intelligence is no longer beholden to the impressions delivered to it, as had been the case in "Philosophicall scribbles." Even less is the soul constrained by the quasi-egalitarian circulation of knowledge from the brain down to "the Body, the Arms, the Fat" and back again that, as noted in chapter 1, Hooke had theorized in *Micrographia*.[152] Instead, knowledge had come to derive from the soul's manipulation of its own quasi-stable order of encoded artifacts representing stimuli constantly vibrating through the material substrate of the human body.[153] In a critical passage from his 1682 lecture, Hooke summarizes his view:

> The Senses . . . are as it were the Collectors or Carriers of the Impressions made by Objects from without, delivering them unto the Repository or Storehouse where they are to be used. Which Impressions being actual Motions . . . become Powers sufficient to effect such Formations of Ideas as the Soul does

> guide and direct them in: For I conceive no Idea can be really formed or stored up in this Repository, without the Directive and Archietonical Power of the Soul; and the Actions or Impressions cease and fail without the concurrent Act of the Soul, which regulates and disposes of such Powers.[154]

Registering in the body's sensorial instruments, vibrating stimuli from objects called impressions are, Hooke allows, necessary to knowing about the world. Yet those mechanical motions only become sufficient to and capable of epistemic function when encrypted into material media by a soul whose hierarchical powers are neither kingly nor imperial. Instead, they are "Directive and Archietonical."

Here we can begin to flesh out the implications of this view for the Repository of artifacts at Gresham College. That is, just as the Royal Society's intellectual activity relied on heterogeneous objects and images transmitted to London from far-flung informants via complex communication infrastructures, so human knowledge in Hooke's lecture of 1682 derived from stimuli received into the mnemonic laboratory of the soul from the body's distributed sensory organs. As with donations such as the putrefying monster flesh of Robert Boyle's untreated colt head, however, the body's brute sensorial impressions were profoundly volatile, unreliable. Delivered to the repository, these fragile transmissions had to be encoded and stabilized into storable forms, whether simply labeled by Hooke and his staff or submitted to more comprehensive transformations. Once translated into appropriate "adapted Matter" by the agency internal to the repository, collected materials—whether at Gresham College or in the repository of the mind—could become tools and cognitive instruments for acts of intelligence that reasoned by comparison, haptic disassembly, compounding, and recombination. Forged from data collected by the experimental body's subservient faculties, these transformations were hardly epiphenomenal to intelligence. Instead, they constituted the very "Action of Thinking, that is, of fixing or darting . . . Radiation more powerfully upon this or that Idea placed in the Repository."[155] And unlike William Petty's Steward, Thomas Willis's rational soul, or even Ned Ward's Keeper, the agent Hooke imagined governing both those bodily collectors and the interior repository was in possession of what he now called an "Archietonical Power." What exactly did he mean?

Architect as Model

In early February 1675, while at work on the architectural plans for his Montagu House in Bloomsbury, Robert Hooke put a down payment on a copy of "J. Dees *Euclid*" with instrument maker Henry Wynn.[156] Nine days later, he was back. Meeting Wynn "at the Painter's Shop on Ludgate Hill against the Old Bayly," he paid the balance of the cost for the volume.[157] What Hooke had in his hands by the end of February 1675 at the very latest was *Elements of Geometrie of the Most Auncient Philosopher Euclide of Megara* (1570), an English translation to which Elizabethan scholar John Dee had added his famous *Mathematicall Praeface*.[158] Hooke read Dee closely and creatively. As noted in the introduction, he dismissed the charges of sorcery and religious enthusiasm contemporaneously leveled against Dee's writings as misapprehensions of those "who understood not the Art and Ingenuity."[159] Like other experimentalists in later Restoration London, moreover, Hooke studied the "revolutionary manifesto" that was Dee's *Praeface* with peculiar attention.[160]

Thumbing through his new purchase in 1675, Hooke would have read how the startling explosion of learning in Renaissance Europe could be traced to the spirit of ancient mathematics. According to Dee, it was because Italy, France, Holland, and Spain had reacquired knowledge of classical calculations that "amongest them do florishe so many cunning and skilfull men, in the inventions of straunge and wonderfull thinges."[161] Not only does mathematics underpin human knowledge for Dee, but it mediates between supernatural realms knowable by reason alone and the terrestrial domains of nature perceived by the senses. Mathematics have "a straunge participation betwene thinges supernaturall, immortall, intellectual, simple and indivisible: and thynges naturall, mortall, sensible, compounded and divisible."[162] While Dee gives short shrift to mechanics who labor without an understanding of the theoretical principles behind their actions, the *Praeface* surveys a capacious field of human arts—from geography to astronomy, painting and fluid mechanics to military strategy—subtended by the mathematician's metaphysical knowing. As one historian has put it, this catalogue attempts to enact "in England the revaluation of the fine arts that was one of the accomplishments of Florentine quattrocento artists and humanists."[163]

Tellingly, architecture and Florentine ideas about the architect figure prominently in Dee's scheme.[164] Foreign to the deeply conservative, guild-bound structure of the arts in Tudor England, Dee's conception of the architect is necessarily articulated through a tissue of quotations. Drawing from Vitruvius, Dee argues, the architect must be a "Storehouse of all workmanship," commanding knowledge of languages, painting, geometry, perspective, arithmetic, history, philosophy, music, medicine, law, and astronomy.[165] Citing Leon Battista Alberti, Dee maintains that the architect is equally the mind that guides skilled, laboring bodies as "*the Chief Master of other Artes . . . the hand of the Carpenter, is the Architectes Instrument*."[166] Possessing full comprehension of the mathematical principles behind action on materials, the architect commands all arts and hence the workers who practice them. Following Plato, he states that the architect is thus

> *Master* ouer all, that make any worke. Whereupon, he is neither Smith, nor Builder; nor, separately, any Artificer; but the Hed, the Prouost, the Directer, and Iudge of all Artificiall workes, and all Artificers. For, the true *Architect*, is hable to teach, Demonstrate, distribute, describe, and Iudge all workes wrought. And he, onely, searcheth out the causes and reasons of all Artificiall thynges.[167]

Now, if developments in early modern English architectural theory and professional identity were modest in comparison to Continental examples, it is also true that Dee was hardly the only source on architecture available to the Restoration experimentalists. Not only did Christopher Wren return from Paris in 1666 with books and prints amounting to "almost all *France* in Paper" (as he claimed to a friend), but he did so having met Gianlorenzo Bernini, then the greatest living architect in Europe.[168] Equally, enthusiasts such as Hooke could acquire encyclopedic collections of Continental architectural treatises and plans without ever leaving London.[169]

That said, it was from Dee's model of the architect that Royal Society Fellows were frequently and creatively pilfering during the early 1680s.[170] In his *Mechanick Exercises on the Whole Art of Printing* (1683–84), printer and future Fellow of the Royal Society Joseph Moxon employed Dee's ideas in an ingenious way.[171] Moxon argued for a more liberal understanding of a figure he called "the Typographer," a conception he drew explicitly from "Dr. Dee, in his Mathematical Preface to Euclid's Elements of Geometrie."[172] This Typographer was not only to be in command of "a Mathematical Science," but he "ought to be equally qualified with all the sciences that becomes an Architect."[173] As with Dee, the Typographer envisioned by Moxon was to possess the ability to direct diverse actions and actors through reasoned thought:

> By a Typographer, I do not mean a Printer, as he is Vulgarly accounted, any more than Dr. Dee means a Carpenter or Mason to be an Architect; But by a Typographer, I mean such a one, who by his own Judgment, from solid reasoning with himself, can either perform, or direct others to perform, from the beginning to the end, all the Handy-works and Physical Operations relating to Typographie.[174]

Architect here denotes the function by which all practical crafts and bodily skills (in Moxon's case, making and composing type, imposing and printing forms, proofreading, and binding) become integrated under the governance of a theoretically informed intelligence. Possessing what Hooke had called "Directive and Archietonical Power," or managerial mastery of all the arts subtended within a field, Moxon's Typographer—like Dee's architect or the soul in its mnemonic repository—could command practical functions distributed among a diverse range of subordinates.

Both bustling with activity, the print shop and the experimental museum seem far from the idealizing realm of John Dee's polymathic architect. So, why had Dee's concepts developed such compelling resonance in experimental-philosophical circles of the early 1680s? Since addressing such questions will be central to the following chapter, I want to conclude here by elucidating how Dee's architectural ideas can clarify our understanding of the practical and imagined life of the Royal Society's Repository in particular. For Robert Hooke, Dee's ideas offered felicitous utility for making sense of both the repository of the mind and the Repository at Gresham College. Because Dee had been keen to distinguish between the intellectual activities of his architect and the humble, mechanical builders who executed that architect's ideas, his model proved useful for negotiating tricky points such as "how the Soul, which is incorporeal, should move and act upon the Ideas which are corporeal."[175] More broadly, Dee's conception of a polymathic intelligence governing and directing diversified laborers mapped well onto how Hooke had actually come to understand the project of collecting. As he was then delegating much of his graphic work to assistants Richard Waller and Henry Hunt, so Hooke looked upon the provisioning of information for the *Philosophical Collections* he edited between 1679 and 1682 as a collective task to be distributed among networks of informants, factors, and far-flung agents—men including John Covel, William Cole, and Henry Jacobs. Equally, gathering objects could be spread advantageously among experimental networks; it could be taken up by Daniel Colwall and other men who could afford to expend what Nehemiah Grew called "the redundant part of their estates" on artifacts.[176] Or this infrastructure of "socially-distributed cognition" could be built simply by collecting donations from experimentalists such as Robert Boyle, Denis Papin, or Gottfried Leibniz, who aimed to publicly advance their own philosophical points through the artifacts they gave to the Royal Society.[177]

What could not be delegated—what were crucial to the institutionalized experimental philosophy practiced by Hooke—were the forging and managing of connections between these geographically distributed agents and the disjointed collections they harvested. When assembling the pages and illustrative plates of *Philosophical Collections* in the early 1680s, Hooke had evolved a range of modestly invasive graphic strategies by which to articulate "*links*, upon which the true Philosophy depends" within materials sourced by the experimental community's self-interested agents and their seemingly atomized programs of research.[178] More literally invasive were the strategies required to stabilize, encode, and integrate the objects gathered by those Fellows in the Repository at Gresham College. And if the fate of Boyle's monstrous colt head shows how fruitfully these haptic transformations could yield insight into a variety of experimental techniques and interests, Hooke's reverse engineering of Gottfried Leibniz's calculating machine (and Zacharias Conrad von Uffenbach's horrified response to such interventions) demonstrates just how quickly they could breed bad feelings.

Cunning and ruthless though they were, these acts were not just idle wickedness. According to Robert Hooke, they were the very fundament of intelligence. After all, the more the Keeper or the soul knew about its own repository—the more it sorted and improved the decaying ideational artifacts made and remade from miscellaneous unstable collections—the more nuanced, elaborate, and ultimately rational its thinking became. "The more steddy and distinct" as the soul was in its attention to its ideas, Hooke wrote, ". . . so is the Idea more compleat, as well as more compounded: And this I conceive to be that Action of the Soul which is commonly called Reasoning."[179] To follow out the terms of the logic that, by the early 1680s, Hooke was sharing with his experimental colleagues, this all-encompassing comportment of wicked intelligence as it liquefied proprietary objects and claims into the collective enterprise of experimental philosophy was positively expected from the soul in its microcosmic repository—or indeed from any archietonical agent acting in its environment. But how did these archietonical powers play in the macrocosm that was Restoration London? And what relation does their peculiar intelligence betray to the modes of wit that notoriously flourished there? With these questions in mind, it is time to turn to the evidence of the built environment.

CHAPTER SIX

The Architecture of Science and the Science of Architecture

FLANKED BY A DRAPED curtain and a fluted column, Christopher Wren stands serenely in the three-quarter-length portrait painted for him by John Closterman around 1695[1] (plate 7). Wren's skin is visible only at its extremities. A ruffled undershirt cinches his wrists, a lace cravat clasps his neck; a mantle of burgundy taffeta shimmers across his left shoulder as it opens to the picture plane. Framed thus, Closterman's portrait becomes a study in orders and equivalences. The natural span of Wren's serpentine left index finger reads luminously against his own meditation on geometrical measure in the analysis of cycloid length, diagrammed on the unfurling scroll of shadowed page at center. A zip of highlight extending his nose, the sitter balances his right wrist on the lip of a bas-relief that has been attributed to Grinling Gibbons, the spectacular sculptural contributor to Wren's architectural projects.[2] Juxtaposed to the stony sporting of that slab's naked, sightless cherub, Closterman's mature, painted sitter faces into the luminous field of the imagined beholder. Describing the arch of Wren's forehead, this pellucid light enables the viewer to analyze relations between bodily nature and mathematical order, dumb sculpture and perceptive painting, and, most of all, the sitter's thoughts and deeds. For, through the crepuscular wedge cut in between the retracted curtain and the parapet at left, Closterman stages a paratactic encounter of Wren and his greatest, then-unfinished creation, St. Paul's Cathedral (circa 1675–1720).

Seen from the southwest, St. Paul's rises from twin towers framing a two-story portico and pediment, much as imagined in the architect's plans from around 1690.[3] This western face of the cathedral was sensitive historically. Turned toward the Court at Westminster, it had been the site of Inigo Jones's celebrated portico, a rare exemplar of Renaissance classicism in English architecture, fatally damaged in the Great Fire of London of 1666, but pulled down only in 1687. Depicting Wren's controversial, coupled-columned response standing in its place, Closterman imagines the massive dome and crowning lantern (still a dozen years from realization) reading in silhouette above the square mile of the City to the east.[4] As royal surveyor to the rebuilding of the City since 1666, Surveyor-General of the King's Works from 1669, and architect of the City's parish churches since 1670, Wren had been central to the transformation of metropolitan London. But that city is invisible in Closterman's rendering. Structured instead by the picture's horizontal axis, the viewer's focus is guided through inferences between the monumental architecture at left and the bodily features of the architect visible at right. Where in 1665 Robert Hooke had only been able to promise Wren's possession of "such a *Mechanical Hand*, and so *Philosophicall* a *Mind*," Closterman could now put paid to the point with St. Paul's as proof.[5]

Sketching causal relations between the present, luminous crown of the architect's skull and the distant, darkened dome of the cathedral, Closterman's portrait also opens a different kind of meditation—almost a metameditation. Among experimentalists in later seventeenth-century London, the leap between assiduous study of a head and speculation on the thoughts inside it was a small one. (See figure 0.7.) Having prepared the graphic images for Thomas Willis's seminal *Cerebri Anatomae* (1664), few could have been more sensitive to the intricate interplays between cerebral tissue and cranial structure than Wren himself. After all, Willis had argued that saline particles heated in the alembic of the brain "being thrust out into the circumference . . . constitute the stony Skull, as it were a bubble covering inclosed wind."[6] If made from brain, the skull's visible registration through skin also signaled the contents and capacities of the mind enclosed within. "*Since the Visible things do manifest the Invisible*," so a philosopher in Robert Boyle's coterie proposed in 1685, "there are no doubt some visible Characters of our Inside, writen by the Hand of God in our outward shape."[7] Drawing on traditions of physiognomic analysis encouraged by fellow medical men such as John Evelyn, this author observed that intelligence could be read in the head's formal structures. He warned, "Narrowness . . . from the Root of the Nose upwards, or from the Right Temple to the Left . . . is but a too visible mark of a very imperfect Soul, and an extraordinary want of Spirits: And those few that such puny heads do lodge, are withal Dormant in a manner, and cannot for want of Room either dilate themselves, or give any light."[8]

High, light, and capacious, the forehead of Christopher Wren visualized by John Closterman pledges the presence of a formidable intellect, an architectural intelligence whose contours and authority commanded significant interest within and beyond experimental-philosophical circles.[9] In the last decades of the seventeenth century, as practices of drawing and picture-collecting exploded through London's social classes, architecture figured as a profoundly mental enterprise, one intricately bound up with the structure of knowledge itself.[10] As one English translator of Vitruvius put it in 1692, "Architecture being an Art that has scarce any other Rule to walk by . . . than what we call a Good Fancy, which truly distinguishes that which is Beautiful and Good from that which is not so."[11] Traditionally, that judgment had been shaped by travel to see the classical ruins of Italy, making architecture by default an accomplishment of the leisured gentleman.[12] This tide was turning in two related ways. First, after the outbreak of the English Civil Wars in 1642 and the ouster of England's premier exponent of this learned architectural tradition, Inigo Jones, increasingly unregulated building proceeded in the hands of London masons. These vernacular builders oscillated between what historians have called "artisan mannerism" and "puritan minimalism," with little learned contact to antiquity.[13] Second, as across later seventeenth-century Europe, the center of architectural gravity was itself moving in space and time. Established by Jean-Baptiste Colbert in 1671 to support Louis XIV's massive building campaigns, France's newly focalizing Académie de l'architecture and its theorists gave increasing credence to the advances of modern Europeans beyond the Mediterranean ancients.[14] Commensurately, placed at the head of London's rebuilding efforts after 1666 were experimental philosophers Christopher Wren and Robert Hooke—neither of whom had ever been farther south than Fontainebleu.

Building on experimental ideas rather than classical ideals appeared to be a precarious business to many English contemporaries. With their air-pumps and bewildering trials "weighing of ayre" (in Charles II's infamous quip), experimental philosophers played a dangerously nugatory game.[15] Constructing "those haughty Edifices they call Systems," so one critic charged, experimentalists ". . . lay their Foundation in the Air, and when they think they are come to solid Ground, the Building disappears, and the Architects tumble down from the Clouds."[16] This charge that experimentalist architecture was too clever by half—condemned by the excess of the very mentalist ideas on which it was built—has found important echo in modern scholarship. "Tyranny of the intellect" was the malady of Wren's architecture diagnosed by John Summerson in a brilliant article of 1937. As Summerson wrote in searing damnation by faint praise, "English architecture, while sharing with other arts the profound disadvantages of intellectual domination, was nevertheless caught up in the ascent of greatness . . . one of whose peaks was the mind of Christopher Wren."[17] If the "basic arbitrariness and lack of emotion" identified by Summerson in Wren's work has characterized Robert Hooke's architecture to an even greater degree, much subsequent scholarship has broken on the shibboleth of that gleaming forehead visualized by Closterman.[18] Either the chasm between the experimentalists' empirical intellectualism and their truly artistic imagination that Summerson detected has been closed by recasting Wren and company as freemasons, natural magicians, and Vitruvian architects in the hermetic *mentalité* of Jones and John Dee.[19] Or, Summerson's formalist critique has been undercut by historicizing experimentalist architecture as an outgrowth of mathematical science.[20] It is by criteria of ingenuity, not aesthetic purity, this view holds, that we should understand the telescopes concealed in public monuments, the puzzling shifts in stylistic vocabulary, and the myriad other cunning quirks that characterize the architecture of Wren and Hooke.[21]

This chapter approaches the problem of experimentalist architecture's intelligence slightly differently. I read it through the collective project of experimental philosophy and the centralizing agency of what, by 1682, Hooke was calling the "Archietonical" mind. My proposition is that the visual practices, organizational concerns, and evolving philosophical methods that had brought experimentalists to ideas of architecture especially in the crucial years of the late 1670s and early 1680s can uniquely—can vitally—illuminate how the Royal Society Fellows not only conceptualized relations between mental life and the built environment, but how they made and understood their architecture. Consequently, my focus is less on their buildings' realized forms than on the interweaving, ever-unfurling ways in which architecture was constructed and thought with. I begin with one of the period's most direct addresses to this problem: Thomas Sprat's account of the causes of experimental-philosophical intelligence. Tracing connections between the capacious thinking valorized by Sprat and the aesthetics of idealizing, urban revitalization planned by Royal Society Fellows after the Great Fire of 1666, I emphasize their discrepancy from the sprawling, mazy, and luridly affluent metropolis that was then actually coming into being. Could architecture bring order to this chaos? And how should an experimental architect do so? These, I argue in the third section, were questions asked pointedly by Christopher Wren in the mid-1670s, from the cockpit of his massive, increasingly complicated architectural practice. If the fabric and fabrication of St. Paul's Cathedral stand as Wren's most powerful answer, then it is there, I argue in conclusion, that the slippery, chiasmic relations between architecture and experimental intelligence can be grasped most clearly.

A hint of this dynamic figures already in John Closterman's portrait. Pairing Wren's physical attributes with those of his architecture, the painting registers an unsettling anthropomorphism that runs through Wren's writing. When enlisted by Charles II to assess the damage done to St. Paul's by the fire of 1666, Wren had prescribed a curious regimen: "Herein wee must imitate the Physician, who when he finds a totall decay of Nature bends his skill to a palliation, to give respite for a better settlement of the estate of the patient."[22] After the decision to raze the old cathedral's charred ruins was made, Wren (whose uncle was Bishop of Ely, whose father had succeeded his brother as Dean of Windsor) shared his condolences with the Dean of St. Paul's: "I must comfort you as I would a freind for the losse of his great Grandfather."[23] What rises behind Wren in Closterman's now-doubled portrait, I think, is not only the product and index of the sitter's intellect but—somehow—a shadowy intelligence of its own.[24] Explaining that claim is the ultimate aim of this chapter.

Wit's Geography

In his officially sponsored and widely read institutional history of the Royal Society of London from 1667, Thomas Sprat set out the causes of what he claimed to be England's eminence in experimental philosophy. Like many contemporaries seeking to account for difference, Sprat approached the problem through a venerable Mediterranean tradition of climatic geography and its power of places.[25] Ancient geographers had explained variations in natural life and human societies as products of the world's division into a five-tier system of frigid, temperate, and tropical zones.[26] According to an authority like Pliny the Elder, peoples in torrid and arctic zones were, respectively, blackened and frosted white—their bodies elongated, their customs made savage—because of excesses of the elements fire and water in their native climates. With elements in blended balance, by contrast, inhabitants of the temperate zones north and south of the equator were "gentle, [with] senses clear, intellects fertile and able to grasp the whole of nature; and they also have governments, which the outer races never have possessed."[27] Climatic zones authorized geopolitical order, naturalizing the expansion of temperate governance over tropical places and offering arguments highly useful to European colonialism.[28]

These ancient ideas found a welcome place in experimental philosophy. In his *Micrographia*, Robert Hooke was contemporaneously proposing that the unusual white coloration of mites discovered in cheese could have been caused by the attributes of their newfound milky environment, just as "*Mores* translated into Northern *European* Climates, after a little time, change both their skin and shape."[29] Sprat deployed such arguments more capaciously. Allowing that "Men's Studies are various according to the different Courses of Life, to which they apply themselves; or the Tempers of the Places, wherein they live," he traced a genealogy of knowledge moving between geography and architecture, from the darkened temples and secretive ways of ancient Egyptians to the public spaces and rational concepts favored by the "hot, earnest and hasty minds" of classical Athenians.[30] Thanks to the printing press, the compass, and other technologies unknown to the ancients, modern philosophy had advanced still further by connecting these climatic zones. Instead of merely compositing the superior features of one local climate, as a paradigmatic ancient like Zeuxis had done in painting his Helen from the most attractive aspects of discrete Crotonian maidens, modern philosophers could compound the industrious, active, and "*inquisitive Humor*" native to northern Europe with the "*cold*, and *circumspect*, and *wary* Disposition" of the Spanish and the Italians.[31] And where Vitruvius himself had credited Roman power to the city's temperate climate, Sprat gave a new name to nature's encouragement of hegemonic, compound thinking: England.[32] "Even the Position of our Climate, the Air, the Influence of the Heaven, the Composition of our *English* Blood; as well as the Embraces of the Ocean," he concluded, "seem to join with the Labours of the *Royal Society*, to render our Country a Land of *experimental Knowledge*."[33]

Whatever other obstacles it might have faced, Sprat's apologetics positively courted challenge in the mid-1660s from the rival Protestant nation just across the North Sea—a nation that appeared to enjoy all the same climatic and geographical advantages as the English. Not only would the Dutch produce their fair share of experimental philosophers in men such as Christiaan Huygens and Antoni van Leeuwenhoek, but they had simultaneously parlayed their climatic circumstances into a commercial empire extending from Brazil to Goa and Japan. To make matters worse, the formidable navy of this wealthiest nation in Europe had been steadily delivering a sequence of crushing defeats to the English just as Sprat was penning his text in the mid-1660s.[34] Beyond bald chauvinism, then, what could possibly sustain Sprat's claims for English supremacy?

Sprat found his answer in England's most populous resource. In the century following antiquarian John Stow's seminal *Survey of London* (1598), London's population had more than doubled, growing from around two hundred thousand to more than half a million by the early eighteenth century. The metropolis expanded not only in numbers of human inhabitants but also in geographical size. From the wealthy, densely populated square mile of the City circumscribed by Roman walls now stretched ever-increasing settlements, thus exacerbating the contrast "between an orderly centre and a disorderly and spatially incoherent inner ring of suburbs."[35] Rebuilt in brick and stone after the Great Fire, London became the model for what historians have dubbed the English urban renaissance.[36] But as home to nearly a tenth of the national population, the capital was also a warren of far-flung commodities and native industries whose intense concentration effectively warped its inhabitants' minds. Londoners, one writer observed, "are equally uncapable both of Attention and Patience . . . they don't allow themselves time either to Hear or See; but like Moles, work in the Dark, and Undermine one another."[37]

Many commentators interpreted this unparalleled metropolitan density of bodies and industries as a veritable breeding ground for the intellect. Conceived in marked opposition to the solitary ruminant or the inspired religious visionary of the Commonwealth era, the fundamentally sociable cleverness known as "wit" was a leading species of metropolitan intelligence.[38] Acknowledging wit as the shiftiest of London's legion crafty wares, poet Abraham Cowley appealed to ancient couplings to elucidate its vital nature:

> In a true piece of *Wit* all things must be,
> Yet all things there *agree*.
> As in the *Ark*, joyn'd without force or strife,
> All *Creatures* dwelt; all *Creatures* that had *Life*.[39]

If philosophers such as Thomas Hobbes took wit to be the quick ability to make connections between ideas, Cowley's poetic ark imagines the product of that wit as discrete parts in sociable union.[40] Urban space provided these conditions of production in spades. As one theorist of wit wrote around 1700, "Men joining from all parts, to do that together which they could not asunder, produce a thousand Things which wholly owe their being but to the Dispositions we reciprocally communicate to each other."[41]

Thomas Sprat himself stands among the most interesting observers of wit and its inscription in London's built environment. As future Anglican bishop and historian of the Royal Society, he also figured as one of the "Town-Wits," those aspirational, semiprofessional literati who emulated in more muted fashion the outrageous aristocratic decadence of the "Court Wits" surrounding (and including) Charles II.[42] For Sprat, the conclusion made obvious by the centrality of sociability to wit was that its greatest practitioners were not men at all. Since wit was "the greatest Art about the smallest Things," then it followed that women were its great masters.[43] Sprat made this point in 1663 to Christopher Wren, friend and fellow alumnus of Oxford's Wadham College: "They have usually their Heads less disturbed with busy Thoughts, their Minds are quicker and readier for new Impressions, they talk more of circumstantial Things, [and] they sit longer together."[44] Making good on the traffic between printed and cognitive impressions favored in the period, Sprat's point is brilliantly underscored in Gerard Valck's contemporary engraving after Peter Lely's portrait of actress and royal mistress Nell Gwyn (fig. 6.1). Known for her intimate sociability with the king, Valck's Gwyn-as-shepherdess mediates between her wooly wards and the implied beholder so effervescently that she outstrips the engraver's very art. As Valck's inscription demurs at least, "The Sculpters part is done [if] the Features hit / of Madam Gwin, No Arte can shew her Witt."[45]

At its apogee among circles of such clever ladies, wit inhered in distinct varieties across the capital's urban landscape, according to commentators such as Sprat. Nurtured by habitual interaction within the occupation-based neighborhoods that continued to dominate London well into the Restoration, witty repartee differed from place to place.[46] As Sprat puts it,

> The Lawyers will laugh at those Jests in the *Temple*, which it may be will not move us at *Charing-cross*. And it is likely that *Tom Killigrew* himself would not seem good Company to a Table of *Benchers*. The Wit beyond *Fleet-bridge* has another Colour from that on this Side. The very Watermen on the *Bank-side* have their Quipps, and their Repartees, which are not intelligible but upon the *Thames*.[47]

Wit was thus to be not only interpreted in terms of climatic physiology and gender but read through grids of local history and social class. Sprat would have known that the Temple and Inns of Court had been central breeding grounds for an influential culture of urbane, witty sociability promulgated during the reigns of the early Stuarts.[48] If this venerable humor of "lawyers" such as Ben Jonson, John Donne, and Inigo Jones departed from more au-courant sensibilities bandied about in the fashionable haunt of Charing Cross or in Thomas Killigrew's nearby Theatre Royal on Drury Lane, then cleverness varied again among the rough workers and rougher trade of London's docks. By the first decades of the eighteenth century, such perceptions had only hardened. Inhabitants of a posh suburb like St. James's Park, so *The Spectator* declared in 1712, "are a distinct *people* from those of Cheapside . . . by several climates and degrees in their way of thinking and conversing together."[49]

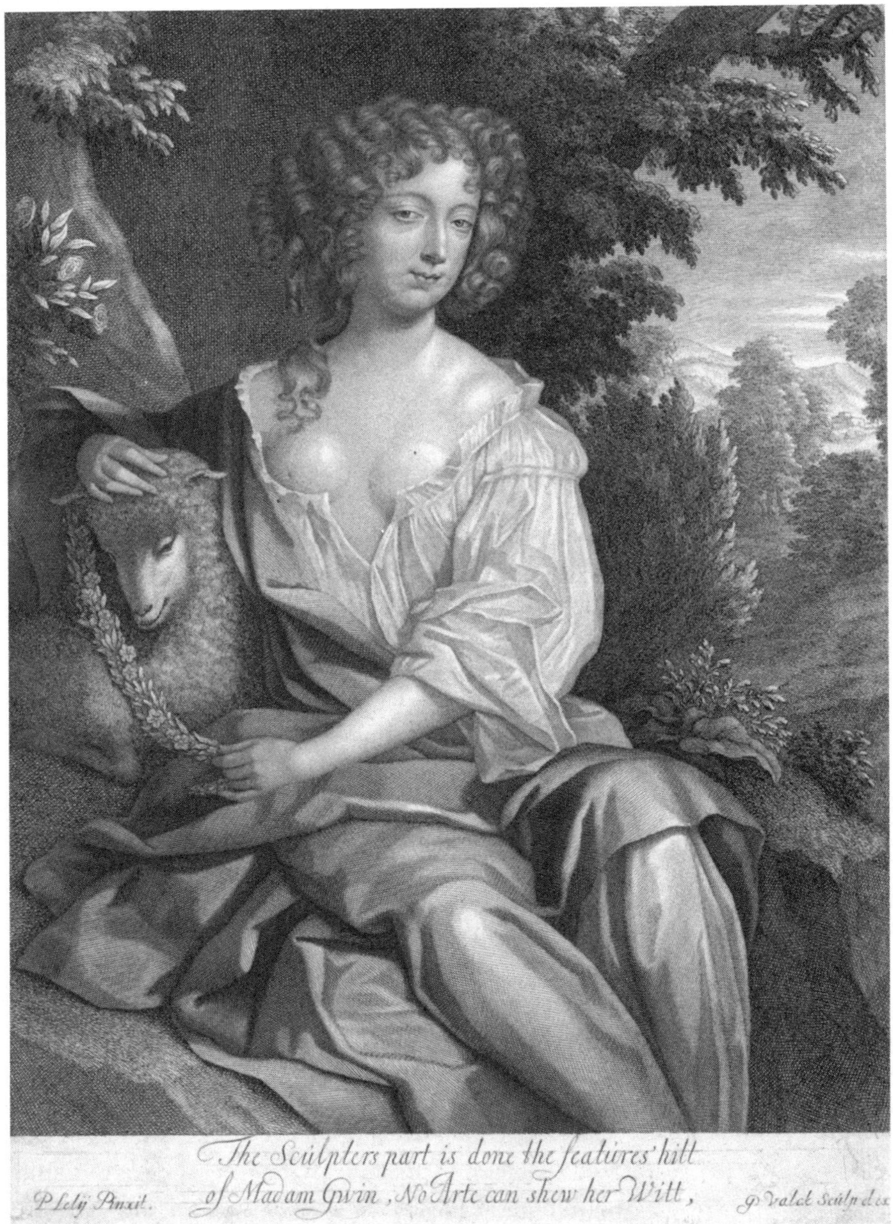

FIGURE 6.1 * Gerard Valck after Peter Lely, *Nell Gwyn* (engraving, circa 1673). © The Trustees of the British Museum (AN514274001).

Within this conversation, the built environment crucially intensified the thinking encouraged by climate and habituated through social mixing. And since introduced into England in Oxford around 1650, few spaces had been more important to intensifying experimental intelligence than the coffeehouse.[50] Exploding through Restoration London, with several hundred cafes in operation by 1700 versus the few dozen boasted by rival cities such as Amsterdam, the coffeehouse was perceived as a keen urban amenity conducive to experimentalism's new social forms. Without "the modern advantage of Coffee-howses in this great Citie," so Royal Society Fellow John Aubrey noted around 1680, ". . . men knew not how to be acquainted, but with their owne Relations, or Societie."[51] A place for business and a salubrious alternative to the pub, the coffeehouse—typically a single large room, occasionally flanked by private chambers—encouraged discourse in a city of strangers.[52] There, one adherent argued, urbanites could "meet *Company*, and by the custome of the house not *such* as at other places *stingy* and reserved to themselves, but *free* and *communicative*, where every Man may modestly begin his *story*, and *propose* to, or *Answer* another as he thinks fit."[53] Coming for the coffee and other exotic commodities sold there, patrons stayed to practice the edifying conversation that was, so this advocate concluded, "the minds best *Dyet*, and the great *Whet-stone* and Incentive of Ingenuity."[54]

From its very earliest incarnations in England, urbane coffeehouse wit stood distinct from the exclusive erudition practiced by scholars within university walls.[55] Consequently, as its critics charged, the public intellectualism spawned by the coffeehouse was just as combustible as it was new. Pamphleteers during the heated years of the Exclusion Crisis (1678–81) imagined the coffeehouses around Charing Cross as meeting points for heterodox partisans contesting the Duke of York's succession.[56] Paired to these internecine politics of religion, poetasters parlayed talk of coffee's medical effects into tales of its diabolical influence. In the play *Knavery in All Trades; or, The Coffee-House* of 1664, audiences heard how coffee is "most pernitious unto the brain, it fires the Pericranium, disorders all the faculties, presents Ideas most delusive; Treason, Murder, (the hand helpes to ambition) twins of villany."[57] The coffeehouse wit—"the *Arch-Devil*, wherewith this *Smoke-hole* is haunted"—embodied that dubious blending.[58] Scoffing at the deliberation required for university learning, the grandstanding manner of this caffeine-fueled jack-of-all-intellectual-trades was inseparable from the heterogeneous ambit of the cafe where social classes would mingle "without regard to degrees or order . . . all blended together, to compose an *Oglio* of Impertinence."[59]

To eager patrons like Hooke, the novel possibility offered by blending aristocrats, minor gentry, mercantile elites, and skilled craftsmen with the coffeehouse's potent chemical reagents was precisely what made it so exciting. Not only did such venues actually *practice* the mixing of social classes merely mouthed by the Royal Society, but Hooke's own splinter "philosophical club" of the mid-1670s met in Garraway's, one of the numerous coffeehouses he visited almost daily.[60] Mapping his geography of experimental excellence in 1667, Thomas Sprat also cited these mixed urban spaces as cause and proof of English advantage over the Dutch. "When they are abroad," he contended, Dutch traders "shew, that they are only a Race of plain *Citizens*, keeping themselves in their own Cells, and Ware-houses; scarce regarding the Acquaintance of any, but those with whom they traffic."[61] Squirreled away in their confined spaces and confining company, Sprat's Dutch merchants were like medieval Scholastics who "shut themselves up in a narrow compass, keeping company with a very few, and . . . addict themselves to some melancholy contemplations."[62] The English were different. While Dutch merchants live "meanly, minding their Gain alone," he proclaimed, "ours converse freely, and learn from all; having in their Behaviour very much of the *Gentility* of the Families, from which so many of them are descended."[63]

Sprat's calculus of experimental intelligence thus boils down to a simple accounting problem. The Dutch and the English shared a range of climatic, confessional, and geographical advantages. The wealth of both societies depended on commerce—"that great *Rudder* of humane affairs," as Hooke was then calling it—inclining each to intellectually advantageous mixing.[64] But what ultimately pushed the English ahead was a liberality or freedom of mind fortified by their broadening social spaces. Menaced by physical enclosure within the "Cells, and Ware-houses" inveterately favored by the mercantile Dutch and hampered among the tight social circles of witty women, experimental intelligence flourished when exposed to blended company, to broad spaces and open air. "To keep themselves free," Sprat concluded, ". . . the true Experimenting has this one thing inseparable from it, never to be a fix'd and settled Art, and never to be limited by constant Rules."[65] By Sprat's math, climate plus commercial mixing minus narrow-mindedness equaled experimental superiority. And one felicitous consequence of London's destructive fire of 1666 was that the Fellows of the Royal Society were given an opportunity to imagine that calculus of experimental intelligence built into the very fabric of the City.

London's Rebuilding and the Builder's London

Ignited by a bakery fire, the medieval, wooden London that combusted so spectacularly and burned for nearly a week in early September 1666 had been a longstanding bugbear to the Restoration intelligentsia. The City was a rainforest, a humid maze of putrid alleys in perpetual downpour thanks to what John Evelyn called the "disposure of the *Spouts* and *Gutters* overhead . . . because it is hereby rendered a *Labyrinth* in its principal passages, and a continual Wet-day after the Storm is over."[66] The sorry state of St. Paul's Cathedral was a particularly sore point. It was bad enough that the cathedral had been used as stables by Parliamentary soldiers during the Civil Wars, but worse that enterprising merchants then converted Inigo Jones's prized portico into a row of shops.[67] One shocked French visitor rued that "this goodly and venerable fabrick . . . be built about and converted into raskally *Ware-Houses*, and so sordidly obscur'd and defac'd."[68] Verging on sacrilege, the cathedral's profane commercial usage revealed the contours of the twisted English mentality only too clearly. "An argument of greater avarice, malice, meanness and deformity of minde cannot possibly be expressed," this anonymous writer concluded.[69]

Along sympathetic lines, several of the plans for urban rebuilding presented to Charles II in the Great Fire's immediate wake envisioned the City made healthy to body and national mind in terms highly favorable to Sprat's account of experimental intelligence. Yet these ideals departed radically from the metropolis actually slithering into existence. In this section, I sketch three key visions for post-Fire London—as seen in turn by experimentalists, merchant-philosophers, and property developers—to illuminate the divisions of interests and aesthetics among London's would-be civic improvers. Why and how the experimentalists were proposing to govern the emerging, menacing cosmologies of metropolitan space they detected among them in larval form will then move into central focus.

When granted a royal audience to propose his plan for rebuilding London in mid-September 1666, Captain Valentine Knight described a city newly riven by connective tissues of thoroughfares and cross streets, its houses set airily, luminously apart. Water coursed through Knight's metropolis, drawing in goods and commodities via a new "Cutt" from the Thames to urban marketplaces and sluicing out waste through a newly rationalized system of sewers.[70] It was the eye that was privileged with this kind of penetrating urban passage in the classicizing plan then proposed to the king by Christopher Wren[71] (fig. 6.2). Opening broad avenues along the crucial east–west axis between City and Court, Wren positioned the grandest of several piazzas just to the west of the Fleet River. Echoing Jones's nearby Covent Garden development, Wren parsed the metropolis into eight sequential views:

> First, straight forward quite through the City: Second, obliquely towards the Right Hand, to the Beginning of the Key, that runs from *Bridewell Dock* to the *Tower*, Third, obliquely on the left to *Smithfield*. Fourth, straight on the Right, to the *Thames*. Fifth, straight on the left, to *Hatton-street*, and *Clarkenwell*.

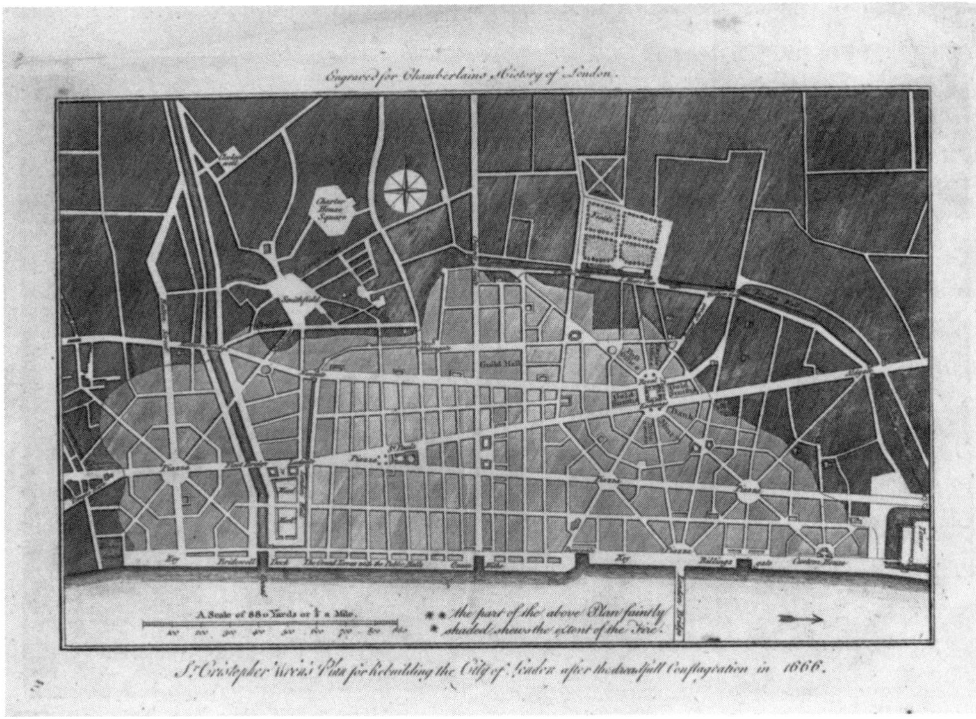

FIGURE 6.2 * Christopher Wren's plan for rebuilding the City of London, published by J. Cooke (etching, late seventeenth century). © The Trustees of the British Museum (AN332550001).

> Sixth, straight backwards, towards *Temple-barr*. Seventh, obliquely on the right, to the Walks of the *Temple*. Eight, obliquely on the left, to *Cursitor's Alley*.[72]

Against Wren's urbanism of the eye, keenly class-conscious John Evelyn encouraged the king to enact truly multisensorial transformations to the built environment.[73] That London should "wrap her stately head in Clowds of Smoake and Sulphur, so full of Stink and Darknesse," he had thundered in 1661, "I deplore with Indignation."[74] Disfiguring warehouses and heavy industries had made the capital more like "the face . . . of *Mount Aetna*, the *Court of Vulcan*, *Stromboli*, or the Suburbs of *Hell*, then an Assembly of Rational Creatures, and the Imperial Seat of our incomparable *Monarch*."[75] Instead, standing as attentive surrogates for the appearance-conscious citizens inside them, Evelyn envisioned houses perched on newly sculpted hills that would "peep over one another successively, with a far better grace, than those do at Genoa."[76] Broad avenues dotted with public spaces and attractive features could mingle polite business with pleasure, seeing with being seen.[77] "In the piazzas should be kept the several markets," Evelyn instructs, "in others . . . should be public fountains placed; not as formerly immured with blind and melancholy walls, but left free to play, and show their crystal waters."[78]

We know less of the plan that Robert Hooke presented to Charles II in September 1666.[79] Apparently, he envisioned London remade as a grid, with "all the chief Streets . . . to lie in an exact strait Line, and all the other cross Streets turning out of them at right Angles."[80] Beyond Richard Waller's claim that it would have "added much to the Beauty and Symmetry of the whole," it is likely that Hooke's rationalizing plan was conversant with ancient ideas about the healthful effects of buildings then prominent among English architectural writers.[81] Equally, Hooke himself soon enacted the Restoration's premier interface between architecture and (disturbed) mind in his new Bethlem Hospital at Moorfields (1674–76; destroyed in 1807), England's only specialized facility for the care of the insane. Although we have frustratingly little writing from him on these points, his single-pile structure and its lengthy, circulating galleries seem to have been designed to amplify the therapeutic advantages of the site in line with classical theories of building situation.[82] Summarizing the available evidence, Christine Stevenson observes that Bethlem's "very form, thin and permeable to the light and air of Moorfields, would help to restore health."[83]

Very different from these therapeutic aesthetics of broad, airy spaces was the plan for London's rebuilding proposed in 1666 by Francis Lodwick (1619–1694).[84] Eldest son of a Dutch merchant and subsequent Fellow of the Royal Society, the largely self-educated Lodwick not only dispensed with the pleasing vistas endorsed by his colleagues but turned Thomas Sprat on his ear, imagining an urbanism redolent with the mercantile interests that were his family's stock in trade. Since they had already lost so much in the Great Fire, Lodwick contended, merchants would be reluctant to return to London without compelling assurances of security to their markets and stores. With the threat of starvation menacing the City, priority had to be placed on efforts to "contrive buildings in relation to trade, that those places for the depositing of goods, may be so contrived, as to save charges to the traders, and loss and damages to his goods."[85] Only by swift, decisive action could London once again be made "the magazine or Storehouse for all sorts of manufactures and other commodities."

Even if we see the broader reconfigurations of urban landscape in late seventeenth-century England as ever pulsating with tensions between private interests and new ideals of public good, what Lodwick proposed was drastic.[86] To ensure that fire could never again leap from structure to structure as it had done in September 1666, warehouses would be built in brick and encased in prophylactic belts "as high as the ridges of the houses." Instead of building cheek-to-jowl as had been disastrously commonplace in the crowded Stuart city, these newly constructed edifices would share no party walls. Nor would commercial warehouses and storage spaces have windows—what Lodwick calls lights—"to the street side, that if any fire happen, it may take no hold through the same, as it did on the churches and other places in the late fire." Amplifying the expanded quays envisioned by Evelyn and Wren, Lodwick also commandeered the banks of the Thames from the "small, close, and noisome dwellings, which harbored for the most [part] poor people." In their place, he proposes "a row of warehouses built facing the river (except that part, appointed for his Majesty's Custom house) of 3 or 4 stories high, strongly built . . . All which storehouses should as before, [have] no doors or lights to the street side, nor any house adjoining to them."

Lodwick's imposing urbanism was literally built around the needs of London's merchants and their commodities. Because porting fresh fruit and vegetables up from the Thames to the City's markets was costly and cumbersome to traders, Lodwick brings the markets to the merchants. Garrisoned amid the impenetrable banks of warehouses with which he flanks the river, designated markets for fruit, vegetable, fish, and other "belly-provisions" would be established "for the conveniency of those gardeners, that came from other parts about the city; as doe fruits they may have the same conveniency with the herb market. And the fruiters, some convenience would also be allowed them close to the river for the landing and disposing of their fruit at easy rates." Rationally organized, inviolably protected, and made supremely convenient for traffickers, Lodwick's vision of London is a veritable hymn to supply-side economics.

In the explosion of building that followed the fire, however, neither the idealizing plans of Royal Society elites nor even the mercantile urbanism advocated by Lodwick held the day. If it is true that by 1670 the square mile of the City was largely rebuilt in brick and stone—overseen by twin teams of surveyors led by Christopher Wren and Robert Hooke—new construction positively erupted further away from the traditional center.[87] Emanating from the vital east–west conduit of the Strand, rows of new terrace houses sprouted in Holborn, Bloomsbury, and Lamb's Conduit Fields to the north and in St. James's Square, Soho, and Piccadilly to the west.[88] Built around a formulaic template of materials and stock floor plans, the few ornamental features of these houses—their stone string courses, gabled roofs, keystones over the new sash windows—were, as Elizabeth McKellar observes, "cheap and easy to produce, thus allowing for changes in fashion and style to be accommodated."[89] Speculatively financed and often of dubious legality, this novel breed of domestic dwelling looked to some just as crooked as the builders who erected them. "New fantasticall perishable Trash" was the assessment of architect and longtime Wren assistant Nicholas Hawksmoor.[90] "Soe far from skill or honesty," Hawksmoor claimed of these contractors, ". . . the Generall part of 'em are more brutall & Stupid then, in the remotest part of Britain and the longer they worke the worse they grow, as you may see in all the Additional Scondrell Streets they are Continuously Cobling up. to Sell by wholesale."[91] Hawksmoor was correct in the shifting scale and market orientation of these constructions. Where Lodwick had eliminated existing housing to accommodate the needs of merchant-suppliers, these new dwellings were designed to expand consumer demand and to explode London's traditional urban space simultaneously.

Few grasped this project with the same savvy and panache as Dr. Nicholas Barbon (circa 1640–1698), late seventeenth-century London's premier property developer. In *An Apology for the Builder* (1685), Barbon squared off against the accusations of greed, excess, and rapine inevitably leveled at the speculative builder when "every year a new Town [is] added to the old one."[92] Rather than causing rents of existing buildings to depreciate, Barbon claimed, expanding housing developments in London would actually increase property values and simultaneously supply an influx of new business to Lodwick's key constituency. Whether in Bloomsbury, St. James's Park, or elsewhere, "the Inhabitants of these places do eat, wear Clothes, and furnish their Houses, and whatsoever Commoditie they use, come first from the Merchants."[93] Nor was this a zero-sum game for the nation; the gain of populations lured to the metropolis by these new dwellings could also benefit the countryside. Creating markets for timber, stone, and other building materials sourced from rural areas, new urban dwellings would accommodate "the Supernumerary and useless Inhabitants of the Country"—those surplus populations that had so troubled elite observers throughout the early modern period.[94] Ultimately, a capital growing in its viable housing stock served the interests of the king, who could profit from an increasing population and the expanding tax revenue it generated.[95]

Barbon's architectural developments and theoretical arguments instantiate what historians have called the "consumer revolution" of the long eighteenth century.[96] Rather than fixing marketplaces into the existing footprint of the ancient City, as Francis Lodwick had done, the profile of that city would *itself* become a speculative market. And instead of courting refined architectural tastes of elites such as John Evelyn—let alone providing the basic shelter necessary for human survival in London's chilly climate—these faddish new dwellings acted as what Barbon called "things necessary to make up the several distinctions of men": they were markers of social status for London's emerging, aspiring middle classes.[97] In this way, the new building Barbon promoted was not only different from the aesthetic values, socio-economic ideology, and clientele profile of the courtly nobles and wealthy citizens for whom experimentalist architects primarily built; it was directly opposed by the Office of the King's Works administered by Surveyor-General Christopher Wren. Pressured from within Charles II's administration to constrain the rampant illegal building in the metropolitan suburbs, Wren himself petitioned the king against these new developments. Construction at Windmill Fields in Soho, he objected, "will prove only receptacles for the poorer Sort, and the Offensive trades, to the annoyance of the better Inhabitants."[98] In aspiration if not fact, as one historian has put it, "the Surveyor and his subordinates at the Office of Works were the 'Police Architectonical.'"[99]

Given the difference of their urban plans and the literal opposition of their interests, it is also telling to note that Barbon was retooling the mentalist tropes proclaimed by Thomas Sprat and so often associated with London's experimentalist architects.[100] For those who imagined the metropolis to be "the Head of the Nation" grown too monstrously large for its body, the physician Barbon corrected the metaphor. "The Metropolis," he claimed in 1685, "is the *heart* of a Nation, through which the Trade and Commodities of it circulate, like the blood through the heart, which by its motion giveth life and growth to the rest of the Body."[101] How could the experimental architects of the Royal Society address this tell-tale reordering of urban anatomy? By what means could they counter the appeal to desire, consumption, and frank irrationality it implied? While the following sections will examine the clever archietonical theories and ingenious architectural strategies they deployed to do so, I conclude by suggesting how their wicked intelligence was divided and effectively defeated.

Pyramid Schemes

In 1638–39, John Greaves, mathematician and subsequent Savilian Professor of Astronomy at Oxford, made an expedition to Egypt. Keen to explore the claim of Proclus that the ancient Egyptian monuments served an astronomical function, Greaves (1602–1652) and a Venetian companion intrepidly descended into the Great Pyramid at Giza. Scaling down steep, lightless stairwells and crawling on their bellies like snakes, Greaves's company came into the heart of a hidden architectural complex. "A very stately peece of work," the mathematician subsequently wrote, "and not inferiour, either in respect of the curiosity of Art, or richnesse of materials, to the most sumptuous, and magnificent buildings."[102] In the faint light of his torch and hidden beneath a film of grime, Greaves discovered corridors, ceilings, and walls of gleaming white marble—stones whose source he later traced to Mount Sinai.[103] Aided by his observations, he could reconstruct how the Pyramids were built. These subterranean chambers, he reckoned, must have stood at the base of a core tower built to the structure's apex, with the sheer, triangular walls of the structure then added as buttresses.[104]

Sharing a fascination with Egyptian culture richly embroidered throughout the early modern learned world, Restoration London's experimental philosophers knew Greaves and his writings well.[105] While Robert Hooke cited Greaves's observations on the Pyramids in several lectures from the 1680s, Wren (like Greaves, a former Savilian Professor of Astronomy at Oxford) offered a very different view of Egypt's spectacular monuments in the years around 1675.[106] Elaborating a line of interpretation from Pliny the Elder dismissed by Greaves, Wren narrated the intertwined relations between agriculture, social life, and architecture in ancient Egypt.[107] Felicitously sited along the banks of the Nile, Wren explained, Egypt had expanded rapidly. Cultivating agriculture in alluvial soils made highly "productive of Corn by the help of the Nile, in a manner without labour," Egypt's people became so numerous that they quickly outstripped the available food supply, causing civil unrest and threatening violent revolt.[108] The Egyptian state had to intervene. Following "the Wisdom of their Ancient Kings and Priests," it commandeered a portion of the annual harvest to be distributed among the hungry and indigent.[109]

Now, in his influential economic writings of the early 1660s, Wren's friend and fellow alumnus of Oxford's philosophical circles, William Petty, had endorsed state-subsidized poor relief as a tool for sustaining and containing a robust modern population. Regarding the nature of work actually *done* by this indigent population to earn their keep, Petty was not dogmatic. "'Tis no matter if it be employed to build a useless Pyramid upon *Salisbury Plain*," he wrote, ". . . for at worst this would keep their mindes to discipline and obedience, and their bodies to a patience of more profitable labours when need shall require it."[110] In the mid-1670s, Wren ran a similar argument by way of architectural-historical explanation. By providing food to its booming populations, the Egyptian state had been able to "support the Popularity, and consequently the Grandure of the Kingdom."[111] Yet what was good for the state as a whole nonetheless posed a risk to the individual. This, according to Wren, was where the Pyramids came in:

> It was not for the Health of the Common people, nor the Policy of the Government, for them to be fed in Idleness; great Multitudes were therefore imploy'd in that which requir'd no great Skill, the Sawing of Stone to square a few different Scantlings; nor was there any need of Scaffolding or Engins, for hands

only would raise them from Step to Step: a little teaching serv'd to make them set by Line.[112]

By Wren's account, the formal simplicity of the great Pyramids conceals the shrewd political function they actually perform. Rising steadily from the unskilled busywork of masses teeming from agricultural overproduction, the Pyramids justified the food rations given by the state while keeping its potentially explosive population in obedient, laboring order. Disavowing the complex construction detailed in Greaves's eyewitness account, Wren located the challenge of building such a "haughty Edifice" less in the skills required for the fabrication or even in the architectural design of the monuments per se. Instead, the challenge was an administrative one, requiring deft management of the materials and working bodies necessary for executing such a massive project. "The difficulty," as Wren put it, "was in Mustering the men to move in order under proper officers . . . to make the men exert their United force in equal time."[113]

I think that Christopher Wren's Egyptian fable can illuminate much about his own contemporaneous architectural practice—particularly about a range of key concerns and practices then consolidating around his monument-building on London's Ludgate Hill.[114] Intensifying in building activity at precisely the moment when this treatise was drafted, St. Paul's Cathedral was, like Wren's Egyptian pyramids, a massive, state-sponsored project.[115] He also regarded that Egyptian-style administration as a defining feature of what he called Europe's greatest "School of Architecture."[116] During his visit to Paris in 1665, he had marveled at Jean-Baptiste Colbert's building works at the Louvre, where "no less than a thousand Hands are constantly employ'd in the Works; some in laying mighty Foundations, some in raising the Stories, Columns, Entablatures, &c., with vast Stones, by great and useful Engines; others in Carving, inlaying of Marbles, Plaistering, Paintings, Gilding, &c."[117] Yet when then asked to identify Paris's finest architectural monument, Wren named the quays along the Seine. This infrastructure had been built, he observed, "with so vast expense and such great quantity of materialls that it exceeded all manner of ways the building of the two greatest pyramids in Egypt."[118] In this section, I read Wren's Egyptian fable against the bureaucratic armature and administrative acumen he crafted in the design and construction of St. Paul's Cathedral—practices that resonate particularly powerfully with the visual and organizational concerns of archietonical experimental philosophy contemporaneously being elaborated by Robert Hooke.

When appointed Surveyor-General of the King's Works in 1669, Wren was placed at the head of a highly complex administrative system. Reinstated with the monarchy at the Restoration in 1660, the Office of Works was as hierarchical an institution as the royal court it served. With the Surveyor-General remunerated officially at over three hundred pounds per annum (to say nothing of the position's other perquisites that could reportedly bring twenty times that sum), subservient "Patent Artisans" were paid on a steeply graded scale.[119] Many officials consequently regarded their posts as sinecures, accepting other incomes and profiting handsomely from the sale of benefits attendant to their offices. Nonetheless, the Office of Works was no trifle. "As the only central building organisation in the kingdom," so its leading historian has written, "the Office of His Majesty's Works was the place where the best architectural advice was to be found."[120] During the forty-nine years in which Wren was at its head, the Office of Works produced formidable results, including the reconstruction of the royal palaces at Hampton Court, St. James, and Whitehall, along with building new hospitals at Chelsea and Greenwich.

St. Paul's Cathedral was the jewel in this crown, visualized by engraver Henry Hulsbergh (died 1729) as the top trophy hung on the pyramid of Wren's prodigious architectural production[121] (fig. 6.3). An adjunct to his main offices at Scotland Yard, Wren's Office of Works at St. Paul's was centered in administrative spaces built into the medieval Convocation House, just to the southwest of the cathedral. If much smaller than the horde of tax clerks who would soon administer the rise of Britain's "fiscal-military state," the organizational structure of the Office of Works was a marvel of architectural administration nonetheless.[122] Wren's office, one scholar has recently proposed, marks "the birth of the modern architectural firm, in which the head office retains overall control but relies on others for the execution."[123] Historian Michael Cooper puts the point even more pithily: "The beginnings of modern construction management can be seen in the way that Wren's office was organised."[124] That organization was sustained by a rich material culture of bureaucracy.[125] Outfitted "with Shelves & lock & hinges to keep bookes, writeings, & paper for the use of the Work," the Convocation House was routinely replenished with office supplies.[126] A typical stationer's delivery from September 1678 included one hundred quills, six pencils, three quarts of ink, numerous skins of vellum, and writing paper in a broad variety of sizes and qualities.[127] The six pounds of candles, two candlesticks, one snuffer, and a pewter inkstand purchased in November 1675 suggest the intensity of work conducted year-round in an office equipped with eleven coal-burning hearths by the end of the decade.[128] This administrative point was incessantly reinforced to the workmen: a work bell set to the sun dial on the Convocation House's wall rang out the roll call at six in the morning, one in the afternoon, and six in the evening, six days per week.[129]

If late seventeenth-century English society was "a series of pyramids . . . governed

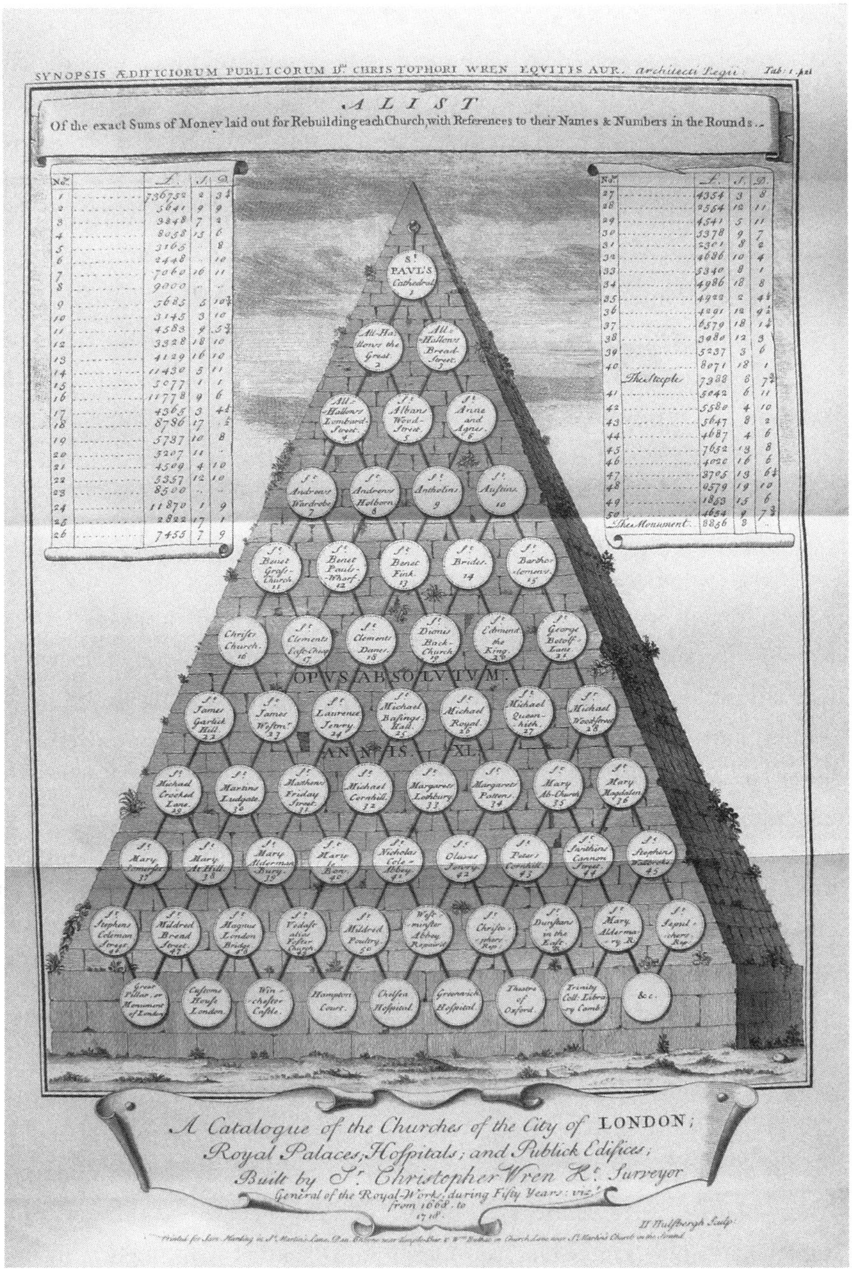

FIGURE 6.3 * Henry Hulsberg's engraving/etching (circa 1720) of *A Catalogue of the Churches of the City of* London . . . *Built by Sr. Christopher Wren Kt. Surveyor General of the Royal-Works, during Fifty Years: vizt. from 1668 to 1718*; published in Christopher Wren Jr., *Parentalia; or, Memoirs of the Family of the Wrens; Viz., of Mathew, Bishop of Ely, Christopher, Dean of Windsor, Etc. but Chiefly of Sir Christopher Wren. . . .* (Farnborough, Hants: Gregg Press, [1750] 1965), 308–9.

by a loose alliance or working agreement between those at the top of each pyramid," then the isosceles triangle of St. Paul's Office of Works branched from headman Christopher Wren to Edward Woodroffe and John Tillison.[130] Made Assistant Surveyor of St. Paul's in 1669, Woodroffe (circa 1622–1675) worked extensively with Wren and Hooke on the City's parish churches in the early 1670s. While acting as draftsman to Wren, Woodroffe's key contribution to this partnership was his extensive know-how in the building trades.[131] As Woodroffe (and then glazier-cum-mason John Oliver) split time between St. Paul's, the parish churches, and other projects, Clerk of the Works Tillison was a veritable fixture on the building site at Ludgate Hill. In charge of keeping accounts and paying the workmen, Tillison also factored in tricky negotiations, such as sourcing the stone from the Isle of Portland that was used so extensively in the cathedral's fabric.[132]

Tillison's greatest responsibility lay in the day-to-day management of the master craftsmen, their individual teams, and the site's swarm of laborers—already over 120 were being paid at Christmas in 1675.[133] Through the Office of Works' prodigious records, we can follow Tillison directing master carpenters John Longland and Israel Knowles in framing and striking a constantly evolving maze of wooden scaffolding to assist master masons Joshua Marshall and Thomas Strong.[134] As the earliest masonry contracts demonstrate, construction work began in June 1675 with the team of Oxfordshire quarryman Strong assigned to the east end of the choir, while royal Master Mason Marshall took the choir's north and south walls.[135] (See plate 10.) Unlike the carpenters and laborers, who were paid by the day, the mason teams—increasing to four in 1679 as work expanded from the choir to the transepts and massive piers of the dome—were paid both by task and by measurement, complicated assessments for which specialist consultants would be periodically recruited.[136] Equally specialized were the lawyers, accountants, and draftsmen such as Nicholas Hawksmoor, who in March 1691 was paid for "assisting the Surveyor in Drawing Designes and other necessary business for the Service of this Work."[137] And while the brilliance of the cathedral's architectural design is most frequently taken as his greatest accomplishment, Wren's contemporaneous Egyptian fable aptly suggests the true nature of his responsibilities. Not only was he charged with the oversight of the cathedral's incredibly complicated structural, legal, aesthetic, and organizational transformations, but he had to wield substantial political influence, thrice standing for Parliament to ensure a steady flow of income for the vastly expensive and deeply controversial project.[138] "I think it is silver upon which the foundation of any worke must be first layd," so Wren had trenchantly observed in 1668, "least it sinke while it is yet rising."[139]

How was this work coordinated? Given what we have seen of Restoration London's archietonical experimental philosophy, it comes as no surprise that Wren's Office of Works relied on an ingenious and variegated array of visual artifacts to communicate within and beyond the office's nerve center on Ludgate Hill. Most prestigious among these visual practices is the extensive body of drawings generated through Wren's collaboration with a shifting coterie of office draftsmen, including Woodroffe, Hooke, and Hawksmoor. Through the roughly three hundred surviving drawings, scholars have been able to identify upwards of seven distinct stages in the cathedral's evolving design.[140] Within this corpus, the drawings change markedly in their function and form (plate 8). As demonstrated by his watercolor renderings from before the Great Fire, many of Wren's early drawings were intended for the eyes of the Rebuilding Commission, the committee of clergy, courtiers, and advisors on whom approval for the building would depend. Once the royal warrant authorizing construction was granted in 1675, the Office of Works' graphic production effectively bifurcated. As his assistants plotted new arrays of scaled working drawings for the craftsmen, the Surveyor himself developed increasingly sketchy pen-and-ink renderings for use within the office[141] (fig. 6.4). Not only do these drawings indicate that Wren saw the authorized plan as highly flexible, but they show how fundamental features of the design remained constantly under reconsideration and adjustment even as the masonry walls rose.[142]

Following a robust tradition of early modern practice, scale models provided the Office of Works with a key mechanism for directing communication between architect and workmen.[143] As early as the spring of 1666, Wren had endorsed architectural modeling "for the Incouragement & Satisfaction of Benefactors that comprehend not Designes & Draughts on paper as well as for the inferior Artificers clearer intelligence of their Business."[144] The most expensive visualization of this kind produced by the office is the enormous object known as the "Great Model" (plate 9). Scaled at one inch to two feet, the model was designed by Wren and Edward Woodroffe, fabricated by master joiner William Clere, and lavished with a coordination of skilled labor surpassed only by the cathedral itself.[145] "Walkt through" by Robert Hooke and then presented to the Rebuilding Commission in 1674, the haptic and optic address of the massive, thirteen-foot-high model also communicated, as it were, too effectively.[146] After the torrent of criticism heaped on the so-called Model Design, Wren's son reported that "the *Surveyor* resolved to make no more Models, or publickly expose his Drawings, which, (as he had found by Experience,) did but lose Time, and subjected his Business many Times, to incompetent Judges."[147] That said, William Clere's team of master woodworkers continued to produce an array of practical models vital to the fabrication of the cathedral's component parts. Paying Richard Clere for "Carveing 4 Capitalls on a pattern piece," Clerk of the Works John

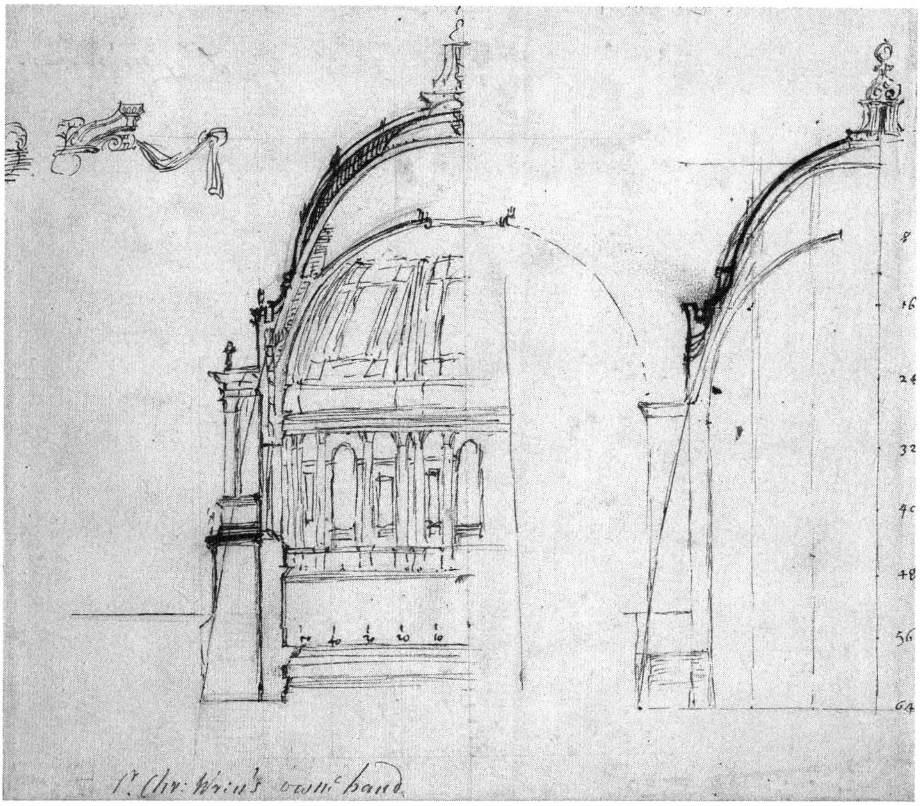

FIGURE 6.4 * Christopher Wren's pen-and-ink study of the design for the dome of St. Paul's Cathedral (circa 1690). © The Trustees of the British Museum (AN14926001).

Tillison dispatched these "Modells of the Capitalls" from the Office of Works to Weymouth, where they were conveyed to stonecutters in the quarries on the Isle of Portland.[148] Designed and fabricated in London, the Cleres' wooden models could thus complement the numerical dimensions and diagrams Wren himself detailed on paper to guide the hands of provincial stoneworkers.[149]

Yet some of the most effective visual artifacts disseminated from the Office of Works were not images in the conventional sense. In 1677, with the coal-tax money allotted by the Rebuilding Act of 1670 nearly exhausted, Henry Compton, Wren's companion in his 1665 journey to France and newly minted bishop of London, penned a plea soliciting donations from the general public. Compton made a brisk case for the national interest in a revitalized St. Paul's while simultaneously countering the complaint—no doubt murmured by the likes of naturalist William Cole—that "so much splendor in the outward circumstances of Gods worship, is . . . not suitable to the spiritual nature and design of the Gospel."[150] Then, utilizing the ready supply of skilled labor available in the print shops flanking St. Paul's churchyard, the Office of Works paid stationer R. Roberts "for printing of Letters to be sent with the Breifs into the Country"; another six thousand copies of these "breifs" were printed in 1679, followed by five thousand more in 1682.[151] Read from pulpits across the country, these documents were returned to London inscribed with the names and financial pledges of individuals in each parish wishing to make a contribution.[152] By inciting a now-national congregation to "act at a distance and make good return hither" (as Hooke would describe his planned experimental periodicals in 1680), the rebuilding effort was able to leverage key donations from the universities and ruling classes.[153]

In his Office of Works at St. Paul's, then, Christopher Wren built an administrative armature worthy of his Egyptian fable. First, like the baroque studio model contemporaneously perfected by Peter Lely, Wren parsed his complex project among a trusted team of deputies centered in the administrative enclave of the cathedral's Convocation House. Second, his office produced, deployed, and archived a prodigious variety of "paperwork"—visual artifacts through which it orchestrated the materials, workmanship, and capital necessary for the massive enterprise. From its bureaucratic metropolitan office, Wren's specialized, salaried team then distributed that field of syntactically and semantically differentiated artifacts, governing their geographical flow and administering the vital materials reciprocally returned by them. It is as if we can watch Bruno Latour's "immutable mobiles" in reverse, as paper and wood are transformed back into stone and money.

Even as they had critiqued the confusions and entrapments caused by artists, Hooke and Wren had been mythologizing—positively mystifying—the managerial strategies requisite to these architectures of mental microcosm and urban macrocosm. Explaining in 1682 how the "Directive and Archietonical" power of the soul could externalize its cogitations, Hooke had appealed to some very strange forces. The immaterial soul, he proposed, gives order to an overwhelming field of sensory impressions constantly collected by the experimental body. To elucidate how that soul could use its ideas to act beyond itself to influence other bodies, though, Hooke repressed the strategic techniques of institutional "Allurements" he had developed so meticulously through the 1670s, appealing instead to musical harmonies and natural magic. Fascination and "Lupus in Fabula" (speak of the devil and the devil comes) were, he proposed, explanatory models for how the soul could "not only influence other Bodies, but be influenc'd by them also."[154] Wren too had conjured musical magic in his Egyptian fable, crediting legendary influences such as Amphion's harp with ancient architects' crucial ability to coordinate labor.[155]

And yet, through the 1670s, Wren had been veritably redefining architectural practice as action-at-a-distance wielded not through occult influence but through hard-headed administration, keen division of labor, and clever visual practices. He had to do so. Overwhelmed with architectural projects in the Office of Works and menaced by the syndicates of private financiers, lawyers, investors, and aspiring contractors commanded by speculative developers such as Nicholas Barbon, Wren's Office of Works might have adopted the organizational motto that Hooke was contemporaneously mandating for his "Cortesian army" of experimental philosophers: to be "well disciplined and regulated though their number be small."[156] But how might that structure of archietonical experimental philosophy enable us to reconsider the experimentalists' architecture itself? Beyond its admirably durable engineering feats, what could this pressured architecture of science teach us about the fragility of the experimentalist cosmology?

St. Paul's Cathedral and the Experimentalist "Art World"

Let us imagine that we are ascending the steps of Wren's west portico circa 1720, just as the cathedral's final flourishes were being completed (plate 10). Underfoot, the lozenges of black-and-white marble pavement laid around 1707 by the team of master mason William Kempster signal the stairs' secular function; exclusive use of white paving stones was confined to the ritual spaces at the eastern end of the choir and the twin chapels flanking the entrance vestibule to the north and south.[157] In the pediment overhead, sculptor Francis Bird's bas relief of the conversion of St. Paul (circa 1706) announces the cathedral's program. Upon our entering the north aisle, the ceiling seems lower than the external elevation has led us to expect. This is because the two-story nave is flanked by single-story side aisles that carry flying buttresses invisible from the street. In what nineteenth-century neo-Gothic architect A. W. N. Pugin would describe as a "miserable expedient . . . worthy only of the debased style in which it has been resorted to," Wren has concealed this medieval load-bearing mechanism behind classicizing screen walls.[158] Such a radical discrepancy between appearance and structure can persuasively be traced to around 1685, when the newly installed regime of Catholic-sympathizing James II doubled the funding for the cathedral, encouraging a new infusion of French architectural influence to be superimposed "as an overlay on Wren's 'basilican' design of 1675–85."[159] Evolving, contested schemes are thus perceptible through the cathedral's classicizing skin.

As we move from the north aisle into the nave and pass eastward along three saucer-domed bays to arrive at the cusp of the crossing and the yawning dome above, these internecine conflicts become all the more pressing (fig. 6.5). Looking eastward

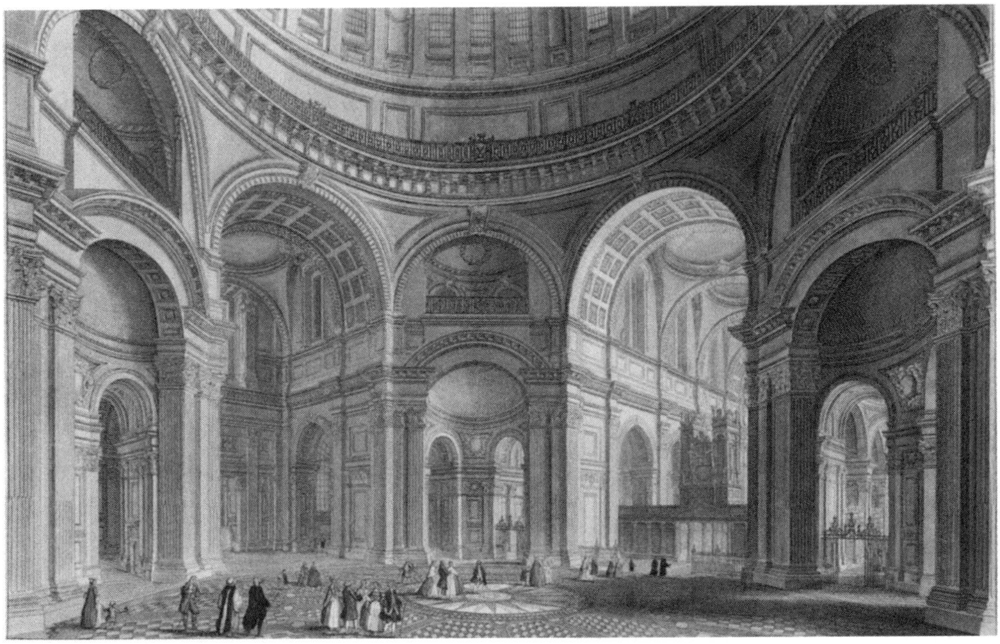

FIGURE 6.5 * Robert Wilkinson, *A Curious Perspective View of the inside of St Paul's Cathedral shewing part of the Dome, the Piers and Arches which support it; and the Entrances into the Choir and Isles* (hand-colored etching and engraving, circa 1770–1810). © The Trustees of the British Museum (AN858730001).

in a view favored by eighteenth-century engravers, Bernard Smith's massive organ supported by Grinling Gibbons's decorative wooden screen blocks our sight into the choir. But as the checkerboard underfoot gives way to the crossing's great marble rosette, we gaze up into the luminous dome and the oculus of its lantern, which had been capped in 1708.[160] Above the metal railing enclosing the "whispering gallery" at the peristyle, we can almost glimpse visualizations of the acts of St. Paul painted (circa 1715–20) in grisaille by Sir James Thornhill—apparently the beneficiary of a political appointment strongly contrary to Wren's wishes.[161] As we exit the crossing via the south transept and walk eastward along the choir's south aisle, the alternating pattern of the floor is now picked up in the milky white of the Portland stone piers juxtaposed to the darkened oak backs of Gibbons's choir stalls, which were in place by 1697.[162] Chalky stone and umber wood then cede to gilded metal at Jean Tijou's elaborate wrought-iron gates, which enclose the choir as if in a floral cage. There, at the eastern end of the choir, in the oldest section of the cathedral whose cornerstone was laid by master mason Thomas Strong in June 1675, Wren envisioned a marble echo of Gianlorenzo Bernini's bronze baldachino at St. Peter's in Rome, with "four Pillars wreathed, of the richest Greek Marbles, supporting a Canopy hemispherical, with proper Decorations of Architecture and Sculpture."[163] Frustrated as he was in this scheme, the architect also met opposition in his design for the choir stalls, which were to bow outward in a circle around a central axis. Unlike Catholics, for whom it is enough "to hear the Murmur of the Mass, and see the Elevation of the Host," Wren explained, the mutual listening and seeing facilitated by this spatial arrangement would uniquely serve Protestant devotion.[164]

Melding materials to serve such site-specific, functional ends had become a focal concern of Christopher Wren's physical architecture in the mid-1670s. In his contemporaneous Trinity College Library at Cambridge (begun in 1676), Wren's structure rises from a colonnade of Tuscan columns that carry a Doric frieze and the upper story's Ionic order.[165] Rather than supporting the floor of the upper story at frieze level, as this classicizing exterior promises, the floor is actually dropped down behind filled lunettes at the springing point of the arches. Through this contrivance, Wren had been able to gain the interior height necessary for soaring arched windows, thus maximizing natural light for reading and for the thirty bookcase-enclosed alcoves lining the library's length. As with the devotional space of seeing and hearing he imagined at St. Paul's, the architect was intensively engaged with crafting this environment conducive to the life of the mind. Designing dimensions for the bookcases and even the library's desks, Wren had quipped, "Wee are scrupulous in small matters and you must pardon us, the Architects are as great pedants as Critics or Heralds."[166]

What I am suggesting is that to construct architectural space down to this level of detail, Wren needed to orchestrate a testy hive of highly skilled, well-paid collaborators. If they were not polite protagonists in the "ordering of the arts" that would come to dominate England's eighteenth century, neither were these agents some nameless craftsmen, but ambitious, ever-innovating players in an emerging art world.[167] Like Lely's painting studio, Grinling Gibbons's sculptural atelier manifests sharp, enterprising efficiencies in the years around 1680: "an increasingly ruthless division . . . along 'horizontal' lines. Instead of each assistant being assigned a carving from start to finish, the composition is subdivided into small tasks, each given to an individual carver."[168] What is more, many master makers administering order from inside these discrete pyramids of artistic labor perceived themselves to be well capable of creative invention—or, at least, more so than their fellow architectural collaborators. Consider the title page to the collection of "Drawings Invented and Desined by John Tijou" published in 1693, where an angelic horde attentively studies a sample of the metalwork Tijou contributed to St. Paul's[169] (fig. 6.6*a*, *b*). Celebrated by celestial patrons of the future "fine arts," Tijou's inventions have banished to a dank pictorial basement the uncouth postures of wood-carvers, whose products bear an uncanny resemblance to the works of Gibbons.[170] Conscious of themselves as artists—competing for claims of "design" and "invention" as proper heirs to the Renaissance *paragone*—these rival teams as much as their diversified materials had to be integrated by Christopher Wren and Robert Hooke to serve the overall architectural fabric.[171]

While this new demand for subsuming parts to whole has served as a veritable definition of the baroque, it is instructive to note how quickly experimentalist architecture would be critiqued precisely for *failing* to achieve that totalizing unity with sufficient centripetal force.[172] In the early 1730s, critic James Ralph found the integration of St. Paul's to be sadly wanting. (See figure 6.5.) The cathedral's choir could have been made "more noble and uniform," he observed, "either compos'd of wood entirely, or marble; for the present mixture of both makes a disagreeable patch-work, that rather disgusts than entertains."[173] Rather than spending excessively on what William Cole would have called "things uselesse" to divine contemplation, the cathedral suffered from a penny-pinching parsimony that diminished its overall effect. "All the intermediate spaces should have been fill'd up," this critic charged, "with the noblest historical paintings; all the majesty of frize-work, cornices, heighten'd with the most costly gildings, should have been lavish'd to adorn it; and one grand flow of magnificent curtain depended from the windows, to finish and adorn the same."[174] These are the failures of artistic nerve that John Summerson cited two hundred years later as wretchedly symptomatic of the experimentalist mentality—its tendency toward "compromise, of disunity of conception . . . [being] an essential part of Wren's mind and of the collective thought of his time."[175] What I want to suggest in conclusion, however, is that Hooke's ongoing thinking about the edifice of scientific knowledge might here be drafted into architectural interpretation. Instead of mentalist failure, Hooke enables us to read this disaggregating, centrifugal tendency of experimentalist baroque architecture as a fundamental, material expression of collectivist philosophical project—even of the shadowy architectural cunning hinted at in John Closterman's portrait of Wren. "Tyranny of the intellect," I think, might better be seen as wicked intelligence.

Around 1675, just as the first stones were being set in the choir's east end, Wren had traced a causal relationship between the easy plentitude of Egyptian agriculture and the sociopolitical necessity of pyramid-building. The human want created by affluence and overpopulation demanded that massive administrative juggernaut instantiated in Egypt's towering monuments. The architect, for Wren, was fundamentally

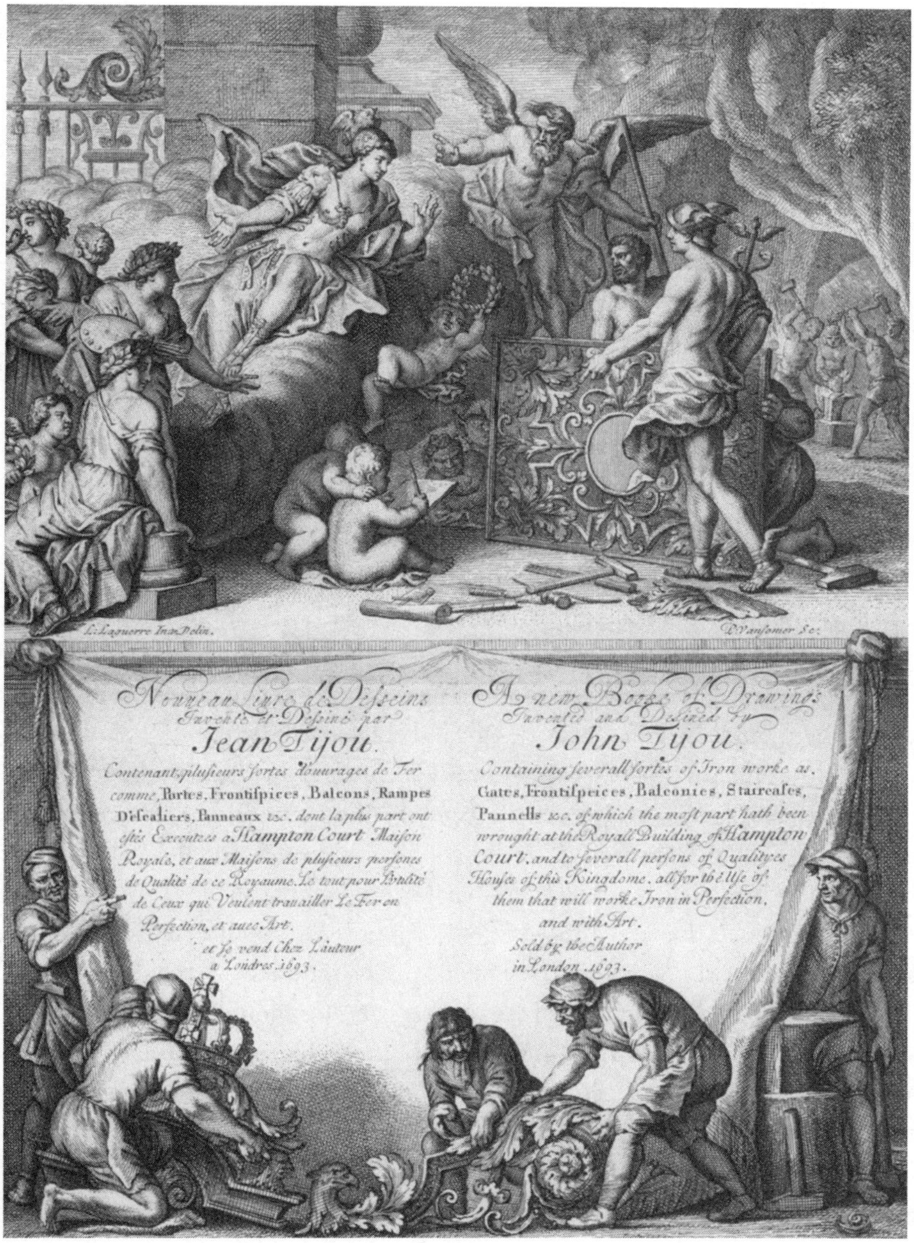

FIGURE 6.6 * Title page to Jean Tijou's *A New Booke of Drawings Invented and Desined by John Tijou* (London: J. Tijou, 1693) designed by Louis Laguerre and engraved by P. Vansomer; Huntington Library, RB 497895. Reproduced by permission of The Huntington Library, San Marino, California.

a governor. "ARCHITECTURE," he claimed, now writing in more current terms, "has its political Use; publick Buildings being the Ornament of a Country; it establishes a Nation, draws People and Commerce; makes the People love their native Country, which Passion is the Original of all great Actions in a Common-wealth."[176] As Wren's architects gave order to economic chaos and stabilizing shape to societies, so Hooke had envisioned architecture in his *General Scheme, or Idea of the Present State of Natural Philosophy* (circa 1666–68) modeling the proper construction of scientific knowledge. Writing in the first blush of his practical involvement with building, Hooke rehearsed classically sanctioned architectural dogma: "The Intellect should first like a skilful Architect, understand what it designs to do, and then consider as near as can be, what things are requisite to be provided in order to this Design, then those Materials are to be carefully sought for and collected, and safely laid up in so convenient an Order."[177] Yet given the limits of human sight and foresight exposed so mercilessly in experimental work, Hooke's philosopher-architect also had to be opportunistic, never refusing "to lay hold on such things, as offering themselves by chance, put him in such a mind that he shall have occasion for them before he can finish his Design."[178]

In the later 1680s, Hooke was sketching a very different conception of relations, not only between parts and whole, design and chance, but between Wren's pairing of agriculture and building. Prompted by the accusation that the Royal Society had "done *just nothing*" in its first quarter century of institutional existence, Hooke acknowledged that all the learned world could see "the Ground designed and set out, the Foundation laid, and the Workmen beginning to raise the Walls, and make use of the Materials that are said to be got in readiness for such a Fabrick."[179] How, though, would the real work of experimental philosophy actually get done? Could it proceed deductively, from natural causes to effects? Or should it work inductively, from effects to causes? Writing on the cusp of the publication of Isaac Newton's *Principia*, which would effectively liquidate the style of philosophy he had long practiced, Hooke appealed to a venerable metaphor:

> An Inquisition by the former Method [that is, deduction] is resembled fitly enough by that Example of an Architect, who hath a full comprehension of what he designs to do and acts accordingly: But the latter is more properly resembled to that of a Husbandman or Gardener, who prepares his Ground and sows his Seed, and diligently cherishes the growing Vegetable, supplying it continually with fitting Moisture, Food, Shelter, & c. observing and cherishing its continual Progression, till it comes to its perfect Ripeness and

> Maturity, and yields him the Fruit of his Labour. Nor is it to be expected that a Production of such Perfection as this is designed, should in an instant be brought to its compleat Ripeness and Perfection; but as all the Works of Nature if it be naturally proceeded with, it must have its due time to acquire its due form and full maturity, by gradual Growth and a natural Progression.[180]

Through their theoretical self-stylings as architects in the later 1670s and early 1680s, London's elite Royal Society Fellows had plotted aggrandizing fictions that effaced much of the messy, complicated work required for actually practicing experimental philosophy *and* for making buildings. Deploying ideas they retrieved from John Dee, Fellows such as Robert Hooke, Christopher Wren, and Joseph Moxon promoted a conception of the managerial, administrative architect-theorist that not only hierarchized the necessarily collaborative projects of experimental knowing and architectural making but elided the critical agency of built artifacts to both enterprises. If it registered at all in these schemes, that tricky traffic from theory to practice figured as one-way directives delivered to workmen by mystified means such as Amphion's harp or "Lupus in Fabula." But in the late 1680s, Hooke now put the deductive philosopher-architect in paratactic juxtaposition to the inductive philosopher-gardener—an epistemological maker who had to rely on the semiautonomous lives and unfurling, unpredictable production of recalcitrant collaborators.

Long crucial to Hooke's own working methods, we can detect an anticipation of that move in his 1682 lecture on the "Directive and Archietonical" powers of the soul. There, by redeploying gravitational attraction to elucidate the reciprocal relations between the soul and those material ideas on which all its knowledge was contingent, Hooke had introduced a mechanism for explaining how things made from embodied collections could also modestly divert, shift, and guide the direction of thinking. In the late 1680s, it was as if those material incursions of chance into and through human design were being felt much more robustly. (See plate 7.) Modeled on a brimming garden that is planned ahead and artfully stimulated but only fructified through nature's cooperation, Hooke had come to suggest, experimental philosophy had to be made both top-down *and* bottom-up; it had to work through deduction *and* induction, by mental design *and* material accident. Moreover, instead of grounding architectural governance in the constraint of nature's excesses, as Wren's Egyptian fable had done, this gardener's intelligence has to be strongly centralized *and* collectively distributed, militantly unified *and* internally divided, actively commanded *and* patiently grown. Yet in such a garden, there is always a snake. That darkened dome rising up as Wren's material companion in

Closterman's portrait is, I think, a shadowy figuration of the wicked intelligence evolving beyond rationalist precept and into the fluxing, wily vitality requisite to the experimental project.

The Exterior View

When satirist Ned Ward has his "London Spy" visit St. Paul's Cathedral around 1700, the experience is positively sobering. Sure, the Spy and his companion do take the opportunity to indulge in some pithy architectural commentary, comparing the size of the cathedral's massive piers to fashionable patches on the face of a passing prostitute.[181] But amid Grinling Gibbons's carvings, the swelling tones of Bernard Smith's organ, and the loose-leaved fruit of the experimentalist art world, the mood of the narrator becomes almost reflective:

> Afternoon Prayers being now ready to begin, we pass'd into the Quire, which was adorn'd with all those Graceful Ornaments, that could any ways add a becoming Beauty and Venerable Decency to so Magnificent a Structure, which indeed consider'd abstractedly from the whole, is so Elegant, Awful, and well Compos'd a part, that nothing but the Glorious Presence of Omnipotence can be worthy of so much Art, Grandure, and Industry, as shines there, to the Honour of God, and Fame of Human Excellence.[182]

A very different scene awaits the Spy outside, where booksellers and vendors of pictorial prints are hawking their wares. He reports, "In our Loitering Perambulation round the outside of *Pauls* . . . were as many Smutty Prints staring the Church in the Face, as a Learned Debauchee ever found in *Aretine*'s Postures."[183]

Anthropomorphizing the cathedral as John Closterman and Christopher Wren had done, the Spy's juxtaposition is more than a collision of the sacred and the profane; it is a faceoff between materially instantiated cosmologies. In his Egyptian fable, Wren had imagined an ancient society whose affluence—its agricultural overproduction, its excessive sexual reproduction—left it dangerously vulnerable. Skillfully coordinating a massive labor force, Egypt's architects and their monuments advanced a political rationale of controlled consumption, civil order, and a stabilized populace. But the impressions leering at Wren's sober pile in Ward's modern telling speak of a different world. Made in multiples and marketed by ranges of third-party retailers, prints are, as Peter Parshall puts it, "the first class of art object that could truly be said to have entered the consumer milieu on a frankly speculative basis, having been fashioned for an anonymous public whose tastes and interests needed to be inferred rather than established through direct negotiation."[184] For many observers assaying these tastes, the most compelling inferences had less to do with the mind's reasons than with its wildest whimsies. Spur of trade and cause of wealth, so Nicholas Barbon put it, "the Wants of the Mind are infinite, Man naturally Aspires, and as his Mind is elevated, his Senses grow more refined, and more capable of Delight; his Desires are inlarged, and his Wants increase with his Wishes, which is for every thing that is rare, can gratifie his Senses, adorn his Body, and promote the Ease, Pleasure, and Pomp of Life."[185] Indeed, that there best might be no governing *head* above this metastasizing hive of desires was, as we will see in this book's conclusion, a proposition sliding into English consciousness at precisely this moment.

Placed before St. Paul's Cathedral, Ned Ward stages that emergent debate in baldly elemental terms. Parting the lusty crowd, Ward's Spy confronts a gawker who is fixated on a saucy print of sexual dalliance. What is so interesting, the Spy queries, in this image of a tryst between a country squire and a milkmaid? "*Why, I'll tell you, Young Man,*" his interlocutor replies, ". . . *I cannot without wonder behold in this Painting, the Madness and vanity of you Young Fellows, with what Confidence you can take a Bear by the Tooth, without the dread of the Danger.*"[186] Our Spy, however, smells a rat: "I rather believe, said I, you gratifie some sensual Appetite, by giving Titilation to your Vitious Thoughts, from the Obscenity of the Action."[187] The terms in which Ward concludes this witty banter salaciously signal the cosmological reordering of cultural anatomy that would quickly turn on the heady intelligence of Robert Hooke and Christopher Wren. Gesturing to the fondling hands and amorous touches of the depicted beau in the naughty number in front of them, the Spy's interlocutor retorts, "*Indeed, Mr.* Inquisitive, *you are much mistaken; but if thy Head had been where his Hand is, I should have view'd it with much more Pleasure, to have thought in what a pretty Condition thy Nose had been.*"[188] In the face of St. Paul's and its experimentalist headmen, Ward plots the degree zero of a new order, burying the Spy's defining vision in the depths of a fantasized, feminine embodiment.

CONCLUSION

IN 1712, NEARLY FOUR decades after its first cornerstones had been laid, St Paul's Cathedral remained troublingly incomplete. One critic complained, "The Iron-Fence has been a long time in the same unfinish'd State, the Church-yard is the same heap of Rubbish it was a year ago; the Chapter-House is at a stand, half built, and . . . the whole Neighbourhood on the Northside are extremely annoy'd by their not taking down the Timber Fences, and removing the Rubbish."[1] Many conniving chiselers and scheming chancers could be blamed for these delays, but responsibility ultimately had to fall on the frail shoulders of Christopher Wren, the octogenarian Surveyor-General of the King's Works. Not only had Wren been slipshod in practicing the management mythologized in his Egyptian fable of the mid-1670s, but the largely absentee Surveyor was willfully flouting rules and regulations. He secretively doled out building contracts to chummy insiders, who then ruled Ludgate Hill in a worryingly "French" (that is, absolutist and arbitrary) manner.[2] Worse, with some fifty thousand pounds' worth of taxpayer-funded work still remaining, the craftsmanship of key cathedral features appeared so conspicuously shoddy that, as this critic imagined, Wren must have been embroiled in some frankly shady dealings. He stated darkly, "When there were most able Bell-Founders, and most skilful Clockmakers who made application to be employ'd, that Sir *Christopher* should use such mean and unskilful Artificers can proceed only from such private Considerations, as every body may reasonably suspect."[3]

Trumped up and partisan though these accusations no doubt were, they offer a salubrious tonic to our perception of Christopher Wren—that "new Archimedes," as Robert Hooke had styled him in 1665. Art historians have often explained this decided turn against Wren and his circles by appealing to a new consolidation of party politics and classicizing aesthetics in the early eighteenth century. Conjoined under the banner of Palladianism, this rising clique of second-generation Whigs reviled "the court taste of the previous half-century, the works of Sir Christopher Wren in particular and anything in the nature of the Baroque."[4] For an influential theorist of that movement such as Anthony Ashley Cooper, third Earl of Shaftesbury (1671–1713), the constitutional monarchy and divided powers achieved at the Glorious Revolution of 1688 modeled conditions keenly favorable to artistic production. Rather than submitting to the whims of the singular "Court-Architect" who had dominated the Surveyorship of Works since 1669, the growth of art demanded driving a wedge between patrons and their artistic protégés.[5] "Without a *Publick Voice*, knowingly guided and directed," Shaftesbury claimed, "there is nothing which can raise a true Ambition in the Artist."[6] In 1718, he got his wish. Wren was summarily dismissed from his post, and the Office of Works (now under command of Richard Boyle, third Earl of Burlington) became an instrument for disseminating a Palladian politics of culture.[7]

Historians of science, meanwhile, have a compelling term for signaling the fate of the freewheeling style of archietonical experimental philosophy crafted with such brio in Restoration London. That term is *Newton*. The rigorously mathematical brand of experimental philosophy announced by Isaac Newton in the *Principia* (1687) not only forbade the kinds of fluid, hypothetical structures routinely deployed by Robert Hooke but reputedly made much of the broader edifice of mechanical philosophy read like a tissue of florid romance.[8] Long bedeviled by his intense personal enmity toward Hooke (that "authorial Antichrist," as Rob Iliffe has described him), Newton wasted little time after his rival's death in 1703.[9] Assuming the presidency of the Royal Society of London in that year, Newton could finally publish his *Opticks* (1704), a work he had withheld until Hooke's death. As the new President continued to inveigh against Hooke, he and his followers successfully defaced the late Curator's reputation—according to legend, destroying all surviving depictions of Hooke.[10] "Penurious, melancholy, mistrustful, and jealous," so one on-message biographer wrote in 1750, Hooke "declared sometimes that he had a great project in his head . . . yet he could never be prevailed with to perfect it; still procrastinating, 'till at last this great design proved an airy phantom, and vanished into nothing."[11] Although the traffic in strange images, monstrous specimens, and curious reports into the Royal Society continued under the guidance of new Secretary Hans Sloane, the experimental philosophy fashioned from them was no longer worthy of that name if not reconciled with mathematical deduction.[12] Restoration styles of philosophizing, so historians tell us, appeared increasingly like Hooke: "very able, very sordid, cynical, wrong headed and whimsical."

Perhaps most instructive for understanding the fate of Restoration wicked intelligence is its place amid eighteenth-century Britain's commercial efflorescence. Although the early Royal Society Fellows considered in this book were hardly averse to trade, their philosophical ventures and visual practices nearly always assumed there would be a backbone of courtly or state patronage. It was from the head of a royally appointed, publicly funded institutional juggernaut that Wren eschewed "Modes and Fashions" to aim his architecture at eternity.[13] And despite his extensive commercial contacts and numerous attempts to monetize experimental inventions, Hooke possessed little sense of a "market" for his exclusive, patron-funded mode of philosophizing.[14] Seen against early eighteenth-century London's grumbling hive of private vices and public benefits detected by theorists such as Bernard Mandeville, these strategies would have looked positively old fashioned. If Britain's visual artists quickly crafted a robust tradition precisely by negotiating and exploiting the possibilities of emerging markets for their work—by "painting for money," in David H. Solkin's phrase—then the abstruse, generally incomprehensible philosophy of Newton equally forged new symbioses with the public lecturers and experimental demonstrators who had been attracting growing audiences since the reign of Queen Anne.[15] Tellingly, within a decade of Newton's ascension as President, the Royal Society had vacated Gresham College and moved into Crane Court, a property in the West End developed and inhabited by none other than Dr. Nicholas Barbon himself.[16] It is as if that merchant tortured by a painter on Snowhill Street and recalled with such *Schadenfreude* by an experimental philosopher had finally found his avenging angel. Barbon wins.

NOTES

ABBREVIATIONS

Birch	Thomas Birch, ed., *The History of the Royal Society of London for Improving of Natural Knowledge* (London: Printed for A. Millar, 1756–57)
BL	British Library
RSCP	Royal Society Classified Papers
RSEL	Royal Society Early Letters
RSLBO	Royal Society Letter Book (Original)
RSRBO	Royal Society Register Book (Original)
TCLC	Trinity College Library, Cambridge
Vertue	The *Notebooks* of George Vertue as published in *Volume of the Walpole Society* 18 (1929–30), 20 (1931–32), 22 (1933–34), 24 (1935–36), 26 (1937–38), 30 (1948–50)
Wren Society	*Wren Society,* ed. A. T. Bolten and H. D. Hendry (Oxford: Oxford University Press, 1923–43)

INTRODUCTION

1. All references to this story are from London Metropolitan Archives CLC/495/MS01758, fol. 131v.

2. On the Beales, see Tabitha Barber, *Mary Beale: Portrait of a Seventeenth-Century Painter, Her Family and Her Studio* (London: Geffrye Museum, 1999).

3. For the Beales' relations with Lely, see Mansfield Kirby Talley, *Portrait Painting in England: Studies in the Technical Literature before 1700* (London: Paul Mellon Centre for British Art, 1981), 270–305.

4. Hooke recorded this story in the famous diary (now in the London Metropolitan Archive), which he kept off and on between 1672 and ca. 1682. Nonetheless, this story and the other tales of confidence schemes and roguish derring-do with which Hooke grouped it were omitted from the published edition of the diary; see *The Diary of Robert Hooke, 1672–1680*, ed. H. W. Robinson and W. Adams (London: Taylor and Francis, 1935). For an insightful discussion of that edition, see Felicity Henderson, "Unpublished

Material from the Memorandum Book of Robert Hooke, Guildhall Library MS 1758," *Notes and Records of the Royal Society of London* 61 (2007): 129–75, esp. 129–35.

5. For Whitchurch, see Henderson, "Unpublished Material," 164, 149, 154. On connections between early modern science and commerce, see Pamela H. Smith and Paula Findlen, eds., *Merchants and Marvels: Commerce, Science and Art in Early Modern Europe* (New York: Routledge, 2002); and Harold J. Cook, *Matters of Exchange: Commerce, Medicine and Science in the Dutch Golden Age* (New Haven, CT: Yale University Press, 2007).

6. On Hooke and Wren as surveyors, see Michael Cooper, *"A More Beautiful City": Robert Hooke and the Rebuilding of London after the Great Fire* (Thrupp: Sutton, 2003). For further bibliography on the architectural work of Hooke and Wren, see especially chapter 6 of the present text.

7. See Joseph Moxon, *Practical Perspective; or, Perspective made Easie* (London: J. Moxon, 1670). For this chronology, compare Ann Bermingham, *Learning to Draw: Studies in the Cultural History of a Polite and Useful Art* (New Haven, CT: Yale University Press, 2000), esp. 33–73; and Kirsti Andersen, *The Geometry of an Art: The History of Mathematical Theory of Perspective from Alberti to Monge* (New York: Springer, 2007), 489–90.

8. Historian of science J. A. Bennett has done much to position the experimental philosophy practiced by Hooke and Wren in a broader tradition of the "mathematical sciences"; see especially J. A. Bennett, *The Mathematical Science of Christopher Wren* (New York: Cambridge University Press, 1982); Bennett, "The Mechanics' Philosophy and the Mechanical Philosophy," *History of Science* 24, no. 1 (March 1986): 1–28; and Anthony Gerbino and Stephen Johnston, *Compass and Rule: Architecture as Mathematical Practice in England 1600–1750* (New Haven, CT: Yale University Press, 2009).

9. See Samuel Y. Edgerton Jr., "Galileo, Florentine 'Disegno,' and the 'Strange Spottednesse of the Moon,'" *Art Journal* 44, no. 3 (Fall 1984): 225–32. For more sympathetic approaches, see Robert Fox, ed., *Thomas Harriot: An Elizabethan Man of Science* (Burlington, VT: Ashgate, 2000).

10. Ellis Waterhouse, *Painting in Britain, 1530–1790* (New Haven, CT: Yale University Press, 1953); and Lucy Gent, ed., *Albion's Classicism: The Visual Arts in Britain, 1550–1660* (New Haven, CT: Yale University Press, 1995).

11. See Jonathan Brown, *Kings and Connoisseurs: Collecting Art in Seventeenth-Century Europe* (Princeton, NJ: Princeton University Press, 1995).

12. Thomas Crow, *Painters and Public Life in Eighteenth-Century Paris* (New Haven, CT: Yale University Press, 1985); and Jonathan Brown and J. H. Elliott, eds., *The Sale of the Century: Artistic Relations between Spain and Great Britain, 1604–1655* (London: Yale University Press, 2002).

13. See Iain Pears, *The Discovery of Painting: The Growth of Interest in the Arts in England, 1680–1768* (New Haven, CT: Yale University Press, 1988). On the broader context, see Steven Pincus, *1688: The First Modern Revolution* (New Haven, CT: Yale University Press, 2009), esp. 49–90.

14. See Carol Gibson-Wood, "Picture Consumption in London at the End of the Seventeenth Century," *Art Bulletin* 84, no. 3 (September 2002): 491–500; and Brian Cowan, "Art and Connoisseurship in the Auction Market of Later Seventeenth-Century London," in *Mapping Markets for Paintings in Europe, 1450–1750*, ed. Neil de Marchi and Hans J. Van Miegroet (Turnhout, Belgium: Brepols, 2006), 263–84.

15. See Ilaria Bignamini, "George Vertue, Art Historian, and Art Institutions in London, 1689–1768," *Walpole Society* 54 (1991): 1–148.

16. On artistic interests in the early Royal Society, see Walter Liedtke, "Pepys and the Pictorial Arts:

'Great Plenty of Good Pictures,'" *Apollo* 133 (1991): 227–37; Antony Griffiths, "The Etchings of John Evelyn," in *Art and Patronage in the Caroline Courts: Essays in Honour of Sir Oliver Millar*, ed. David Howarth (Cambridge: Cambridge University Press, 1993), 51–67; Carol Gibson-Wood, "Classification and Value in a Seventeenth-Century Museum: William Courten's Collections," *Journal of the History of Collections* 9, no. 1 (1997): 61–77; Craig Ashley Hanson, *The English Virtuoso: Art, Medicine, and Antiquarianism in the Age of Empiricism* (Chicago: University of Chicago Press, 2009), esp. 58–92; and Celina Fox, *The Arts of Industry in the Age of Enlightenment* (New Haven, CT: Yale University Press, 2009), 13–44.

17. Steven Shapin and Simon Schaffer, *Leviathan and the Air-Pump: Hobbes, Boyle and the Experimental Life* (Princeton, NJ: Princeton University Press, 1985), 60.

18. See Svetlana Alpers, *The Art of Describing: Dutch Art in the Seventeenth Century* (Chicago: University of Chicago Press, 1983), esp. 72–118. I discuss Alpers's influential account of Hooke and various interpretations it has informed in chapters 1 and 2.

19. Perhaps we might count this tale among the legions of artistic deceptions catalogued by early modern Europeans—an artistic ruse achieved, in this case, by an ingenious feint of medical authority. On conjunctions between art and medicine in this period, see Hanson, *English Virtuoso*; and Barbara Maria Stafford, *Body Criticism: Imaging the Unseen in Enlightenment Art and Medicine* (Cambridge, MA: MIT Press, 1991). On artistic mythology more broadly, see Ernst Kris and Otto Kurz, *Legend, Myth, and Magic in the Image of the Artist* (New Haven, CT: Yale University Press, 1979).

20. Simon Dickie, "Hilarity and Pitilessness in the Mid-Eighteenth Century: English Jestbook Humor," *Eighteenth-Century Studies* 37, no. 1 (Fall 2003): 2. As in many early modern manuscripts, Hooke flipped his diary around and inscribed entries at both ends of the volume. Thus, this collection of "Storys" can be said to appear either at the conclusion or at the beginning of his narrated life.

21. In modeling this narrative, I've drawn key inspiration from Joseph Leo Koerner, *The Moment of Self-Portraiture in German Renaissance Art* (Chicago: University of Chicago Press, 1993).

22. For the key statements of these arguments, see Shapin and Schaffer, *Leviathan and the Air-Pump*, esp. 49–79, 283–344; and Steven Shapin, *A Social History of Truth: Civility and Science in Seventeenth-Century England* (Chicago: University of Chicago Press, 1994). This position has received recent complement and endorsement from studies such as Deborah E. Harkness, *The Jewel House: Elizabethan London and the Scientific Revolution* (New Haven, CT: Yale University Press, 2007), esp. 211–53.

23. Steven Shapin, "Who Was Robert Hooke?," in *Robert Hooke: New Studies*, ed. Michael Hunter and Simon Schaffer (Woodbridge: Boydell Press, 1989), 255–85. See also Stephen Pumfrey, "Ideas above His Station: A Social Study of Hooke's Curatorship of Experiments," *History of Science* 29 (1991): 1–44.

24. Robert Hooke, *Micrographia; or, Some Physiological Descriptions of Minute Bodies made by Magnifying Glass* (London: John Martyn and James Allestry, 1665), sig. Gii.

25. See especially Mordechai Feingold, "Robert Hooke: Gentleman of Science," in *Robert Hooke: Tercentennial Studies*, ed. Michael Cooper and Michael Hunter (Aldershot: Ashgate, 2006), 203–17.

26. On the deformation of Hooke's body, see Nick Wilding, "Graphic Technologies," in Cooper and Hunter, *Robert Hooke: Tercentennial Studies*, 123–34. On his occult connections, see John Henry, "Robert Hooke: The Incongruous Mechanist," in Hunter and Schaffer, *Robert Hooke: New Studies*, 149–80; and Penelope Gouk, *Music, Science and Natural Magic in Seventeenth-Century England* (New Haven, CT: Yale University Press, 1999), 194.

27. Quotation from Anne Goldgar, *Impolite Learning: Conduct and Community in the Republic of*

Letters, 1680–1750 (New Haven, CT: Yale University Press, 1995), 9. See Michael Hunter, *Science and Society in Restoration England* (London: Cambridge University Press, 1981); Hunter, *Establishing the New Science: The Experience of the Early Royal Society* (Woodbridge: Boydell Press, 1989); Adrian Johns, *The Nature of the Book: Print and Knowledge in the Making* (Chicago: University of Chicago Press, 1998); Rob Iliffe, "'In the Warehouse': Privacy, Property and Priority in the Early Royal Society," *History of Science* 30 (1992): 29–68; Iliffe, "Material Doubts: Hooke, Artisan Culture and the Exchange of Information in 1670s London," *British Journal of the History of Science* 28 (1995): 285–318; and Ofer Gal, *Meanest Foundations and Nobler Superstructures: Hooke, Newton, and the "Compounding of the Celestiall Motions of the Planetts"* (London: Kluwer, 2002).

28. J. R. Jones, *Country and Court: England 1658–1714* (Cambridge, MA: Harvard University Press, 1978), 1.

29. A brilliant treatment of this theme is Marshall Sahlins, "The Sadness of Sweetness: The Native Anthropology of Western Cosmology," *Current Anthropology* 37, no. 3 (June 1996): 395–415.

30. See especially J. A. Bennett, "Christopher Wren: Astronomy, Architecture, and the Mathematical Sciences," *Journal for the History of Astronomy* 6 (1975): 149–84; Hentie Louw, "The 'Mechanick Artist' in Late Seventeenth Century English and French Architecture: The Work of Robert Hooke, Christopher Wren and Claude Perrault Compared as Products of an Interactive Science/Architecture Relationship," in Cooper and Hunter, *Robert Hooke: Tercentennial Studies*, 181–99; and Robert Williams, *Art, Theory, and Culture in Sixteenth-Century Italy: From Techne to Metatechne* (New York: Cambridge University Press, 1997).

31. John Flamsteed to Richard Towneley, November 1686; cited in J. L. E. Dreyer, "Flamsteed's Letters to Richard Towneley," *Observatory* 580 (September 1922): 294.

32. John Summerson, "The Tyranny of the Intellect," *Journal of the Royal Institute of British Architects* (February 20, 1937): 373–90. More recently, Lydia Soo has similarly described her interpretive project as reconciling Wren's "mind-set as an experimental scientist and as a practicing architect"; Lydia Soo, ed., *Wren's "Tracts" on Architecture and Other Writings* (New York: Cambridge University Press, 1998), 196.

33. Tellingly, the lone section of Kerry Downes's catalogue for his 1982 exhibition on Wren that dealt with his scientific images was written by J. A. Bennett (whose work has long been an exception to this pattern); see Downes, ed., *Sir Christopher Wren: An Exhibition Selected by Kerry Downes at the Whitechapel Art Gallery* (London: Trefoil Books, 1982), 44–49. This pattern is changing quickly; see Lisa Jardine, "Monuments and Microscopes: Scientific Thinking on a Grand Scale in the Early Royal Society," *Notes and Records of the Royal Society of London* 55, no. 2 (2001): 289–308; and James W. P. Campbell, *Building St. Paul's* (New York: Thames and Hudson, 2007).

34. Influential histories of English art that, respectively, conclude by gesturing to the Royal Society or begin at the Glorious Revolution in 1688, neglecting the Royal Society entirely, include David Howarth, *Lord Arundel and His Circle* (New Haven, CT: Yale University Press, 1985); and David H. Solkin, *Painting for Money: The Visual Arts and the Public Sphere in Eighteenth-Century England* (New Haven, CT: Yale University Press, 1992).

35. Horace Walpole, *Anecdotes of Painting in England* (Strawberry Hill: Walpole, 1763), 3:20. Interestingly, French histories of Hooke in the eighteenth century were often far more favorable than British contemporaries; see Louis de Jaucourt, "Wight L'ile de," *Encyclopédie, ou Dictionnaire Raisonné* (Neufchastel: Samuel Faulche, 1765), vol. 26, *Venerien–Z*, 613–15.

36. See John Stoye, *The Siege of Vienna: The Last Great Trial between Cross and Crescent* (New York: Pegasus Books, 2000); and Zdenko Zlator, *Between the Double Eagle and the Crescent: The Republic of Dubrovnik and the Origins of the Eastern Question* (New York: Columbia University Press, 1992), esp. 128–49.

37. Robert Hooke, "Dr. Hook's Discourse to the Royal Society, May 21. 1684. Shewing a Way How to communicate one's Mind at great Distances," in *Philosophical Experiments and Observations of the Late Eminent Dr. Robert Hooke*, ed. William Derham (London: W. and J. Innys, 1726), 143.

38. See Steven Pincus, "'Coffee Politicians Does Create': Coffeehouses and Restoration Political Culture," *Journal of Modern History* 67 (December 1995): 807–34.

39. On Hooke's relationship with Wilkins, see Lisa Jardine, *The Curious Life of Robert Hooke: The Man Who Measured London* (London: Harper Perennial, 2003), esp. 66–77, 108–13.

40. See John Wilkins, *Mercury; or, The Secret and Swift Messenger* (London: I. Norton, 1641), 90–92.

41. Howard Gardner, *Frames of Reference: The Theory of Multiple Intelligences* (New York: Basic Books, 1983), 60–61.

42. A classic account is Theodore K. Rabb, *The Struggle for Stability in Early Modern Europe* (New York: Oxford University Press, 1975).

43. On seventeenth-century language schemes and their connections to science, see M. M. Slaughter, *Universal Languages and Scientific Taxonomy in the Seventeenth Century* (London: Cambridge University Press, 1982); and Richard Markeley, *Fallen Languages: Crises of Representation in Newtonian England, 1660–1740* (Ithaca, NY: Cornell University Press, 1993).

44. Recent interpretations, particularly in art history, have claimed that Wilkins's characters are "natural signs" rather than conventional symbols. While acknowledging potential advantages of signs that signify "*Naturally*," Wilkins clearly stipulates that his characters refer "by *Institution*"; John Wilkins, *An Essay towards a Real Character and a Philosophical Language* (London: S. Gellibrand and J. Martyn, 1668), 385–86. See, for counterexample, Michael Leja, "Peirce, Visuality and Art," *Representations* 72 (Autumn 2000): 97–122.

45. Robert Hooke to Gottfried Leibniz, July 12, 1680; Royal Society Early Letters (henceforth RSEL), vol. H3, fol. 64. For Hooke's views on Mandarin, see his "Some Observations, and Conjectures Concerning the *Chinese* Characters," *Philosophical Transactions* 16 (1686): 63–78.

46. Michel Foucault, *The Order of Things: An Archaeology of the Human Sciences* (London: Routledge Classics, 2004), 141.

47. Peter Galison, *Image and Logic: A Material Culture of Microphysics* (Chicago: University of Chicago Press, 1997), 19.

48. For a recent study that also frames epistemological shifts around juxtapositions of pictorial and diagrammatic forms, see John Bender and Michael Marrinan, *The Culture of Diagram* (Stanford, CA: Stanford University Press, 2010).

49. David Freedberg, *The Eye of the Lynx: Galileo, His Friends, and the Beginnings of Modern Natural History* (Chicago: University of Chicago Press, 2002), 4.

50. Ibid., 476n1.

51. Hooke's best-known use of Wilkins's philosophical language, for example, was a personal attack on Henry Oldenburg, then Secretary of the Royal Society—an act that scandalized the philosophical community. For a concise account of this episode, see Johns, *Nature of the Book*, 521–31.

52. Dr. William Durston to Lord Brouncker, July 22, 1669; RSEL D1:11.

53. Durston to Henry Oldenburg, November 2, 1669; RSEL D1:14.

54. Wilkins, *Mercury*, 11.

55. For discussion of and bibliography on recent philosophical studies of modeling, see especially chapter 2.

56. This is a simplification of Elkins's complex taxonomic scheme, which subdivides the picture-writing-notation triad into an additional seven categories. In his more recent work, Elkins has argued for and performed even further levels of specification; see the following by James Elkins: *The Domain of Images* (Ithaca, NY: Cornell University Press, 1999); *Visual Practices across the University* (Munich: Wilhelm Fink Verlag, 2007); and *Six Stories from the End of Representation: Images in Painting, Photography, Astronomy, Microscopy, Particle Physics, and Quantum Mechanics, 1980–2000* (Stanford, CA: Stanford University Press, 2008). In chapters 1 through 3, I cite several examples of the rich literature documenting how artistic conventions have informed scientific visual practices.

57. Michael Cole, *Cellini and the Principles of Sculpture* (New York: Cambridge University Press, 2002), esp. 15–44.

58. Pamela H. Smith, *The Body of the Artisan: Art and Experience in the Scientific Revolution* (Chicago: University of Chicago Press, 2004), 59.

59. Svetlana Alpers and Michael Baxandall, *Tiepolo and the Pictorial Intelligence* (New Haven, CT: Yale University Press, 1994).

60. Jeff Wall, "Photography and Liquid Intelligence," in *Jeff Wall*, ed. T. de Duve et al. (London: Phaidon, 1996), 90–93.

61. Here I am adapting Hooke's famous description of Wren as a new Archimedes in whom "such a *Mechanical Hand*, and so *Philosophical* a *Mind*" had been conjoined; see Hooke, *Micrographia*, sig. Giv.

62. An interesting index of the idiosyncrasy of Hooke's reading is a note scrawled on the flyleaf of a copy of Meric Casaubon's *A True and Faithful Relation* (1659) now in the British Library. In a cursive hand dating to the last quarter of the eighteenth century, the note reads: "This Book made a great noise upon its first publication and many years after. The character of it was secured by Dr. Hooke, who believed that Casaubon, Archbishop Usher & other learned men, were mistaken in their notions about it & that in reality, the author never fell under such delusions, but being a Man of great Art & intrigue, made use of this strange method of writing to conceal things of a political nature and instead of a pretended Enthusiast, was a real Spy, but that interpretation is not received for very solid reasons, which are shown in his life"; see British Library 31.g.8. I thank Giles Mandelbrote for his suggestions on the dating of this note.

63. Meric Casaubon, *A True & Faithful Relation of what Passed for Many Yeers between Dr. John Dee and Some Spirits* (London: D. Maxwell, 1659), fol. 18. Important recent accounts of Dee include William H. Sherman, *John Dee: The Politics of Reading and Writing in the English Renaissance* (Amherst: University of Massachusetts Press, 1995); and Deborah E. Harkness, "Managing an Experimental Household: The Dees of Mortlake and the Practice of Natural Philosophy," *Isis* 88, no. 2 (1997): 247–62.

64. Robert Hooke, "Of Dr. Dee's Book of Spirits," in *The Posthumous Works of Robert Hooke*, ed. Richard Waller (London: Sam. Smith and Benj. Walford, 1705), 205.

65. See ibid., 205.

66. Ibid., 207.

67. Ibid., 206.

68. Marcel Detienne and Jean-Pierre Vernant, *Cunning Intelligence in Greek Culture and Society*, trans. Janet Lloyd (Atlantic Highlands, NJ: Humanities Press, 1978), 11.

69. Ibid., 18.

70. In addition to the sources cited above, this phrase has also drawn from classic studies of poetic intelligence, trickster's intelligence, and scientific procedure such as Stanley Fish, *Surprised by Sin: The Reader in "Paradise Lost"* (Cambridge, MA: Harvard University Press, 1997); Lewis Hyde, *Trickster Makes This World: Mischief, Myth and Art* (New York: North Point, 1998); and Paul Feyerabend, *Against Method* (London: Verso, 1978).

71. G. W. Leibniz to Henry Oldenburg, March 8, 1673, in *The Correspondence of Henry Oldenburg*, ed. and trans. A. Rupert Hall, Marie Boas Hall, and Eberhard Reichmann (Madison: University of Wisconsin Press, 1965–75), 9:493.

72. For Hooke's flying abilities, see Richard Waller, "The Life of Dr. Robert Hooke," in Waller, *Posthumous Works of Robert Hooke*, iv; and Henry, "Robert Hooke," 174.

73. For a concise statement of this position, see James Elkins, "Visual Practices across the University," in *Beyond Mimesis and Convention: Representation in Art and Science*, ed. Roman Frigg and Matthew C. Hunter (New York: Springer, 2010), 169–92.

74. Ann Bermingham, *Landscape and Ideology: The English Rustic Tradition, 1740–1860* (Los Angeles: University of California Press, 1986), 2.

CHAPTER ONE

1. For a concise statement, see J. A. Bennett, "Hooke's Instruments," in *London's Leonardo: The Life and Work of Robert Hooke*, ed. J. A. Bennett et al. (New York: Oxford University Press, 2003), 63–104.

2. For a still-useful chronology of Hooke's publications, see Geoffrey Keynes, *A Bibliography of Robert Hooke* (Oxford: Clarendon Press, 1960). The drawings of the 1680s are now to be found in the Royal Society Classified Papers (henceforth RSCP), vol. 24, fols. 282–310.

3. See RSCP vol. 24, fol. 308. On the observatory, see Derek Howse, *Greenwich Observatory: The Royal Observatory at Greenwich and Herstmonceux, 1675–1975* (London: Taylor and Francis, 1975).

4. For example of this usage, see RSCP vol. 24, fol. 298.

5. See, for example, Ann Bermingham, *Learning to Draw: Studies in the Cultural History of a Polite and Useful Art* (New Haven, CT: Yale University Press, 2000), esp. 64–73. This chronology of Hooke's draftsmanship is discussed in chapter 3.

6. These drawings appear in Royal Society MS 131 ("Waller Scrapbook"), fols. 111A–112C. See also Roberta J. M. Olson, ". . . And They Saw Stars: Renaissance Representations of Comets and Pretelescopic Astronomy," *Art Journal* 44, no. 3 (1984): 216–24.

7. For Huygens's drawing (dated January 2, 1665), see Royal Society Early Letters (henceforth RSEL), vol. H1, fol. 43. For Hevelius's contemporaneous images of comets, see Johannes Hevelius, *Cometographia* (Danzig: Simon Reiniger, 1668).

8. Birch 1:511.

9. For Hooke's (ultimately successful) appointment, see Michael Hunter, *Establishing the New Science: The Experience of the Early Royal Society* (Woodbridge: Boydell Press, 1989), esp. 282–97.

10. See Walter Liedtke, "Pepys and the Pictorial Arts: 'Great Plenty of Good Pictures,'" *Apollo* 133

(1991): 227–37; and Carol Gibson-Wood, "Classification and Value in a Seventeenth-Century Museum: William Courten's Collections," *Journal of the History of Collections* 9, no. 1 (1997): 61–77.

11. Historian of science Lisa Jardine makes this attribution in her biography of Christopher Wren, while also endorsing it in her biography of Hooke. However, these drawings were engraved and published in 1707, attributed to Royal Society Fellow John Ray (1627–1705), and credited to his time in Rome—an attribution that clarifies several inconsistencies with Jardine's claim. Compare Lisa Jardine, *On a Grander Scale: The Outstanding Career of Sir Christopher Wren* (London: HarperCollins, 2003), 268–73; Jardine, *The Curious Life of Robert Hooke: The Man Who Measured London* (London: Harper Perennial, 2003), 115–16; and [John Ray], "Observations made at Rome, by the late Reverend Mr. John Ray, of the Comet which appeared Anno 1664. . . . ," *Philosophical Transactions* 25 (1707): 2350–52.

12. The key positions on Hooke's social status are Steven Shapin, "Who Was Robert Hooke?," in *Robert Hooke: New Studies*, ed. Michael Hunter and Simon Schaffer (Woodbridge: Boydell Press, 1989), 255–85; and Mordechai Feingold, "Robert Hooke: Gentleman of Science," in *Robert Hooke: Tercentennial Studies*, ed. Michael Cooper and Michael Hunter (Aldershot: Ashgate, 2006), 203–17.

13. See *The Posthumous Works of Robert Hooke*, ed. Richard Waller (London: Sam. Smith and Benj. Walford, 1705), 149.

14. Heinrich Wölfflin, *Principles of Art History: The Problem of the Development of Style in Later Art*, trans. M. D. Hottinger (New York: Dover, 1950), 14; Heinrich Wölfflin, *Kunstgeschichtliche Grundbegriffe: Das Problem der Stilentwicklung in der Neuren Kunst* (Munich: Hugo Bruckmann, 1920), 15.

15. Svetlana Alpers, *The Art of Describing: Dutch Art in the Seventeenth Century* (Chicago: University of Chicago Press, 1983); Stuart Clark, *Vanities of the Eye: Vision in Early Modern European Culture* (New York: Oxford University Press, 2007). More broadly, see Hal Foster, ed., *Vision and Visuality* (Seattle: Bay Press, 1988).

16. Robert Hooke, *Micrographia; or, Some Physiological Descriptions of Minute Bodies made by Magnifying Glass* (London: John Martyn and James Allestry, 1665), sig. Aiv.

17. Robert Hooke, *A Discourse on the Nature of Comets*, in Waller, *Posthumous Works of Robert Hooke*, 152.

18. See particularly Steven Shapin and Simon Schaffer, *Leviathan and the Air-Pump: Hobbes, Boyle and the Experimental Life* (Princeton, NJ: Princeton University Press, 1985), esp. 60–79.

19. Classic statements of this position are Edgar Zilsel, "The Origins of William Gilbert's Scientific Method," *Journal of the History of Ideas* 2, no. 1 (January 1941): 1–32; and Erwin Panofsky, "Artist, Scientist, Genius: Notes on the 'Renaissance-Dämmerung,'" in *The Renaissance: Six Essays*, ed. W. K. Ferguson et al. (New York: Harper and Row, 1962), 121–82. More recently, see Pamela H. Smith, *The Body of the Artisan: Art and Experience in the Scientific Revolution* (Chicago: University of Chicago Press, 2004). A useful survey of relevant scholarship can be found in John Henry, *The Scientific Revolution and the Origins of Modern Science* (New York: Palgrave, 2002).

20. See Mario Biagioli, *Galileo Courtier: The Practice of Science in the Culture of Absolutism* (Chicago: University of Chicago Press, 1993); Deborah E. Harkness, *The Jewel House: Elizabethan London and the Scientific Revolution* (New Haven, CT: Yale University Press, 2007); and Antonio Barrera-Osorio, *Experiencing Nature: The Spanish American Empire and the Early Scientific Revolution* (Austin: University of Texas Press, 2006).

21. Francis Bacon, preface to *Magna Instauratio*, in *Francis Bacon: Essays*, ed. R. F. Jones (New York: Odyssey Press, 1937), 254. On this point, see Antonio Perez-Ramos, "Bacon's Forms and the Maker's

Knowledge Tradition," in *The Cambridge Companion to Bacon*, ed. M. Peltonen (New York: Cambridge University Press, 1996), 99–120.

22. For examples of works that have revised the traditional opposition between Scholasticism and experimentalism, see Mordechai Feingold, *The Mathematician's Apprentice: Science, Universities and Society in England, 1560–1640* (London: Cambridge University Press, 1984); and William R. Newman, *Atoms and Alchemy: Chymistry and the Experimental Origins of the Scientific Revolution* (Chicago: University of Chicago Press, 2006).

23. Classic studies of the foment of experimental philosophy in England include Allen G. Debus, *Science and Education in the Seventeenth Century: The Webster-Ward Debate* (New York: American Elsevier, 1970); Charles Webster, *The Great Instauration: Science, Medicine, and Reform, 1626–1660* (New York: Holmes and Meier, 1976); and Robert G. Frank Jr., *Harvey and the Oxford Physiologists: A Study of Scientific Ideas* (Berkeley: University of California Press, 1980).

24. Recent readings of *Micrographia* include John T. Harwood, "Rhetoric and Graphics in *Micrographia*," in Hunter and Schaffer, *Robert Hooke: New Studies*, 119–47; Michael Aaron Dennis, "Graphic Understanding: Instruments and Interpretation in Robert Hooke's *Micrographia*," *Science in Context* 3, no. 2 (1989): 309–65; Janice Neri, "Some Early Drawings by Robert Hooke," *Archives of Natural History* 32, no. 1 (2005): 41–47; Neri, "Between Observation and Image: Representation of Insects in Robert Hooke's *Micrographia*," in *The Art of Natural History: Illustrated Treatises and Botanical Paintings, 1400–1850*, ed. Teresa O'Malley and Amy R. W. Meyers (New Haven, CT: Yale University Press, 2008), 83–108; and Domenico Bertoloni Meli, "The Representation of Insects in the Seventeenth Century: A Comparative Approach," *Annals of Science* 67, no. 3 (2010): 405–29.

25. Hooke, *Micrographia*, sig. Aii.

26. Ibid., sig. Biii.

27. On the disjunction between the Society's inclusive rhetoric and actual practice, see Michael Hunter, *Science and Society in Restoration England* (London: Cambridge University Press, 1981).

28. Hooke, *Micrographia*, sig. Aiv.

29. See ibid., sig. Bi.

30. Ibid., plate 2.

31. For different readings of this important plate, compare Barbara Maria Stafford, *Body Criticism: Imaging the Unseen in Enlightenment Art and Medicine* (Cambridge, MA: MIT Press, 1991), 352–56; Adrian Johns, *The Nature of the Book: Print and Knowledge in the Making* (Chicago: University of Chicago Press, 1998), 430–34; and Angela Fischel, "Sehen, Darstellen, Beschreiben: Mikroskopische Beobachtung in den Kupferstichen der *Micrographia*," *Kunsttexte.de* 1 (2002): 1–10.

32. Hooke, *Micrographia*, plate 3.

33. Ibid.

34. For this phrase, see Robert Hooke, "An Instrument of Use to Take the Draught, or Picture of Any Thing," in *Philosophical Experiments and Observations of the Late Eminent Dr. Robert Hooke*, ed. William Derham (London: W. and J. Innys, 1726), 294. On Hooke's attitude toward prints more broadly, see Matthew C. Hunter, "The Theory of the Impression according to Robert Hooke," in *Printed Images in Early Modern Britain: Essays in Interpretation*, ed. Michael Hunter (Aldershot: Ashgate, 2010), 164–93.

35. On didactic dimensions, see Dennis, "Graphic Understanding," esp. 330–35.

36. For extensive discussions of instruments and technique, see Hooke, *Micrographia*, sigs. Div–Fiii.

37. Ibid. sig. Biv.

38. For this measurement technique and Hooke's famous depiction of his microscopic instruments, see ibid., sig. Fii and scheme 1.

39. See R. H. Nuttall, "That Curious Curiosity: The Scotoscope," *Notes and Records of the Royal Society of London* 42 (1988): 133–38.

40. Hooke, *Micrographia*, 126.

41. In spring and summer 2008, I was able to observe with and draw from two late seventeenth-century microscopes produced by London instrument makers with whom Robert Hooke was closely connected. Now in the collection of the Science Museum, London, these instruments are John Mann's 45× magnification compound microscope with a tripod support (ca. 1700; Inv. 1928–772) and an anonymous compound microscope of approximately 60× magnification in the "Hooke style" (ca. 1675; Inv. 1928–786). Without using a laboratory-quality fiber-optic light source, I was completely incapable of perceiving my observational target through either instrument.

42. Hooke, *Micrographia*, 168.

43. Ibid.

44. On these points, see Alfred Gell, "The Technology of Enchantment and the Enchantment of Technology," in *The Art of Anthropology: Essays and Diagrams*, ed. Eric Hirsch (London: Athlone Press, 1999), 159–86.

45. See J. A. Bennett, "Hooke's Instruments for Astronomy and Navigation"; and A. D. C. Simpson, "Robert Hooke and Practical Optics: Technical Support at a Scientific Frontier," in Hunter and Schaffer, *Robert Hooke: New Studies*, 21–32; 33–62.

46. Hooke, *Micrographia*, 176.

47. Ibid., sig. Fiv.

48. Henry Power described the eyes of "The Gray, or Horse-Fly" as "an incomparable pleasant spectacle: 'tis of a semisphaeroidal figure; black and waved, or rather indented all over with a pure Emerauld-green, so that it looks like green silk Irish-stitch, drawn upon a black ground, and all latticed or chequered with dimples like Common Flyes, which makes the Indentures look more pleasantly"; Henry Power, *Experimental Philosophy, in Three Books* (London: T. Roycroft, 1664), 6–7. For more on Power, see Charles Webster, "Henry Power's Experimental Philosophy," *Ambix* 14, no. 3 (1967): 150–78.

49. Hooke, *Micrographia*, sig. Fiv.

50. Ibid.

51. Ibid.

52. Ibid., 175, 176, scheme 23.

53. For Hooke's use of drone fly eyes as a model, see ibid., 182–83.

54. Ibid., 176, scheme 24.

55. Comparison might be made between Hooke's visualizations of "true form" and what historians of science Lorraine Daston and Peter Galison have influentially called "truth to nature"; see Lorraine Daston and Peter Galison, "The Image of Objectivity," *Representations* 40 (Autumn 1992): 81–120, esp. 84–96.

56. August 13, 1664, entry, in *The Diary of Samuel Pepys*, ed. Robert Latham and William Matthews (Los Angeles: University of California Press, 1971), 5:240. See also Marjorie Hope Nicolson, *Pepys' Diary and the New Science* (Charlottesville: University of Virginia Press, 1965).

57. See Cole's letter (dated Bristol, September 30, 1669) in Robert Hooke, *Lectiones Cutlerianæ* (London: J. Martyn, 1679), 48.

58. See Keynes, *Bibliography of Robert Hooke*, 97–108.

59. See [Christopher Wren], "The Description of an Instrument Invented Divers Years Ago by Dr. Wren, for Drawing the Out-lines of Any Object in Perspective," *Philosophical Transactions of the Royal Society of London* 4 (1669): 898–99. For more on this instrument, see J. A. Bennett, *The Mathematical Science of Christopher Wren* (New York: Cambridge University Press, 1982), 71–76.

60. Among other things, Alpers opposes the activity of the Albertian beholder to what she sees as the passive, receptive, disembodied eye organizing northern depiction; see Alpers, *Art of Describing*, 36–43. Although the responses to Alpers's influential work are legion, two instructively opposing views are Eddy de Jongh, "The Art of Describing; Dutch Art in the Seventeenth Century," *Simiolus* 14, no. 1 (1984): 51–59; and Louis Marin, "In Praise of Appearance," *October* 37 (Summer 1986): 98–112.

61. Leon Battista Alberti, *"On Painting" and "On Sculpture": The Latin Texts of "De pictura" and "De statua,"* ed. and trans. Cecil Grayson (London: Phaidon, 1972), 39, 37; for Latin, see 38, 36.

62. Ibid., 43; for Latin, see 42. For Alberti's relationship to medieval visual theory, see David Lindberg, *Theories of Vision from Al-Kindi to Kepler* (Chicago: University of Chicago Press, 1976), 149–54.

63. Alberti, *"On Painting" and "On Sculpture,"* 69.

64. See Giovanni Battista Armenini, *On the True Precepts of the Art of Painting* [1586], ed. and trans. E. J. Olszewski (New York: Burt Franklin, 1977), 123. More broadly, see David Winton Bell Gallery, *Children of Mercury: The Education of Artists in the Sixteenth and Seventeenth Century* (Providence, RI: Brown University Press, 1984).

65. Armenini, *On the True Precepts of the Art of Painting*, 124.

66. Ibid.

67. John Evelyn, *Sculptura; or, The History, and Art of Chalcography and Engraving in Copper* (London: J. C. for G. Beedle and T. Collins, 1662), 107–8.

68. See Alexander Browne, *Ars pictoria; or, An academy Treating of Drawing, Painting, Limning, and Etching* (London: J. Redmayne, 1669). See also Kim Sloane, *"A Noble Art": Amateur Arts and Drawing Masters c. 1600–1800* (London: British Museum Press, 2001), 15.

69. On Hooke's apprenticeship, compare Richard Waller, "The Life of Dr. Robert Hooke," in Waller, *Posthumous Works of Robert Hooke*, ii; John Aubrey, *Brief Lives*, ed. J. Buchanan-Brown (New York: Penguin, 2000), 394–95; and Matthew C. Hunter, "Hooke's Figurations: A Figural Drawing Attributed to Robert Hooke," *Notes and Records of the Royal Society of London* 64 (2010): 251–60.

70. See, for example, Diana Dethloff, "Patterns of Drawing Collecting in Late Seventeenth and Early Eighteenth Century England," in *Drawing: Masters and Methods, Raphael to Rodin*, ed. Diana Dethloff (London: Royal Academy of Art, 1992), 197. Connections between experimental philosophy and Lely's studio practice are discussed at length in chapter 3 of the present text.

71. Dethloff, "Patterns," 200.

72. For the provenance of this album, see Giles Waterfield, *Mr. Cartwright's Pictures: A Seventeenth Century Collection* (London: Dulwich Picture Gallery, 1987), 14.

73. See Ilaria Bignamini, "The Artist's Model from Lely to Hogarth," in *The Artist's Model: Its Role in British Art from Lely to Etty*, ed. Ilaria Bignamini and Martin Postle (Nottingham: Nottingham University Art Gallery, 1991), 8–10.

74. A contemporaneous depiction of these studio conventions can be seen in Johann Heinrich Schönfeld's *A Class in the Academy* (ca. 1669, Alte Galerie, Universalmuseum Joanneum, Graz, Austria); see Ulrich Becker, ed., *Alte Galerie: Masterpieces* (Graz, Austria: Landesmuseum Joanneum, 2005), 146.

75. Hooke, "An Instrument of Use," 296, 295.

76. Robert Hooke, untitled lecture from July 14, 1697; RSCP vol. 20, fol. 89.

77. Hooke, *Micrographia*, 1.

78. Ibid., 91.

79. Ibid.

80. Historian of science Peter Heering has been able to re-create projected images using eighteenth-century solar microscopes with (reportedly) incredible clarity and resolution. Although the invention of the solar microscope has repeatedly been assigned to the end of the first quarter of the eighteenth century, Hooke was extremely interested in using projected images for pictorial purposes; compare Peter Heering, "The Enlightened Microscope: Re-enactment and Analysis of Projections with Eighteenth-Century Solar Microscopes," *British Journal for the History of Science* 41, no. 3 September 2008): 345–67; and [Robert Hooke], "A Contrivance to make the Picture of any thing appear on a Wall . . . ," *Philosophical Transactions of the Royal Society of London* 4, no. 38 (August 17, 1668): 741–43. For the chronology of the solar microscope, see Barbara Maria Stafford and Frances Terpak, *Devices of Wonder: From the World in a Box to Images on a Screen* (Los Angeles: Getty Research Institute, 2001), 215–20.

81. For the controversial circumstances under which Hooke actually did obtain this post in 1665, see Stephen Inwood, *The Man Who Knew Too Much: The Strange and Inventive Life of Robert Hooke 1635–1703* (London: Macmillan, 2002), esp. 30–33. More broadly, see Anthony Gerbino and Stephen Johnston, *Compass and Rule: Architecture as Mathematical Practice in England 1600–1750* (New Haven, CT: Yale University Press, 2009).

82. Hooke had access to many narratives of Columbus's life and work, but the account given in Sir Ferdinando Gorges's 1659 *America Painted to the Life* is particularly suggestive. Hooke's reading of Columbus is highly sympathetic to that provided by Gorges, an English colonist who acquired the rights to develop Maine during the reign of Charles I. Hooke was friendly with Gorges's grandson; see *The Diary of Robert Hooke, 1672–1680*, ed. H. W. Robinson and W. Adams (London: Taylor and Francis, 1935), 28, 76, 82, 88, 227, and 260. On Hooke's broader contacts with merchants and voyagers, see Jardine, *Curious Life*, esp. 247–87. For a magisterial reading of Columbus's natural philosophy, see Nicolás Wey Gómez, *The Tropics of Empire: Why Columbus Sailed South to the Indies* (Cambridge, MA: MIT Press, 2008).

83. Robert Hooke, *A General Scheme, or Idea of the Present State of Natural Philosophy*, in Waller, *Posthumous Works of Robert Hooke*, 20.

84. Ibid., 21.

85. Ibid., 20. Compare with Gorges: Columbus, threatened with mutiny by his fearful crew, was able to placate them "by his observation of the heavens, the air, and the clouds, [showing] that they could not now be far off from land, he promised them that if within three days they did not discover land, he would not prosecute his design any farther . . . but it fell out happily according to his expectation and foresight"; Sir Ferdinando Gorges, *America painted to the life. The history of the Spaniards proceeding in America* (London: T. J. for N. Brook, 1659), 1–2.

86. See Steven Shapin, "'A Scholar and a Gentleman': The Problematic Identity of the Scientific Practitioner in Early Modern England," *History of Science* 29 (1991): 279–327; Adrian Johns, "Prudence

and Pedantry in Early Modern Cosmology: The Trade of Al Ross," *History of Science* 25 (1997): 23–59; and Simon Schaffer, "Regeneration: The Body of Natural Philosophers in Restoration England," in *Science Incarnate: Historical Embodiments of Natural Knowledge*, ed. Christopher Lawrence and Steven Shapin (Chicago: University of Chicago Press, 1998), 83–120.

87. Hooke, *General Scheme*, 20.

88. Ibid., 21.

89. On the meanings of registration and the practices associated with the Royal Society's Register Books (in which many of Hooke's drawings were then archived), see Johns, *Nature of the Book*, esp. 477–531.

90. Cited in Anthony Pagden, *Lords of All the World: Ideologies of Empire in Spain, Britain, and France c.1500–c.1800* (New Haven, CT: Yale University Press, 1995), esp. 73–86; see also John T. Juricek, "English Territorial Claims in North America under Elizabeth and the Early Stuarts," *Terrae Incognitae* 7–8 (1976): 9.

91. See Gary W. Kronk, *Cometography: A Catalog of Comets*, vol. 1, *Ancient–1799* (New York: Cambridge University Press, 1999), 369.

92. December 12, 1680, entry, in *The Diary of John Evelyn*, ed. A. Dobson (London: Macmillan, 1906), 3:66.

93. Edmond Halley, *Astronomical Tables with Precepts in both English and Latin* (London: Printed for William Innys, 1752), Tttt4. More broadly, see Simon Schaffer, "Halley's Atheism and the End of the World," *Notes and Records of the Royal Society of London* 32, no. 1 (July 1977): 17–40; and Sara Schechner Genuth, *Comets, Popular Culture and the Birth of Modern Cosmology* (Princeton, NJ: Princeton University Press, 1997).

94. See David W. Hughes, "Edmond Halley: His Interest in Comets," in *Standing on the Shoulders of Giants: A Longer View of Newton and Halley*, ed. Norman Joseph William Thrower (Los Angeles: University of California Press, 1990), 325.

95. As elsewhere, Hooke identified *Cometographia*, the capacious study of comets engraved and published by Polish astronomer Johannes Hevelius, as especially infected with this "prejudicate and prepossest Poetic Fancy"; see Hooke, *Discourse on the Nature of Comets*, 152.

96. See Robert Hooke, *Cometa*, in *Lectures and Collections* (London: J. Martyn, 1678), 6. An authoritative study of premodern comet theories is Tofigh Heidarzadeh, *A History of Physical Theories of Comets, from Aristotle to Whipple* (New York: Springer, 2008). For more on Hooke's comet theory, see Richard S. Westfall, "Hooke and the Law of Universal Gravitation: A Reappraisal of a Reappraisal," *British Journal for the History of Science* 3, no. 3 (June 1967): 245–61; and J. A. Bennett, "Hooke and Wren and the System of the World: Some Points toward an Historical Account," *British Journal for the History of Science* 8, no. 1 (March 1975): 32–61.

97. See René Descartes, *Principles of Philosophy*, trans. Valentine Rodger Miller and Reese P. Miller (Boston: D. Reidel, 1984), esp. 60–61.

98. Hooke, *Cometa*, 11–14.

99. Ibid., 47–53.

100. Hooke, *Discourse on the Nature of Comets*, 152.

101. Ibid., 151.

102. Ibid., 152.

103. Ibid.

104. Ibid.

105. See RSCP vol. 24, fols. 283–283v (December 29, 1680), and fols. 288–288v (January 9, 1681).

106. See ibid., fol. 286 (January 5, 1681), fol. 287 (January 7, 1681), fol. 288 (January 9, 1681), and fol. 290 (January 13, 1681).

107. See ibid., fol. 291v (January 10, 1681) and fol. 293 (January 24, 1681).

108. Hooke, *Discourse on the Nature of Comets*, 152.

109. Ibid., 149.

110. For an influential narrative of this kind, see Lorraine Daston and Peter Galison, *Objectivity* (New York: Zone Books, 2007).

111. RSCP vol. 24, fol. 281. For more on Woodward and his relations with Hooke, see Joseph M. Levine, *Dr. Woodward's Shield: History, Science, and Satire in Augustan England* (Berkeley: University of California Press, 1977).

112. See Hooke, *Discourse on the Nature of Comets*, 149.

113. According to the 1703 sale catalogue of his library, Hooke had a 1603 Latin folio edition of *Uranometria* with figures (which sold for 17 s. 6 d.) and a 1654 Latin quarto edition, apparently without figures; see Leona Rostenburg, *The Library of Robert Hooke: The Scientific Book Trade of Restoration England* (Santa Monica, CA: MODOC Press, 1989), 6, 18.

114. See Deborah J. Warner, *The Sky Explored: Celestial Cartography 1500–1800* (New York: Alan R. Liss, 1979), 18–19; and George Sergeant Snyder, *Maps of the Heavens* (London: André Deutsch, 1984), 93.

115. Warner cites reprints of the plates alone (without Bayer's text) at Augsburg in 1624 and at Ulm in 1639, 1641, 1648, 1655, 1661, 1666, and 1689; see Warner, *The Sky Explored*, 19.

116. This connection to Dutch mannerism is not coincidental. Several of Mair's constellation schemata were adapted directly from prints made by Jacques de Gheyn II for a humanist emblem book published in 1600 by Hugo Grotius. Compare Hugo Grotius, *Syntagma Arateorum* (Lyon: Ex Officina Plantiniana, 1600); and I. Q. Van Regteren Altena, *Jacques de Gheyn: Three Generations* (London: Martinus Nijhoff, 1983), 1:45.

117. See Nelson Goodman, *Languages of Art: An Approach to a Theory of Symbols* (New York: Bobbs-Merrill, 1968), 21–26.

118. On Newton's use of Bayer, see Eric G. Forbes, "The Comet of 1680–1681," in Thrower, *Standing on the Shoulders of Giants*, 317. Although it is beyond the scope of this chapter, the relationship between the comet drawings made independently by Hooke and Newton in the early 1680s is a fascinating topic worthy of further investigation. For a reproduced example of Newton's comet drawings (which Hooke would not have known), see Richard S. Westfall, *Never at Rest: A Biography of Isaac Newton* (New York: Cambridge University Press, 1980), 396.

119. On August 20, 1682, Hooke observed, "Through the telescope it [the comet] appeared of this shape, supposing the black white and the white black"; RSCP vol. 24, fol. 302.

120. January 16, 1681; ibid., fol. 292.

121. January 30, 1681; ibid., fol. 296.

122. Hooke, *Cometa*, 2.

123. Hooke, *Discourse on the Nature of Comets*, 156.

124. For a preliminary account of this model, see Matthew Hunter, "Experiment, Theory, Representation: Robert Hooke's Material Models," in *Beyond Mimesis and Convention: Representation in Art and Science*, ed. Roman Frigg and Matthew C. Hunter (New York: Springer, 2010), 193–219.

125. Hooke, *Discourse on the Nature of Comets*, 158.

126. This is not to say that my account precludes telling a more determinedly disciplined art-historical story here. No doubt one might press on the fact that Hooke and his circle would have had access to Bayer both in the 1660s and in the 1680s. One might then stress that *only* following a change in the structure of, say, the *Kunstwollen* was Bayer then redeployed to visualize ephemeral, optical events. My point is simply that we can craft an argument more compellingly keyed to the tone and objects of Hooke's images by tactically resisting this impulse. I thank Caroline Arscott for pressing me on this point.

127. James William Grant, "Remarks on Hooke's Observations of the Comets of 1680 and 1682," *Monthly Notices of the Royal Astronomical Society* 14 (January 1854): 77.

128. Ibid., 77–78.

129. I draw this phrase from James Elkins, "On Visual Desperation and the Bodies of Protozoa," *Representations* 40 (Autumn 1992): 33–56.

130. Leonardo da Vinci, *Leonardo on Painting*, ed. M. Kemp, selected and trans. Martin Kemp and Margaret Walker (New Haven, CT: Yale University Press, 1989), 193.

131. Compare Smith, *Body of the Artisan*; and Louis Marin, *To Destroy Painting* [1977], trans. Mette Hjort (Chicago: University of Chicago Press, 1995).

CHAPTER TWO

1. Royal Society Classified Papers (henceforth RSCP), vol. 13:1. In the mid-eighteenth century, this paper fish was bound together with measurements of a "Gyant-Child" from Scotland and John Locke's illustrated report of a French child with catastrophically elongated fingernails. For Locke's image, see RSCP 13:5. For the published form of Locke's report without the image, see John Locke, "An Account of One Who Had Horny Excrescencies or Extraordinary Large Nails on His Fingers and Toes, by Mr. Locke," *Philosophical Transactions* 19, no. 230 (July 1697): 694–96.

2. See Christopher Wren Jr., *Parentalia; or, Memoirs of the Family of the Wrens* (London: T. Osborn and R. Dodsley, 1750), 227.

3. See, respectively, Adam Smyth, "'Rend and Teare in Peeces': Textual Fragmentation in Seventeenth-Century England," *Seventeenth Century* 19, no. 1 (2004): 36–52; Lucia Nuti, "'To Make the Whole Progress a Lineall Visible Demonstration': The Journal of Edward Dummer," *Word & Image* 15, no. 3 (1999): 292–305; Ann Blair, "Reading Strategies for Coping with Information Overload ca. 1550–1700," *Journal of the History of Ideas* 64, no. 1 (2003): 11–28; Martin Rudwick, "George Cuvier's Paper Museum of Fossil Bones," *Archives of Natural History* 27, no. 1 (2000): 51–68; and Kristel Smentek, "The Collector's Cut: Why Pierre-Jean Mariette Tore Up His Drawings and Put Them Back Together Again," *Master Drawings* 46, no. 1 (Spring 2008): 36–60.

4. Ursula Klein, "Techniques of Modelling and Paper-Tools in Classical Chemistry," in *Models as Mediators: Perspectives on Natural and Social Science*, ed. Mary S. Morgan and Margaret C. Morrison (Cambridge: Cambridge University Press, 1999), 146–67; and Bruno Latour, "Drawing Things Together," in *Representation in Scientific Practice*, ed. Michael Lynch and Steve Woolgar (London: MIT Press, 1990), 22.

5. John Evelyn's collection of early issues of *Philosophical Transactions* (now British Library EVE.A.149), for example, contains this assembled instrument; I thank Meghan Doherty for kindly sharing this information with me. However, repression of the paper micrometer's patched and pasted form is not new. In etchings of the newly opened Royal Observatory at Greenwich made around 1676–77, for

example, Francis Place provided an exploded view of the equipment in the observatory's well telescope that replicates Hooke's micrometer design *minus* the patch. For a reproduction of this rare etching, see Derek Howse, *Francis Place and the History of the Greenwich Observatory* (New York: Science History Publications, 1975), 58–61.

6. See Adrian Johns, "Martyn, John (1617/18–1680)," in *Oxford Dictionary of National Biography*, ed. H. C. G. Matthew and Brian Harrison (New York: Oxford University Press, 2004), 37:37–38.

7. On the history of the micrometer, compare Clare Vincent, "A Beam Compass by Christoph Trechsler the Elder and the Origin of the Micrometer Screw," *Metropolitan Museum Journal* 24 (1989): 209–22; and Randall C. Brooks, "The Development of Micrometers in the Seventeenth, Eighteenth and Nineteenth Centuries," *Journal for the History of Astronomy* 22 (1991): 127–73. I thank Nick Dew for pushing me on this point.

8. Robert Hooke, *Animadversions on the first part of the Machina Coelestis of the Honourable, Learned and deservedly Famous Astronomer Johannes Hevelius, &c.* (London: T. R. for John Martyn, 1674), in *Lectiones Cutlerianae* (London: John Martyn, 1679), 8.

9. Ibid., 9. J. Z. Buchwald explains how Hooke's experiment imagines a test subject with roughly 20/20 vision who can distinguish one minute of arc such that the stripes could be differentiated if seen from some 290 feet; see J. Z. Buchwald, "Discrepant Measurements and Experimental Knowledge in the Early Modern Era," *Archive for History of Exact Sciences* 60, no. 6 (2006): 565–649, esp. 619–21.

10. See, for example, Bruno Latour, *We Have Never Been Modern*, trans. Catherine Porter (Cambridge, MA: Harvard University Press, 1993); Arthur C. Clarke, *2001: A Space Odyssey* (New York: Roc, 1999), 202.

11. See [Adrien Auzout], "An Extract of a Letter Written Decemb. 28. 1666. By M. Auzout to the Publisher, Concerning a Way of His, for Taking the Diameters of the Planets . . . ," *Philosophical Transactions* 21, no. 1 (1667): 373–75.

12. For more on Auzout's micrometer, see Robert M. McKeon, "Établissement de l'astronomie de précision et œuvre d'Adrien Auzout" (PhD diss., École des hautes études en sciences sociales, Paris, 1965), esp. 1–81.

13. [Richard Towneley], "An Extract of a Letter, Written by Mr. Richard Towneley to Dr. Croon, Touching the Invention of Dividing a Foot into Many Thousand Parts, for Mathematical Purposes," *Philosophical Transactions* 25, no. 2 (May 6, 1667): 457. For the broader political context, see Steven Pincus, *Protestantism and Patriotism: Ideologies and the Making of English Foreign Policy, 1650–1668* (New York: Cambridge University Press, 1996).

14. [Towneley], "An Extract," 458.

15. Allan Chapman, "Jeremiah Horrocks, the Transit of Venus, and the 'New Astronomy' in Early Seventeenth-Century England," *Quarterly Journal of the Royal Astronomical Society* 31 (1990): 344.

16. For these drawings, see RSCP 2:13. Towneley's inscription is transcribed in Matthew C. Hunter, "Robert Hooke Fecit: Making and Knowing in Restoration London" (PhD diss., University of Chicago, 2007), 172.

17. For the Royal Society context, see Rob Iliffe, "'In the Warehouse': Privacy, Property and Priority in the Early Royal Society," *History of Science* 30 (1992): 29–68; and Adrian Johns, *The Nature of the Book: Print and Knowledge in the Making* (Chicago: University of Chicago Press, 1998). More broadly, see Mario Biagioli and Peter Galison, eds., *Scientific Authorship: Credit and Intellectual Property in Science* (New York: Routledge, 2003).

18. See J. A. Bennett, "Hooke's Instruments," in *London's Leonardo: The Life and Work of Robert Hooke*, ed. J. A. Bennett et al. (New York: Oxford University Press, 2003), 63–104.

19. See Robert Hooke, "More Wayes For the same Purpose, Intimated by M. Hook," *Philosophical Transactions* 25, no. 2 (May 6, 1667): 459.

20. [Richard Towneley and Robert Hooke], "A Description of an Instrument for dividing a foot into many thousand parts, and thereby measuring the diameters of planets to a great exactness, &c. The instrument itself and being contrived and sent by Mr. Richard Towneley to the R. Society, and this following description of it being made by Mr. Hook was read before the Society Novemb. 14, 1667," Royal Society Register Book (Original) (henceforth RSRBO), vol. 3, fol. 227.

21. [Robert Hooke], "A Description of an Instrument for dividing a foot into many thousand parts. . . . ," *Philosophical Transactions* 29, no. 2 (November 11, 1667): 542.

22. For the images, see RSRBO vol. 3, fols. 227–30.

23. [Hooke], "A Description," 543. This innovation was quickly implemented. John Flamsteed wrote the following to Henry Oldenburg in 1671: "Not findeing yt Mr. Townly's micrometer is so devised as to be convenient for some observations I intend, I have devised a new one in which I can have thread sights or other as I please and parallel threads disposed perpendicular to ye ostensors like a kind of rete [i.e. a net] which is very convenient"; John Flamsteed to Henry Oldenburg, December 2, 1671, in *The Correspondence of Henry Oldenburg*, ed. and trans. A. Rupert Hall, Marie Boas Hall, and Eberhard Reichmann (Madison: University of Wisconsin Press, 1971), 8:390.

24. Beyond the omission of Hooke's amended indexes from the copy of his drawing in the Register Book, an offprint of the micrometer engraving from the *Philosophical Transactions* now bound into that Register Book (and included here as my fig. 2.2) shows Hooke's enhanced pointers crossed out with three inky slashes; see RSRBO vol. 3, fols. 227–30.

25. For a roughly contemporaneous example in which Hooke also extended drawings of machines with wax, see his "The way to sound the Depths of the Sea" (September 30, 1663) in RSCP 20:23.

26. See Françoise Viatte, "Weaving a Rope of Sand," *Yale French Studies* 89 (1996): 85–102.

27. [Hooke], "A Description," 543.

28. Ibid.

29. For an authoritative account of these traditions, see Robert G. Frank Jr., *Harvey and the Oxford Physiologists: A Study of Scientific Ideas* (Los Angeles: University of California Press, 1980), esp. 140–63. Influential studies of early modern anatomical practices and ideas include Barbara Maria Stafford, *Body Criticism: Imaging the Unseen in Enlightenment Art and Medicine* (Cambridge, MA: MIT Press, 1991); Jonathan Sawday, *The Body Emblazoned: Dissection and the Human Body in Renaissance Culture* (New York: Routledge, 1996); and Roger French, *Dissection and Vivisection in the European Renaissance* (Brookfield: Ashgate, 1999).

30. Robert Boyle, *New Experiments Physico-Mechanicall, Touching the Spring of the Air* (1660), in *The Works of Robert Boyle*, ed. Michael Hunter and E. B. Davis (London: Pickering and Chatto, 1999), 1:278–79. More broadly, see Steven Shapin and Simon Schaffer, *Leviathan and the Air-Pump: Hobbes, Boyle and the Experimental Life* (Princeton, NJ: Princeton University Press, 1985).

31. For examples of this London-based program as developed by Dr. William Croone, see Birch 1:433, 444.

32. Ibid., 1:475.

33. Although they should not be considered an exhaustive record, six related trials are listed in the Royal Society's Register Books; see RSRBO 3:64 (November 9, 1664), 65–66 (November 23, 1664), 203–5 (October 24, 1667), 265–67 (December 19, 1667), 297 (May 14, 1668), and 307–8 (July 4, 1668).

34. Henry Oldenburg to Robert Boyle, November 10, 1664, in *The Correspondence of Henry Oldenburg*, ed. and trans. A. Rupert Hall and Marie Boas Hall, and Eberhard Reichmann, vol. 2, *1663–1665* (Madison: University of Wisconsin Press, 1966), 296–97. This experiment was also reported in Thomas Sprat's widely read *History of the Royal-Society of London, For the Improving of Natural Knowledge* (London: T. R. for J. Martyn, 1667), 232.

35. Francis Bacon, *The Advancement of Learning* [1605] (Reprint, Chicago: Encyclopaedia Britannica, 1952), bk. 2, §2.1.34. A classic account of this "Baconian tradition" of experiment is Thomas Kuhn, "Mathematical Versus Experimental Traditions in the Development of Physical Science," in *The Essential Tension: Selected Studies in Scientific Tradition and Change* (Chicago: University of Chicago Press, 1977), 31–65. For an important critique of Kuhn's scheme, see J. A. Bennett, "The Mechanics' Philosophy and the Mechanical Philosophy," *History of Science* 24, no. 1 (March 1986): 1–28.

36. On relations between experiment and representation, see Matthew Hunter, "Experiment, Theory, Representation: Robert Hooke's Material Models," in *Beyond Mimesis and Convention: Representation in Art and Science*, ed. Roman Frigg and Matthew C. Hunter (New York: Springer, 2010), 193–219.

37. While the original manuscript of this letter has been lost, an eighteenth-century copy survives in the British Library; see BL Add. MS 6194, fols. 27–28.

38. Ibid., fol. 27v.

39. For this opposition of dry and wet visual technologies, see Jeff Wall, "Photography and Liquid Intelligence," in *Jeff Wall*, ed. Thierry de Duve et al. (London: Phaidon, 1996), 90–93.

40. BL Add. MS 6194, fol. 27v.

41. Ibid.

42. Ibid., fol. 28r.

43. Other contributions that have noted this include Andreas Holger-Maehle and Ulrich Tröhler, "Animal Experimentation from Antiquity to the End of the Eighteenth Century: Attitudes and Arguments," in *Vivisection in Historical Perspective*, ed. Nicolaas Rupke (New York: Routledge, 1987), 14–47; and Janice Neri, *The Insect and the Image: Visualizing Nature in Early Modern Europe, 1500–1700* (Minneapolis: University of Minnesota Press, 2011), 105–38.

44. A widely read statement of Descartes's view of animals is part 5 of *Discours de la Méthode* (1637); see in *The Philosophical Writings of Descartes*, trans. John Cottingham et al., vol. 1 (New York: Cambridge University Press, 1985), 131–40. On the infrequency with which the animal/machine doctrine was actually used to justify vivisection, see Anita Guerrini, "The Ethics of Animal Experimentation in Seventeenth-Century England," *Journal of the History of Ideas* 50, no. 3 (1999): 391–407. A compelling revision is Claus Zittel, "Conflicting Pictures: Illustrating Descartes' *Traité de l'homme*," in *Silent Messengers: The Circulation of Material Objects of Knowledge in the Early Modern Low Countries*, ed. Sven Dupré and Christoph Lüthy (Berlin: LIT, 2011), 217–60.

45. See R. T. Gunther, ed., *Early Science in Oxford*, vol. 6, *The Life and Work of Robert Hooke* (Oxford: Oxford University Press, 1930), 210.

46. On "violent literalizations," see Jeremy Biles, "I, Insect, or Bataille and the Crush Freaks," *Janus*

Head 7, no. 1 (2004): 115–31; and Mark Seltzer, *Serial Killers: Death and Life in America's Wound Culture* (New York: Routledge, 1998).

47. For Boyle, see Malcolm R. Oster, "The 'Beame of Diuinity': Animal Suffering in the Early Thought of Robert Boyle," *British Journal for the History of Science* 22, no. 2 (July 1989): 151–79. More broadly, see William F. Bynum, "The Anatomical Method, Natural Theology, and the Functions of the Brain," *Isis* 64, no. 4 (December 1973): 445–68.

48. Robert Hooke, *Micrographia; or, Some Physiological Descriptions of Minute Bodies made by Magnifying Glass* (London: John Martyn and James Allestry, 1665), 186.

49. Ibid.

50. For Wren's innovative work on intravenous drug injection and its application to vivisection, see W. C. Gibson, "The Bio-Medical Pursuits of Christopher Wren," *Medical History* 14, no. 4 (October 1970): 331–41.

51. BL Add. MS 6194, fol. 28r.

52. Svetlana Alpers, *The Art of Describing: Dutch Art in the Seventeenth Century* (Chicago: University of Chicago Press, 1983), 91; and Stafford, *Body Criticism*, 47. See also Catherine Wilson, "Visual Surface and Visual Symbol: The Microscope and the Occult in Early Modern Science," *Journal of the History of Ideas* 49, no. 1 (1988): 85–108.

53. On the importance of the rupture in this chronology, see Guerrini, "Ethics of Animal Experimentation," 400–402.

54. A classic account is Marcel Detienne and Jean-Pierre Vernant, *Cunning Intelligence in Greek Culture and Society*, trans. Janet Lloyd (Atlantic Highlands, NJ: Humanities Press, 1978).

55. The account of Collins's donation reads as follows: "Mr. Collins presented the society with an excellent double horizontal dial, containing some new additions to that instrument; as also several books of which he was the author, viz. 1. *A Treatise of Geometrical Dialing* 2. *The Sector on a Quadrant: or, a Treatise containing the description and use of four several quadrants, &c.* 3. *The Mariners plan Scale new planed.* 4. *An Introduction to merchants Accounts. The Doctrine of Decimal Arithmetic, Simple Interest, &c. as also of Compound Interest and Annuities, generally preferred for any time of payment, &c.* in one printed sheet in 8vo. Besides these he presented Remmelini's anatomical cuts"; Birch 2:202.

56. For more on Remellin's prints, see Andrea Carlino, *Paper Bodies: A Catalogue of Anatomical Fugitive Sheets 1538–1687*, trans. Noga Arikha (London: Wellcome Institute for the History of Medicine, 1999), esp. 71–73.

57. On Hooke's relationship to the Royal Society Library, see Linda Levy Peck, "Uncovering the Arundel Library at the Royal Society: Changing Meanings of Science and the Fate of the Norfolk Donation," *Notes and Records of the Royal Society of London* 52, no. 1 (1998): 3–24.

58. I am not asserting that Remellin's volume was the only source Hooke knew in which flaps were utilized. Indeed, many books related to architecture (including a text by John Dee that we will see Hooke reading in chapter 5) used this technique. My point is that Remellin's volume arrived precisely and publicly at this crucial moment in Hooke's experimental activities, enabling him to redeploy the flap technique to solve a puzzle. For more on flaps in architectural publications, see Anthony Gerbino and Stephen Johnston, *Compass and Rule: Architecture as Mathematical Practice in England 1600–1750* (New Haven, CT: Yale University Press, 2009), esp. 57–59. More broadly, see Suzanne Karr Schmidt and Kimberly Nichols, *Altered and Adorned: Using Renaissance Prints in Daily Life* (Chicago: Art Institute of Chicago, 2011), esp.

73–91; and Susan Dackerman, *Prints and the Pursuit of Knowledge in Early Modern Europe* (Cambridge, MA: Harvard Art Museums, 2011).

59. For various approaches to this problem, see Michael Lynch, "Discipline and the Material Form of Images: An Analysis of Scientific Visibility," *Social Studies of Science* 15, no. 1 (February 1985): 37–66; Gordon Fyfe and John Law, eds., *Picturing Power: Visual Depiction and Social Relations* (London: Routledge, 1988); and Frigg and Hunter, *Beyond Mimesis and Convention*.

60. Thomas Kuhn, *The Structure of Scientific Revolutions* (Chicago: University of Chicago Press, 1965), 38. For interesting readings of Kuhn's thinking on puzzle-solving, see Thomas Nickles, "Kuhnian Puzzle Solving and Schema Theory," *Philosophy of Science* 67, pt. 2 (September 2000): S242–55; and Thomas Leddy, "Is the Creative Process a Form of Puzzle Solving?," *Journal of Aesthetic Education* 24, no. 3 (Autumn 1990): 83–97.

61. Although it is beyond the scope of this chapter, examination of relations between Hooke's paper micrometer and early modern games might well draw out additional features through which to expand the meanings of this puzzle analogy. For an approach that goes toward that direction, see James Elkins, *Why Are Our Pictures Puzzles? On the Modern Origins of Pictorial Complexity* (New York: Routledge, 1999). I thank Lia Markey and especially Jason La Fountain for this point.

62. For a particularly compelling reading of the force of "elegance" in contemporary physics, see Peter Galison, "Theory Bound and Unbound: Superstrings and Experiment," in *Laws of Nature: Essays on the Philosophical, Scientific and Historical Dimensions*, ed. Fridel Weinert (New York: Walter de Gruyter, 1995), 269–408.

63. By its experimental uselessness, I mean that making and using the paper micrometer would do nothing to resolve the intolerable suffering of animals in Hooke's actual vivisections.

64. Various appeals to "plain style" include Richard Foster Jones, "Science and English Prose Style in the Third Quarter of the Seventeenth Century," *Proceedings of the Modern Language Association* 45 (1930): 977–1009; Laura Lunger Knoppers, "The Politics of Portraiture: Oliver Cromwell and the Plain Style," *Renaissance Quarterly* 12, no. 22 (1998): 1281–1319; and Ann Bermingham, *Learning to Draw: Studies in the Cultural History of a Polite and Useful Art* (New Haven, CT: Yale University Press, 2000), esp. 64–73.

65. Hooke, *Micrographia*, 190.

66. Ibid.

67. On technologized images as traps, see the account of scientists "cornered" and "swamped" by cascades in Latour, "Drawing Things Together," esp. 48, 42; and Alfred Gell's scintillating essays "The Technology of Enchantment and the Enchantment of Technology" and "Vogel's Net: Traps as Artworks and Artworks as Traps," in *The Art of Anthropology: Essays and Diagrams*, ed. Eric Hirsch (London: Athlone Press, 1999), 129–214. Relevant art-historical sources will be discussed below.

68. For a broader account of Hooke's anxieties about visual art and its relation to his theories of perception and cognition, see Matthew C. Hunter, "The Theory of the Impression according to Robert Hooke," in *Printed Images in Early Modern Britain: Essays in Interpretation*, ed. Michael Hunter (Aldershot: Ashgate, 2010), 164–93.

69. Robert Hooke, "An Instrument of Use to Take the Draught, or Picture of Any Thing," in *Philosophical Experiments and Observations of the Late Eminent Dr. Robert Hooke*, ed. William Derham (London: W. and J. Innys, 1726), 294.

70. Ibid.

71. William B. MacGregor, "The Authority of Prints: An Early Modern Perspective," *Art History* 22, no. 3 (September 1999): 389–420.

72. Robert Hooke, *A General Scheme, or Idea of the Present State of Natural Philosophy*, in *The Posthumous Works of Robert Hooke*, ed. Richard Waller (London: Sam. Smith and Benj. Walford, 1705), 64.

73. On this point, see Johns, *Nature of the Book*, 380–443.

74. Robert Hooke, preface to Robert Knox, *An Historical Relation of the Island of Ceylon in the East Indies* (London: Richard Chiswell, 1681), sig. A6.

75. John Evelyn, *Sculptura; or, The History, and Art of Chalcography and Engraving in Copper* (London: J. C. for G. Beedle and T. Collins, 1662), 107.

76. Robert Hooke, "A Curious Dissertation Concerning the Causes of the Power & Effects of Music," RSCP 2:31 (fol. 1). For a transcription of and commentary on this lecture, see Penelope Gouk, "The Role of Acoustics and Music Theory in the Scientific Work of Robert Hooke," *Annals of Science* 37 (1980): 573–605.

77. RSCP 2:31 (fol. 1v).

78. [Robert Hooke], "A Contrivance to make the Picture of any thing appear on a Wall. . . . ," *Philosophical Transactions of the Royal Society of London* 4, no. 38 (August 17, 1668): 743. This article was published by Henry Oldenburg (and possibly also written by Hooke) in response to a reputed miracle of a transformation of the host in a Catholic church near the French Protestant academy at Saumur in 1668. For a discussion of the propagandistic context and philosophical debate surrounding this "miracle" and its technological debunking, see Matthew C. Hunter, "Robert Hooke Fecit," 226–34. More broadly, see Laurent Mannoni, *The Great Art of Light and Shadow: Archaeology of the Cinema*, trans. Richard Crangle (Exeter: Exeter University Press, 2000).

79. [Hooke], "A Contrivance," 742.

80. See Henri Justel to Henry Oldenburg, September 29, 1668, in *The Correspondence of Henry Oldenburg*, vol. 5, *1668–1669*, ed. and trans. A. Rupert Hall and Marie Boas Hall, and Eberhard Reichmann (Madison: University of Wisconsin Press, 1968), 74.

81. For a recent history of intellectuals' responses to these problems, see Stuart Clark, *Vanities of the Eye: Vision in Early Modern European Culture* (New York: Oxford University Press, 2007).

82. Michelangelo, quoted in Giovanni Battista Armenini, *On the True Precepts of the Art of Painting* [1586], ed. and trans. E. J. Olszewski (New York: Burt Franklin, 1977), 138. For the sinister dimensions of artistic creativity in the northern tradition, see Claudia Swan, *Art, Science and Witchcraft in Early Modern Holland: Jacques de Gheyn II (1565–1629)* (Cambridge: Cambridge University Press, 2005).

83. Michael Cole, "Cellini's Blood," *Art Bulletin* 81, no. 2 (June 1999): 215–35.

84. Louis Marin, *To Destroy Painting* [1977], trans. Mette Hjort (Chicago: University of Chicago Press, 1995), esp. 160–64; Michael Fried, *Realism, Writing, Disfiguration: On Thomas Eakins and Stephen Crane* (Chicago: University of Chicago Press, 1987), esp. 64–65.

85. See Pamela H. Smith, *The Body of the Artisan: Art and Experience in the Scientific Revolution* (Chicago: University of Chicago Press, 2004), esp. 228–36.

86. I owe the term *lightness* to Katie Scott.

87. Hooke, *Micrographia*, sig. Ai.

88. Compare Claude Levi-Strauss, *The Savage Mind* (Chicago: University of Chicago Press, 1968), esp. 16–19; and Michele and Robert Root-Bernstein, "Thinkering," *Psychology Today* (August 21, 2008): http://www.psychologytoday.com/blog/imagine/200808/thinkering.

89. See Kendall Walton, *Mimesis as Make-Believe: On the Foundations of the Representational Arts* (Cambridge, MA: Harvard University Press, 1990), 11–69.

90. I would stress that while members of the Royal Society may have objected to the way in which Hooke represented discrete features of Richard Towneley's micrometer, they had no problem in understanding his contrivance as a representation of that machine. On the definition of these terms, see ibid., 38–39.

91. The description and dating to February 21, 1666, comes from Samuel Pepys's diary; see *The Diary of Samuel Pepys*, ed. Robert Latham and William Matthews (Los Angeles: University of California Press, 1972), 7:51. For more on the dating of Hooke's felting lecture, see Rosemary Weinberg, *The History of the Worshipful Company of Feltmakers 1604–2004* (Chichester: Phillimore, 2004), xvii. For an argument that links Hooke's craft knowledge directly to his appointment as Cutlerian Lecturer in the history of trades, see Michael Hunter, *Establishing the New Science: The Experience of the Early Royal Society* (Woodbridge: Boydell Press, 1989), 279–338.

92. In seventeenth-century England, paper was made not from wood pulp but, like felt, from rags; see Philip Gaskell, *A New Introduction to Bibliography* (Oxford: Clarendon Press, 1974), esp. 57–66.

93. All citations that follow regarding this image are from Robert Hooke, "The Way of Making Felts," RSCP 20:96 (fol. 223). However, Hooke's copious notations on felting extend over several heavily corrected folio pages; see RSCP 20:96 (fols. 223r–227v).

94. See Sheila McTighe, "Abraham Bosse and the Language of Artisans: Genre and Perspective in the Académie royale de peinture et de sculpture, 1640–1670," *Oxford Art Journal* 21, no. 1 (1998): 1–26.

95. Although Hooke explicitly gives instructions to an engraver here (i.e. "This Draught must not be inverted"), I have been unable to locate any evidence suggesting that a print was ever actually made from this drawing.

96. Walton, *Mimesis as Make-Believe*, 22.

97. Hooke, *General Scheme*, 57. The dating of Hooke's *General Scheme* is a notoriously tricky problem. The text was posthumously published in 1705 by Richard Waller, who failed to supply a specific date for it; Hooke's original manuscripts for this text are now lost. While dates of ca.1666–ca.1668 have generally been cited, leading historian of the early Royal Society, Michael Hunter, has recently followed P. J. Pugliese in dating the lectures comprising the *General Scheme* to ca. 1668. I find that date convincing, especially given the fact that absolutely no discussion of organic tissue growth or bodily interiors appeared anywhere in Hooke's exhaustive notes on felting/papermaking from 1666. For various accounts of the dating, compare Michael Hunter, "Hooke the Natural Philosopher," in Bennett et al., *London's Leonardo*, 105–62, esp. 117; Mary B. Hesse, "Hooke's Philosophical Algebra," *Isis* 57, no. 1 (Spring 1966): 67–83, esp. 68; P. W. Wood, "Methodology and Apologetics: Thomas Sprat's *History of the Royal Society*," *British Journal for the History of Science* 13, no. 43 (1980): 1–26, esp. 7n45; and Stephen Inwood, *The Forgotten Genius: The Biography of Robert Hooke 1635–1703* (San Francisco: Macadam/Cage, 2003), 112.

98. Hooke, *General Scheme*, 57.

99. See Erin O'Connor, "Embodied Knowledge in Glassblowing: The Experience of Meaning and the Struggle towards Proficiency," *Sociological Review* 55, Suppl. 1 (May 2007): 126–41.

100. Although I am not aware of further instances in which Hooke used paper to pursue this kind of conceptual modeling per se, he was clearly interested in other aspects of its physical properties, such as its ability to absorb fluids. In *Micrographia*, he explains the deficiencies of printed images in part by noting "the *uneven surface* of the *paper*, which at best appears no smother then a very course piece of *shag d*

cloth." Like most early modern intellectuals, moreover, Hooke also collected recipes for making ink. See Hooke, *Micrographia*, 3; and "Ink for the Rolling-Press," in *Philosophical Experiments and Observations*, 188–90. See also [Henry Oldenburg], "An Experiment of a Way of Preparing a Liquor, That Shall Sink into, and Colour the Whole Body of Marble, Causing a Picture, Drawn on a Surface, to Appear Also in the Inmost Parts of the Stone," *Philosophical Transactions* 1 (1665–66): 125–27.

101. See especially Morgan and Morrison, *Models as Mediators*.

102. Roman Frigg, "Models and Fiction," *Synthese* 172 (2010): 258.

103. Tarja Knuuttila and Atro Voutilainen, "A Parser as an Epistemic Artifact: A Material View on Models," *Philosophy of Science* 70 (December 2003): 1489.

104. Compare Hans-Jörg Rheinberger, *Toward a History of Epistemic Things: Synthesizing Proteins in the Test Tube* (Stanford, CA: Stanford University Press, 1997); David Davies, "Learning through Fictional Narratives in Art and Science," in Frigg and Hunter, *Beyond Mimesis and Convention*, 51–70; and Tarja Knuuttila and Mieke Boon, "How Do Models Give Us Knowledge? The Case of Carnot's Ideal Heat Engine," *European Journal for Philosophy of Science* 1, no. 3 (2011): 309–34. Here it might be technically correct in Walton's sense to describe the paper micrometer as functioning both as a prompter and as an "object of imagining"; see Walton, *Mimesis as Make-Believe*, 25–28.

105. Giorgio Vasari, *The Lives of the Artists*, trans. J. C. Bondanella and P. Bondanella (New York: Oxford University Press, 1991), 425.

106. On Hooke's scientific methodology, see Hesse, "Hooke's Philosophical Algebra"; D. R. Oldroyd, "Robert Hooke's Methodology of Science as Exemplified in his 'Discourse of Earthquakes,'" *British Journal for the History of Science* 6, no. 22 (1972): 109–30; and William Lynch, *Solomon's Child: Method in the Early Royal Society of London* (Stanford, CA: Stanford University Press, 2001), esp. 70–115.

107. For readings of Hooke's philosophical methods along these lines, see Mark E. Ehrlich, "Mechanism and Activity in the Scientific Revolution: The Case of Robert Hooke," *Annals of Science* 52 (1992): 127–51; and Michael Hunter, "Hooke the Natural Philosopher." See also Domenico Bertoloni Meli, *Thinking with Objects: The Transformation of Mechanics in the Seventeenth Century* (Baltimore: Johns Hopkins University Press, 2006).

108. For studies that emphasize the role of magic in Hooke's mechanical philosophy, see John Henry, "Robert Hooke: The Incongruous Mechanist," in *Robert Hooke: New Studies*, ed. Michael Hunter and Simon Schaffer (Woodbridge: Boydell Press, 1989), 149–80; and Penelope Gouk, *Music, Science and Natural Magic in Seventeenth-Century England* (New Haven, CT: Yale University Press, 1999), esp. 193–223.

109. Birch 2:91.

110. René Descartes, *Principles of Philosophy* [1647], trans. Valentine Rodger Miller and Reese P. Miller (Boston: D. Reidel, 1984), 60–61.

111. Birch 2:91.

112. Ibid., 2:92. For Hooke's various mathematical errors in these contentions, see Alexandre Koyré, "La gravitation universelle de Kepler à Newton," *Archives Internationales d'Histoire des Sciences* 4, no. 14 (January 1951): 638–53, esp. 649.

113. Robert Hooke, *An Attempt to Prove the Motion of the Earth by Observations* (London: J. Martyn, 1674), 28. For an interesting speculation on the possible sources for Hooke's conception of this "sphere of Activity," see J. A. Bennett, "Hooke and Wren and the System of the World: Some Points toward an Historical Account," *British Journal for the History of Science* 8, no. 1 (March 1975): 32–61, esp. 39–40.

114. Hooke, *An Attempt*, 28. On this difference of Hooke's theory from that of Borelli, see Alexandre Koyré, "An Unpublished Letter of Robert Hooke to Isaac Newton," *Isis* 43, no. 4 (December 1952): 312–37, esp. 318–19.

115. Hooke, *An Attempt*, 28.

116. Robert Hooke to Isaac Newton, January 6, 1680, in *The Correspondence of Isaac Newton*, vol. 2, *1676–1687*, ed. H. W. Turnbull (Cambridge: Cambridge University Press for the Royal Society, 1960), 309.

117. In addition to the works cited above, important statements in this literature include Louise Diehl Patterson, "Hooke's Gravitation Theory and Its Influence on Newton. I: Hooke's Gravitation Theory," *Isis* 40, no. 4 (November 1949): 327–41; Johannes Lohne, "Hooke *versus* Newton," *Centaurus* 7, no. 1 (1960): 6–52; Richard S. Westfall, "Hooke and the Law of Universal Gravitation: A Reappraisal of a Reappraisal," *British Journal for the History of Science* 3, no. 3 (June 1967): 245–61; Patri J. Pugliese, "Robert Hooke and the Dynamics of Motion in a Curved Path," in Hunter and Schaffer, *Robert Hooke: New Studies*, 181–205; and Ofer Gal, *Meanest Foundations and Nobler Superstructures: Hooke, Newton, and the "Compounding of the Celestiall Motions of the Planetts"* (London: Kluwer, 2002).

118. Compare Domenico Bertoloni Meli, "Who Is Afraid of Centrifugal Force?," *Early Science and Medicine* 10, no. 4 (2005): 535–43; and Michael Nauenberg, "Robert Hooke's Seminal Contribution to Orbital Dynamics," in *Robert Hooke: Tercentennial Studies*, ed. Michael Cooper and Michael Hunter (Aldershot: Ashgate, 2006), 3–32.

119. Robert Hooke, "Sect. VII," *Lectures of Light*, in Waller, *Posthumous Works of Robert Hooke*, 146. See chapter 5 for further discussion and secondary sources of this lecture.

120. Hooke goes to some length to stress that ideas are "material and bulky" and, "being material . . . subject to decay"; see ibid., 142, 144. For the much longer tradition on which Hooke's thinking drew and departed, see Erwin Panofsky, *Idea: A Concept in Art Theory*, trans. J. S. Peake (New York: Harper and Row, 1968).

121. Hooke, "Sect. VII," 140.

122. This use of the inverse-square law has also been noted by Douwe Draaisma, "Hooke on Memory and the Memory of Hooke," in Cooper and Hunter, *Robert Hooke: Tercentennial Studies*, 116.

123. Hooke, "Sect. VII," 145–46.

124. See ibid., 146.

125. For this episode, see Sachiko Kusukawa, "The *Historia Piscium* (1686)," *Notes and Records of the Royal Society of London* 54, no. 2 (2000): 179–97.

126. Wren to Lord Brouncker [ca. 1662?], in Wren Jr., *Parentalia*, 221.

127. John Flamsteed to Richard Towneley, November 1686; cited in J. L. E. Dreyer, "Flamsteed's Letters to Richard Towneley," *Observatory* 580 (September 1922): 294.

CHAPTER THREE

1. [Thomas Povey], "An Account of A Secret in the Use of Painting in Answer to the Command of the R. Society Brought in by Mr. Povey, and read before the Society Dec. 19, 1667," Royal Society Register Book Original (henceforth RSRBO) vol. 3, fols. 259–64. This text is reproduced in Birch 2:227–30.

2. July 1, 1664, entry, in *The Diary of John Evelyn*, ed. A. Dobson (London: Macmillan, 1906), 2:211.

3. For the relationship between Povey and Hoogstraten, see Celeste Brusati, *Artifice and Illusion: The Art and Writing of Samuel van Hoogstraten* (Chicago: University of Chicago Press, 1995).

4. September 21, 1664, entry, in *The Diary of Samuel Pepys*, ed. Robert Latham and William Matthews (Los Angeles: University of California Press, 1971), 5:277.

5. [Povey], "An Account," fol. 261.

6. On the appetite for news, see Steven Pincus, "'Coffee Politicians Does Create': Coffeehouses and Restoration Political Culture," *Journal of Modern History* 67 (December 1995): 807–34. For comparable egg tempera recipes, see Cennino Cennini, *The Craftsman's Handbook* [ca. 1437], trans. Daniel V. Thompson (New York: Dover, 1960), 51.

7. [Povey], "An Account," fol. 261.

8. On sheen, compare Roland Barthes, "The World as Object," in *Calligram: Essays in New Art History from France*, ed. Norman Bryson (New York: Cambridge University Press, 1988), 106–15; and Krista Thompson, "The Sound of Light: Reflections on Art History in the Visual Culture of Hip-Hop," *Art Bulletin* 91, no. 4 (December 2009): 481–505.

9. [Povey], "An Account," fol. 263.

10. See Karel van Mander, *The Lives of the Illustrious Netherlandish and German Painters* [1609], ed. and trans. H. Miedema (Doornspijk: Davaco, 1999), 1:56–57. Van Mander's text was clearly known by leading English critics: "Whilest great *Vasari* and *Vermander* shall / Interpret the deep mystery of all, / And I unto our modern Picts shall show, / What due renown to thy fair Art they owe"; see Richard Lovelace, "Peinture. A Panegyrick to the best Picture of Friendship Mr. Pet. Lilly," in *Lucasta: Posthume Poems of Richard Lovelace Esq.* (London: William Godbid for Clement Darby, 1659), 65–66. For evidence of Povey's broader chymical interests, see Thomas Povey, "The Method, Manner and Order of the Transmutation of Copper into Brass, etc.," *Philosophical Transactions of the Royal Society of London* 22 (1700–1701): 474–75.

11. [Povey], "An Account," fol. 263.

12. For a reading of Povey's activities as a case of science advancing art, see Craig Ashley Hanson, *The English Virtuoso: Art, Medicine, and Antiquarianism in the Age of Empiricism* (Chicago: University of Chicago Press, 2009), 98. More broadly, see Iain Pears, *The Discovery of Painting: The Growth of Interest in the Arts in England, 1680–1768* (New Haven, CT: Yale University Press, 1988); and Ann Bermingham, *Learning to Draw: Studies in the Cultural History of a Polite and Useful Art* (New Haven, CT: Yale University Press, 2000), esp. 64–73.

13. [Povey], "An Account," fols. 263, 262.

14. See Birch 2:230.

15. Although the literature on cross-pollinations between painting and science in early modern Europe is now vast, distinguishable traditions of interpretation are perceptible. One such tradition has emphasized the direct historical agency exerted by painters on the natural sciences. See classic studies such as William M. Ivins Jr., *On the Rationalization of Sight* (New York: Metropolitan Museum of Art, 1938); Erwin Panofsky, "Artist, Scientist, Genius: Notes on the 'Renaissance-Dämmerung,'" in *The Renaissance: Six Essays*, ed. W. K. Ferguson et al. (New York: Harper and Row, 1962), 121–82; and Samuel Y. Edgerton Jr., "Galileo, Florentine 'Disegno,' and the 'Strange Spottednesse of the Moon,'" *Art Journal* 44, no. 3 (Fall 1984): 225–32. A second approach has explored how optical sciences and instruments specifically informed so-called naturalistic strains of painting in Renaissance and baroque Europe. For example: Martin Kemp, *The Science of Art: Optical Themes in Western Art from Brunelleschi to Seurat* (New Haven, CT:

Yale University Press, 1990); and Philip Steadman, *Vermeer's Camera: Uncovering the Truth behind the Masterpieces* (New York: Oxford University Press, 2002). A third, more self-consciously interdisciplinary literature has seen painting and science as privileged iterations of broader epistemic concerns: Svetlana Alpers, *The Art of Describing: Dutch Art in the Seventeenth Century* (Chicago: University of Chicago Press, 1983); and David Freedberg, *The Eye of the Lynx: Galileo, His Friends, and the Beginnings of Modern Natural History* (Chicago: University of Chicago Press, 2002). For a recent synthetic overview, see Pamela H. Smith, *The Body of the Artisan: Art and Experience in the Scientific Revolution* (Chicago: University of Chicago Press, 2004).

16. For a concise statement of this position, see James Elkins, "Visual Practices across the University," in *Beyond Mimesis and Convention: Representation in Art and Science*, ed. Roman Frigg and Matthew C. Hunter (New York: Springer, 2010), 169–92.

17. See Diana Dethloff, "Lely, Sir Peter (1618–1680)," in *Oxford Dictionary of National Biography*, ed. H. C. G. Matthew and Brian Harrison (New York: Oxford University Press, 2004), 33:303–9.

18. For the very limited scholarship we have on this painting, see Jacques Foucart, "Peter Lely, Dutch History Painter," *Hoogsteder-Naumann Mercury* 2 (1985): 17–26, esp. 18; and Udo Kultermann, "Woman Asleep and the Artist," *Artibus et Historiae* 11, no. 22 (1990): 129–61.

19. [Richard Graham], *A Short Account of the most Eminent Painters, both Ancient and Modern. . . .*, in *De Arte Graphica. The Art of Painting by C.A. Du Fresnoy with Remarks*, trans. John Dryden (London: J. Heptinstall for W. Rogers, 1695), 343–44.

20. Richard Flecknoe, "To *LILY* on his excellent Painting," in *A Treatise of the Sports of Wit* (London: Printed for Flecknoe, 1675), 17.

21. Recent influential accounts of scientific observation in early modern Europe include Brian W. Ogilvie, *The Science of Describing: Natural History in Renaissance Europe* (Chicago: University of Chicago Press, 2006), esp. 87–138; and Lorraine Daston and Peter Galison, *Objectivity* (New York: Zone Books, 2010).

22. Although the provenance of Lely's subject pictures is extremely patchy, we know that his *Susanna and the Elders*, now at Burghley House in Cambridgeshire, was likely commissioned or acquired by William Cavendish, third Earl of Devonshire, a founding member of the Royal Society, and scion of a leading Royalist family. Strong support of Charles II's regime runs like a red thread among the other owners of these subject pictures, who include Henry Bennet, Earl of Arlington, Secretary of State to Charles II, and member of the infamous Cabal (an acronym designating the cadre of the king's leading ministers in the 1670s); Restoration actor William Cartwright, a leading member of the King's Company at the Theatre Royal, Drury Lane; John Grenville, who was created first Earl of Bath among numerous other titles by Charles II; and William Fermor, Baron Leominster, a Tory MP. For the sources of this information, see Giles Waterfield, *Mr. Cartwright's Pictures: A Seventeenth Century Collection* (London: Dulwich Picture Gallery, 1987), 58–59; Appendix C (a), The Tate Gallery Report 1961–2 (London: H.M. Stationery Office, 1962), 25–26; "Editorial: Sir Peter Lely's Collection," *Burlington Magazine* 83, no. 485 (1943): 188; and Matthew C. Hunter, "Robert Hooke Fecit: Making and Knowing in Restoration London" (PhD diss., University of Chicago, 2007), 274–78.

23. On the importance of the court presence in the early Royal Society, see Michael Hunter, *Science and Society in Restoration England* (New York: Cambridge University Press, 1981), 113–35; and Hunter, *The Royal Society and Its Fellows 1660–1700: The Morphology of an Early Scientific Institution* (Oxford: BSHS, 1994), 25–34.

24. Joseph Leo Koerner, *The Moment of Self-Portraiture in German Renaissance Art* (Chicago: University of Chicago Press, 1993), 322.

25. "Few artistic themes have offered so satisfying an opportunity for legitimized voyeurism as Susanna and the Elders"—Mary Garrard, "Artemisia and Susanna," in *Feminism and Art History: Questioning the Litany*, ed. Norma Broude and Mary D. Garrard (New York: Harper and Row, 1982), 149. See also John Berger, *Ways of Seeing* (New York: Viking Press, 1972), 50–51.

26. Svetlana Alpers, *Rembrandt's Enterprise: The Studio and the Market* (Chicago: University of Chicago Press, 1988), 45.

27. On the purchase of this painting from the sale of Charles's collections, see Jonathan Brown and J. H. Elliott, eds., *The Sale of the Century: Artistic Relations between Spain and Great Britain, 1604–1655* (London: Yale University Press, 2002), esp. 94–96. On the sale more broadly, see Jerry Brotton, *The Sale of the Late King's Goods: Charles I and His Art Collection* (London: Macmillan, 2006).

28. Lely bought this painting (valued at seven pounds) on April 24, 1650; see Oliver Millar, ed., "The Inventories and Valuations of the King's Goods, 1649–51," *Walpole Society* 43 (1972): 276. Although I have been unable to determine which of Tintoretto's numerous treatments of the Susanna theme Lely might have owned, various candidates are discussed in Rodolfo Pallucchini and Paola Rossi, *Tintoretto: Le Opere sacre e profane* (Milan: Electa, 1994), 1:160, 171, 173, 210.

29. [Graham], *A Short Account*, 343. For a nearly identical statement, see the claims of artist-historian George Vertue, whose *Notebooks* were published by the Walpole Society in six volumes between 1930 and 1950: Vertue 18, 20, 22, 24, 26, and 30); Vertue 20:148.

30. [Charles Le Brun], *Conference of Monsieur Le Brun . . . Upon Expression, General and Particular* (London: John Smith, 1701), 7–8. For the original French, see *Conférence de monsieur le Brun, premier peintre du roi . . . sus l'expression générale et particulière* (Amsterdam: J. L. De Lorme, 1698), 10–11.

31. Lorraine Daston, "Curiosity in Early Modern Science," *Word & Image* 11, no. 4 (1995): 391.

32. Robert Aylett, *Svsanna; or, The Arraignment of the Two Vnjvst Elders* (London: John Teage, 1620), 17.

33. Ibid., 20.

34. Ibid., 20–21.

35. For an instructive account of Aylett's poetry and religious views, see Frederick M. Padelford, "Robert Aylett," *Huntington Library Bulletin* 10 (October 1936): 1–48; and Padelford, "Robert Aylett: A Supplement," *Huntington Library Quarterly* 2, no. 4 (July 1939): 471–78.

36. Giovanni Boccaccio, *The Decameron: Containing An Hundred Pleasant Novels* (London: James Iaggard, 1620), 179.

37. Ibid., 179v.

38. On relations between chiaroscuro and blinding, see Louis Marin, *To Destroy Painting* [1977], trans. Mette Hjort (Chicago: University of Chicago Press, 1994), 150–64.

39. See Victor I. Stoichita, *The Self-Aware Image: An Insight into Early Modern Meta-Painting*, trans. A.-M. Glasheen (New York: Cambridge University Press, 1997), esp. 30–63.

40. Boccaccio, *Decameron*, 179v. On this approach to artistic education, see David Winton Bell Gallery, *Children of Mercury: The Education of Artists in the Sixteenth and Seventeenth Century* (Providence, RI: Brown University Press, 1984).

41. Boccaccio, *Decameron*, 179v.

42. Ibid.

43. Ibid., 180.

44. See Claire Pace, "'Delineated Lives': Themes and Variations in Seventeenth-Century Poems about Portraits," *Word & Image* 2, no. 1 (1986): 1–17.

45. See Raymond A. Anselment, "'Clouded Majesty': Richard Lovelace, Sir Peter Lely, and the Royalist Spirit," *Studies in Philology* 86, no. 3 (Summer 1989): 367–87.

46. Lovelace's claim here is especially rich given the complex pictorial quotations of Anthony van Dyck (especially his *Charles I and Henrietta Maria*, ca. 1632, now in the Archiepiscopal Castle and Gardens, Kromeríž, Czech Republic) through which Lely composed this double portrait. For Van Dyck's portrait, see Susan Barnes et al., *Van Dyck: A Complete Catalogue of the Paintings* (New Haven, CT: Yale University Press, 2004), 6.46, 460–62.

47. Richard Lovelace, "To my Worthy Friend Mr. Peter Lilly: on that excellent Picture of his Majesty, and the Duke of York, drawne by him at Hampton-Court," in *Lucasta: Epodes, Odes, Sonnets, Songs, &c.* (London: Thomas Harper, 1649), 62.

48. Ibid.

49. Ibid., 61.

50. Ibid., 62.

51. Ibid.

52. Seneca, "On Gathering Ideas," in *Ad Lucilium Epistulae Morales*, trans. R. H. Gummere (London: Heinemann, 1920), 2:281.

53. Lovelace, "To my Worthy Friend Mr. Peter Lilly," 62.

54. For example, Simon Schaffer, "Regeneration: The Body of Natural Philosophers in Restoration England," in *Science Incarnate: Historical Embodiments of Natural Knowledge*, ed. C. Lawrence and S. Shapin (Chicago: University of Chicago Press, 1998), 83–120.

55. See Royal Society Early Letters, vol. L1, fol. 9. Leeuwenhoek's image and accompanying text were then printed as [Antoni van Leeuwenhoek], "Microscopical Observations of Mr. Leewenhoeck, Concerning the Optic Nerve. . . . ," *Philosophical Transactions* 10 (1675): 378–80. On Descartes's experiment, which is recounted in Discourse Five of *Optics* (1637), see *The Philosophical Writings of Descartes*, trans. John Cottingham et al. (New York: Cambridge University Press, 1985), 1:166–67.

56. [Antoni van Leeuwenhoek], "A Specimen of Some Observations Made by a Microscope, Contrived by Mr. Leewenhoeck in Holland, Lately Communicated by Dr. Regnerus de Graaf," *Philosophical Transactions* 8 (1673): 6038. On Dutch microscopy, see Edward G. Ruestow, *The Microscope in the Dutch Republic: The Shaping of Discovery* (New York: Cambridge University Press, 1996).

57. See Immanuel Kant, *Critique of Judgment* [1790], trans. Werner S. Pluhar (Cambridge: Hackett, 1987), esp. §43, p. 170.

58. Ibid., §44, p. 173.

59. Hooke, of course, also made observations through insect eyes; see Robert Hooke, *Micrographia; or, Some Physiological Descriptions of Minute Bodies made by Magnifying Glass* (London: John Martyn and James Allestry, 1665), esp. 175–80.

60. This claim was recorded in the diary of Samuel Pepys on August 8, 1666; see E. N. da C. Andrade, "Samuel Pepys and the Royal Society," *Notes and Records of the Royal Society of London* 18, no. 2 (1963): 86.

61. Robert Hooke, "Sect. VI," *Lectures of Light*, in *The Posthumous Works of Robert Hooke*, ed. Richard Waller (London: Sam. Smith and Benj. Walford, 1705), 134.

62. Robert Hooke, *A General Scheme, or Idea of the Present State of Natural Philosophy*, in Waller, *Posthumous Works of Robert Hooke*, 8. On scaling with reference to Hooke, see Paul Emmons, "Size Matters: Virtual Scale and Bodily Imagination in Architectural Drawing," *Architectural Research Quarterly* 9, nos. 3–4 (2005): 227–35.

63. Hooke, *General Scheme*, 9.

64. Lely was also literally elevated by the Stuarts, knighted by Charles II in 1679; see Bainbrigg Buckeridge, *An Essay Towards an English School*, in *The Art of Painting . . . : Third edition* [1754] (London: Cornmarket Press, 1969), 403.

65. On the "moral economy" of scientific observation, see Lorraine Daston, "The Moral Economy of Science," *Osiris* 10 (1995): 2–24; and Daston, "On Scientific Observation," *Isis* 99, no. 1 (March 2008): 97–110.

66. [Robert Boyle], "Mr. Boyle's Account of the above-mentioned *Invention*, and the *Experiments* thereon," in *Parentalia; or, Memoirs of the Family of the Wrens*, ed. Christopher Wren Jr. (London: T. Osborn and R. Dodsley, 1750), 230.

67. See Sidney W. Mintz, *Sweetness and Power: The Place of Sugar in Modern History* (New York: Viking, 1985); Brian Cowan, *The Social Life of Coffee: The Emergence of the British Coffeehouse* (New Haven, CT: Yale University Press, 2005); Iain Gately, *Tobacco: A Cultural History of How an Exotic Plant Seduced Civilization* (New York: Grove Press, 2001); and Richard Davenport-Hines, *The Pursuit of Oblivion: A Social History of Drugs* (London: Phoenix Press, 2001).

68. On this trade in "ophium," see BL MSS Ewr D 1076, fols. 45, 59; for Bengal, see Davenport-Hines, *Pursuit of Oblivion*, 11.

69. See Frank Dikötter, Lars Laamann, and Zhou Xun, *Narcotic Culture: A History of Drugs in China* (London: Hurst, 2004), esp. 24–36.

70. See Barbara Hodgson, *In the Arms of Morpheus: The Tragic History of Laudanum, Morphine and Patent Medicines* (Vancouver: Greystone, 2001), esp. 45–77.

71. [Boyle], "Mr. Boyle's Account," 230.

72. For this assessment of Willis, see Carl Zimmer, *Soul Made Flesh: The Discovery of the Brain—and How It Changed the World* (New York: Free Press, 2004), 7.

73. Andreas Holger-Maehle, "Pharmacological Experimentation with Opium in the Eighteenth Century," in *Drugs and Narcotics in History*, ed. Roy Porter and Mikulás Teich (New York: Cambridge University Press, 1995), 53; and Thomas Willis, *Pharmaceutica Rationalis; or, An Exercitation of the Operations of Medicines in Humane Bodies: Shewing the Signs, Causes and Cures of most Distempers incident Thereunto* (London: J. B. for T. Dring, 1684), §7, p. 126.

74. Willis, *Pharmaceutica Rationalis*, 126.

75. Ibid., 126, 127.

76. All references to this report are from Dr. Edward Smith, "Of the Use of Opium amongst the Turks," RSRBO vol. 7, fols. 97–99.

77. The experiment was effectively compromised from the beginning as, Smith concedes, Shatoor "came to me the next morning, but excused his having taken halfe a dram before, because he wanted strength to rise out of bed without it."

78. Moses Pitt, ed., *The English Atlas* (Oxford: M. Pitt, 1682), 5:81. For Hooke's involvement with this atlas project, see E. G. R. Taylor, "'The English Atlas' of Moses Pitt, 1680–83," *Geographical Journal* 95, no. 4 (April 1940): 292–99.

79. Pitt, *English Atlas*, 5:81.

80. See Trinity College Library, Cambridge (henceforth TCLC), MS o.11a.1 (10a).

81. For an overview of Linschoten's voyages and publications, see Ernest van den Boogaart, *Civil and Corrupt Asia: Image and Text in the "Itinerario" and "Icones" of Jan Hugyen van Linschoten* (Chicago: University of Chicago Press, 2003), esp. 1–10. For a general account of Hooke's notes, see Lisa Jardine, *The Curious Life of Robert Hooke: The Man Who Measured London* (London: Harper Perennial, 2003), 82–85.

82. Further contemporaneous evidence of elite experimentalist interest in "that stupefying Herb *Datura*" can be found in the correspondence between Sir Robert Moray and Sir Philberto Vernatti ("Resident in Batavia in Java Major") published in Thomas Sprat, *The History of the Royal-Society of London, For the Improving of Natural Knowledge* (London: T. R. for J. Martyn, 1667), 158–72, esp. 161–62.

83. TCLC o.11a.1 10b. For the passage in Wolfe's translation that Hooke's notes follow very closely, see Jan Huygen van Linschoten, *Iohn Huighen van Linschoten his Discourse of Voyages into the Easte & West Indies. . . .*, trans. John Wolfe (London: John Wolfe, 1598), chap. 31, p. 60.

84. TCLC o.11a.1 10b; compare Wolfe's *Iohn Huighen van Linschoten his Discourse of Voyages*, chap. 31, p. 60. For more on the relationship between Linschoten and Paludanus, see Harold J. Cook, *Matters of Exchange: Commerce, Medicine and Science in the Dutch Golden Age* (New Haven, CT: Yale University Press, 2007), 120–32.

85. See Harold J. Cook, "The Cutting Edge of a Revolution? Medicine and Natural History near the Shores of the North Sea," in *Renaissance and Revolution: Humanists, Scholars, Craftsmen and Natural Philosophers in Early Modern Europe*, ed. J. V. Field and Frank A. J. L. James (Cambridge: Cambridge University Press, 1993), 45–61; Charles Webster, "Alchemical and Paracelsian Medicine," in *Health, Medicine and Mortality in the Sixteenth Century*, ed. Charles Webster (New York: Cambridge University Press, 1979), 301–34; and Hanson, *English Virtuoso*, esp. 8–11.

86. On Hooke's drug-taking, see Linda McCray Beier, "Experience and Experiment: Robert Hooke, Illness and Medicine," in *Robert Hooke: New Studies*, ed. Michael Hunter and Simon Schaffer (Woodbridge: Boydell Press, 1989), 235–52; Lisa Jardine, "Hooke The Man: His Diary and His Health," in *London's Leonardo: The Life and Work of Robert Hooke*, ed. J. A. Bennett et al. (New York: Oxford University Press, 2003), 214–47; and Stephen Inwood, *The Man Who Knew Too Much: The Strange and Inventive Life of Robert Hooke 1635–1703* (London: Macmillan, 2002), 398–99.

87. Hooke, *Micrographia*, 203.

88. Ibid.

89. Ibid., 204. In a contemporaneous letter, Hooke also explained how "with a thin Cable of glass I have often imprisoned small flying insects so as to view them with a microscope"; Robert Hooke to John Beale, June 24, 1664; Royal Society Classified Papers (hereafter RSCP) 20:30.

90. For recent studies of this series, see Michael Wenzel, "The Windsor Beauties by Sir Peter Lely and the Collection of Paintings at St. James's Palace, 1674," *Journal of the History of Collections* 14, no. 2 (2002): 205–13; and Catharine MacLeod and Julia Marciari Alexander, "The 'Windsor Beauties' and the Beauties Series in Restoration London," in *Politics, Transgression, and Representation at the Court of Charles II*,

ed. Julia Manciari Alexander and Catharine MacLeod (New Haven, CT: Yale University Press, 2007), 81–120.

91. See Diana de Marly, "Undress in the Oeuvre of Lely," *Burlington Magazine* 120, no. 908 (November 1978): 749–50.

92. Alexander Pope, *The First Epistle of the Second Book of Horace Imitated*, in *Poems of Alexander Pope*, vol. 4, *Imitations of Horace*, ed. J. Butt (London: Methuen, 1939), 207–9.

93. See William Pietz, "The Problem of the Fetish, I," *RES* 9 (1995): esp. 8–9; and Pietz, "The Problem of the Fetish, II: The Origins of the Fetish," *RES* 12 (1987): esp. 42–43.

94. BL Add. 22950, fol. 41.

95. Compare, for example, the accounts given in Margaret Whinney and Oliver Millar, *English Art 1625–1714* (Oxford: Clarendon Press, 1957), 173; and Julia Manciari Alexander, "Beauties, Bawds and Bravura: The Critical History of Restoration Portraits of Women," in *Painted Ladies: Women at the Court of Charles II*, ed. Catharine MacLeod and Julia Manciari Alexander (London: National Portrait Gallery, 2001), 62–71.

96. See Cyril Hughes Hartmann's introduction to Anthony Hamilton, *Memoirs of the Comte de Gramont*, trans. Peter Quennell (London: Routledge, 1932), 1–15.

97. Hamilton, *Memoirs of the Comte de Gramont*, 112.

98. Ibid., 110.

99. Here, Elizabeth Hamilton (1641–1708), wife of the Comte de Gramont and sister of the author, is the paradigm. "She did not attempt to shine in conversation," Hamilton observes approvingly. ". . . still more did she avoid that sort of affected deliberation in her discourse, which is so heavy that it sends the hearer to sleep"; ibid., 118.

100. Ibid., 170.

101. Ibid., 190.

102. See Michael Fried, *Absorption and Theatricality: Painting and Beholder in the Age of Diderot* (Los Angeles: University of California Press, 1980), esp. 66.

103. On the careers and reputations of these women, see Alexander, "Beauties, Bawds."

104. On the king's reputation, see James Grantham Turner, "Pepys and the Private Parts of Monarchy," in *Culture and Society in the Stuart Restoration: Literature, Drama, History*, ed. Gerald MacLean (Cambridge: Cambridge University Press, 1995), 95–110; and Paul Hammond, "The King's Two Bodies: Representations of Charles II," in *Culture, Politics and Society in Britain, 1660–1800*, ed. Jeremy Black and Jeremy Gregory (Manchester: Manchester University Press, 1991), 13–48.

105. See Kevin Sharpe, "'Thy Longing Country's Darling and Desire': Aesthetics, Sex, and Politics in the England of Charles II," in Alexander and MacLeod, *Politics, Transgression, and Representation*, 1–32.

106. See *The Diary of Robert Hooke, 1672–1680*, ed. H. W. Robinson and W. Adams (London: Taylor and Francis, 1935), 430–31.

107. For useful calculation of prices, see Jocelyn Powell, *Restoration Theatre Production* (London: Routledge and Kegan Paul, 1984), 12.

108. For the location of this coffeehouse, see Hooke, *Diary*, 466.

109. For discussion of depicted items, see Edward Tyson, *Phocæna; or, the Anatomy of a Porpess, dissected at Gresham College. . . .* (London: Benj. Tooke, 1680), 31, 34–35, 19–23.

110. My rough working calculation is that at the very least, Hooke produced something on the order of 70–80 drawings in the 1660s; roughly 30–50 drawings in the 1670s (a figure made especially vague given the largely unknown extent of Hooke's contribution to John Ogilby's mapping projects); and approximately 30–40 drawings in the 1680s. For an expanded account of the sources for these estimates, see Matthew C. Hunter, "Robert Hooke Fecit," 258–59. On Hooke's work with Ogilby, see E. G. R. Taylor, "Robert Hooke and the Cartographical Projects of the Late Seventeenth Century (1666–1696)," *Geographical Journal* 90, no. 6 (December 1937): 529–40.

111. For the latest drawing by Hooke I have been able to locate (dated June 2, 1686), see his depiction of "a waterwork newly finished by Mr. Aldersey at Hackney," RSCP vol. 20, fol. 159.

112. A brilliant reading of the meanings of abandoning picture-making on which I have drawn is Thierry de Duve, *Pictorial Nominalism: On Marcel Duchamp's Passage from Painting to the Readymade*, trans. Dana Polan (Minneapolis: University of Minnesota Press, 1991); and de Duve, "The Readymade and the Tube of Paint," *Artforum* 24 (1986): 110–21.

113. For these artists, see Alpers, *Art of Describing*, 112.

114. On this arc of Velázquez's career, see Jonathan Brown, *Velázquez, Painter and Courtier* (New Haven, CT: Yale University Press, 1986).

115. See John Aubrey, *Brief Lives*, ed. J. Buchanan-Brown (New York: Penguin, 2000), 394. For the broader, rhetorical trope of the rustic youth who makes images through incredible, autodidact ingenuity (exemplified par excellence by Vasari's Giotto), see Ernst Kris and Otto Kurz, *Legend, Myth, and Magic in the Image of the Artist* (New Haven, CT: Yale University Press, 1979), esp. 26–38. For the political context of Hooke's "discovery" by Charles I's court artists, see Jardine, *Curious Life*, 21–56.

116. Sir Robert Moray to Christopher Wren, August 13, 1661, in Christopher Wren Jr., *Parentalia; or, Memoirs of the Family of the Wrens* (London: T. Osborn and R. Dodsley, 1750), 211.

117. For this commission, see A. D. C. Simpson, "Robert Hooke and Practical Optics: Technical Support at a Scientific Frontier," in Michael Hunter and Simon Schaffer, *Robert Hooke: New Studies*, 33–62, esp. 41–43. More broadly, see Sachiko Kusukawa, "Picturing Knowledge in the Early Royal Society: The Examples of Richard Waller and Henry Hunt," *Notes and Records of the Royal Society* 65, no. 3 (2011): 273–94.

118. See Mary Ezell, "Richard Waller S.R.S: 'In Pursuit of Nature,'" *Notes and Records of the Royal Society of London* 38, no. 2 (1984): 216–17.

119. Ibid., 221.

120. For Waller's claims on his visual contribution, see Waller, *Posthumous Works of Robert Hooke*, 149.

121. For this claim about Hunt, see Robert Hooke to John Aubrey, August 24, 1675, Bodleian Library MS Aubrey vol. 12, fol. 186. For more on Hunt, see Jardine, *Curious Life*, esp. 249–51, 288–323; and Inwood, *Man Who Knew Too Much*, 383–440.

122. For the change of Wren's drawing in the 1680s, see Anthony Geraghty, *The Architectural Drawings of Sir Christopher Wren at All Souls College, Oxford: A Complete Catalogue* (Burlington, VT: Lund Humphries, 2007), 11.

123. See Steven Shapin, "Who Was Robert Hooke?" in Michael Hunter and Simon Schaffer, *Robert Hooke: New Studies*, 255–85; and Shapin, *A Social History of Truth: Civility and Science in Seventeenth-Century England* (Chicago: University of Chicago Press, 1994), esp. 355–407.

124. For various examples of Lely's generosity, see Vertue 20:134; 18:108, 125; 24:29; and 26:33.

125. On baroque workshop practices north and south of the Alps, compare Anne T. Woollett et al., *Rubens and Brueghel: A Working Friendship* (Los Angeles: Getty Research Institute, 2006); and Patrizia Cavazzini, *Painting as Business in Early Seventeenth-Century Rome* (University Park: Pennsylvania State University Press, 2008).

126. On the "Academy of 1673," see Ilaria Bignamini and Martin Postle, eds., *The Artist's Model: Its Role in British Art from Lely to Etty* (Nottingham: Nottingham University Art Gallery, 1991), 8.

127. Mary Bryan H. Curd, *Flemish and Dutch Artists in Early Modern England: Collaboration and Competition, 1460–1680* (Burlington, VT: Ashgate, 2010), 111.

128. For a characteristic statement of connoisseurial critique of Lely's late, studio-reliant manner, see R. B. Beckett, *Lely* (London: Routledge and Kegan Paul, 1951), 23–25. However, while the modern analysis has tended to disparage Lely's late work, some period sources present a different view. According to Richard Graham (and reiterated in full by George Vertue), "as to his *Art*, certain it is, that his *last Pieces* were his *best*, and that he gain'd ground, and improv'd himself every day, even to the very *Moment* in which *Death* snatch'd his *Pencil* out of his *hand* in an *Apoplectic Fit*, *Anno* 1680"; [Graham], *A Short Account*, 344; compare Vertue 20:148.

129. See Ella Hendriks and Karen Groen, "Lely's Studio Practice," *Bulletin of the Hamilton Kerr Institute* 2 (1994): 21–38.

130. Catharine MacLeod, "'Good but Not Like': Peter Lely, Portrait Practice and the Creation of a Court Look," in *Painted Ladies: Women at the Court of Charles II*, ed. Catharine MacLeod and Julia Marciari Alexander (London: National Portrait Gallery, 2001), 52.

131. See Diana Dethloff, "The Executors' Account Book and the Dispersal of Sir Peter Lely's Collection," *Journal of the History of Collections* 8, no. 1 (1996): 15–51, esp. 17–22.

132. For Lely's employment of mezzotint engraver Jan Van Somer, see Vertue 18:35; on his re-use of postures and plates, see MacLeod, "'Good but Not Like,'" 56–57.

133. Judith Hook, *The Baroque Age in England* (London: Thames and Hudson, 1976), 93.

134. For this characterization, see Curd, *Flemish and Dutch Artists*, 118.

135. Conservator Nicole Ryder draws this conclusion based on the period evidence of paint applied on top of natural resin varnish; see Nicole Ryder, *Report on the Treatment of DPG.555* (May 2006), Dulwich Picture Gallery, London. I thank Ms. Ryder for discussing these points with me and allowing me to watch her conservation work on the picture.

136. For an overview of Caravaggio's technique, see Cavazzini, *Painting as Business*, 64–74. For the contest between Lely and Kneller, see J. Douglas Stewart, "Kneller, Sir Godfrey, baronet (1646–1723)," in Matthew and Harrison, *Oxford Dictionary of National Biography*, 31:877–85.

137. For a specific case of Lely recycling motifs from his collection into his pictorial oeuvre, see David Ekserdjian, "A Portrait by Lely and a Drawing after Correggio: An Artist's Use of His Collection of Drawings," *Apollo* (May 1998): 28–29. More broadly, see Maria H. Loh, "New and Improved: Repetition as Originality in Italian Baroque Practice and Theory," *Art Bulletin* 86, no. 3 (September 2004): 477–504; and Paul Duro, "'The Surest Measure of Perfection': Approaches to Imitation in Seventeenth-Century French Art and Theory," *Word & Image* 25, no. 4 (October–December 2009): 363–83.

138. Vertue 24:22.

139. On this point, see Celina Fox, *The Arts of Industry in the Age of Enlightenment* (New Haven, CT: Yale University Press, 2009), 294–95.

140. Tyson, *Phocæna*, 3–4.

141. Ibid., 4.

CHAPTER FOUR

1. For a brief biography, see Elisabeth Leedham-Green, "Covel [Colvill], John (1638–1722)," in *Oxford Dictionary of National Biography*, ed. H. C. G. Matthew and Brian Harrison (Oxford: Oxford University Press, 2004), 13:726–27. On the Levant Company and its textile trade in Turkey, see Ralph Davis, *Aleppo and Devonshire Square: English Traders in the Levant in the Eighteenth Century* (Toronto: Macmillan, 1967), esp. 26–42.

2. Steven Pincus, *1688: The First Modern Revolution* (New Haven, CT: Yale University Press, 2009), 52. For Cambridgeshire, where industrial growth was slower, see A. Sarah Bendall, *Maps, Land and Society: A History, with a Carto-Bibliography of Cambridgeshire Estate Maps, c. 1600–1836* (New York: Cambridge University Press, 1992), esp. 13–23.

3. Pincus, *First Modern Revolution*, 73.

4. See John Spurr, *England in the 1670s: "This Masquerading Age"* (Malden, MA: Blackwell, 2000).

5. All references are to BL Add. MS 57495. For logic and religion, see fols. 5v–18; fols. 24–27; fol. 27v. For floral drawings, see fols. 29–73. For the glued leaves, see fol. 75. For wax impressions, see fols. 80v–87. For drawings of medals, see fols. 87v–89; fol. 112; fol. 114. The attribution of these drawings to Hooke was first made by Janice L. Neri; see Janice Neri, *The Insect and the Image: Visualizing Nature in Early Modern Europe, 1500–1700* (Minneapolis: University of Minnesota Press, 2011), 105–38; Neri, "Some Early Drawings by Robert Hooke," *Archives of Natural History* 32, no. 1 (2005): 41–47; and Neri, "Between Observation and Image: Representation of Insects in Robert Hooke's *Micrographia*," in *The Art of Natural History: Illustrated Treatises and Botanical Paintings, 1400–1850*, ed. Teresa O'Malley and Amy R. W. Meyers (New Haven, CT: Yale University Press, 2008), 83–108.

6. On commonplacing, see Adam Smyth, "'Rend and Teare in Peeces': Textual Fragmentation in Seventeenth-Century England," *Seventeenth Century* 19, no. 1 (2004): 36–52; and Ann Blair, "Reading Strategies for Coping with Information Overload ca. 1550–1700," *Journal of the History of Ideas* 64, no. 1 (2003): 11–28.

7. For identifications of these insects, see Neri, "Some Early Drawings," 41–47.

8. BL Add. MS 22911, fol. 213.

9. For the correspondence regarding the manuscript's request and return, see BL Sloane 4065, fol. 281; and BL Sloane 3340, fols. 323–24.

10. See Charles A. Rivington, "Early Printers to the Royal Society, 1663–1708," *Notes and Records of the Royal Society of London* 39, no. 1 (Sept. 1984): 1–27; and David A. K. Kronick, "Notes on the Printing History of the Early *Philosophical Transactions*," *Libraries & Culture* 25, no. 2 (1990): 243–68.

11. On the frequency of priority disputes, see Rob Iliffe, "'In the Warehouse': Privacy, Property and Priority in the Early Royal Society," *History of Science* 30 (1992): 29–68.

12. See Marie Boas Hall, *Henry Oldenburg: Shaping the Royal Society* (New York: Oxford University Press, 2002), esp. 144–48, 157.

13. For broader discussions, see David Landau and Peter Parshall, *The Renaissance Print, 1470–1550* (New Haven, CT: Yale University Press, 1996), esp. 237–59; Sachiko Kusukawa, "Leonhart Fuchs on the

Importance of Pictures," *Journal of the History of Ideas* 58, no. 3 (1997): 403–28; and Matthew C. Hunter, "The Theory of the Impression according to Robert Hooke," in *Printed Images in Early Modern Britain: Essays in Interpretation*, ed. Michael Hunter (Aldershot: Ashgate, 2010), 164–93.

14. Royal Society Classified Papers (henceforth RSCP), 2:22.

15. For this phrase, see Henry Oldenburg to R. F. Sluse, April 2, 1669, in *The Correspondence of Henry Oldenburg*, vol. 5, *1668–1669*, ed. and trans. A. Rupert Hall, Marie Boas Hall, and Eberhard Reichmann (Madison: University of Wisconsin Press, 1968), 468–70.

16. For the context of this arrangement, see Marie Boas Hall, *Henry Oldenburg*, 106–8.

17. See Adrian Johns, *The Nature of the Book: Print and Knowledge in the Making* (Chicago: University of Chicago Press, 1998), 519–34.

18. On the appetite for news, see Steven Pincus, "'Coffee Politicians Does Create': Coffeehouses and Restoration Political Culture," *Journal of Modern History* 67 (December 1995): 807–34.

19. Peter Nelson to Henry Oldenburg, August 22, 1668; Royal Society Early Letters (henceforth RSEL), N:31.

20. On this point, see Johns, *Nature of the Book*, esp. 535–36.

21. For a recent account of the main disputes between Hooke and Oldenburg, see Lisa Jardine, *The Curious Life of Robert Hooke: The Man Who Measured London* (London: Harper Perennial, 2003), 191–213.

22. Stephen Inwood, *The Forgotten Genius: The Biography of Robert Hooke 1635–1703* (San Francisco: MacAdam/Cage, 2003), 302.

23. On the volatility of this moment, compare Lotte Mulligan and Glenn Mulligan, "Reconstructing Restoration Science: Styles of Leadership and Social Composition of the Early Royal Society," *Social Studies of Science* 11 (1981): 327–64; and Michael Hunter and Paul B. Wood, "Towards Solomon's House: Rival Strategies for Reforming the Early Royal Society," *History of Science* 24 (1986): 49–108.

24. See Bruno Latour, "Drawing Things Together," in *Representation in Scientific Practice*, ed. Michael Lynch and Steve Woolgar (London: MIT Press, 1990), 19–68.

25. William Cole, "General Proposalls of the Termes and Conditions for Surrender of my Collections"; MS Ashmole 1820 a, fols. 65–69; reproduced in A. J. Turner, "A Forgotten Naturalist of the Seventeenth Century: William Cole of Bristol and His Collections," *Archives of Natural History* 11, no. 1 (1982): 35.

26. Ibid., in Turner, "Forgotten Naturalist," 35.

27. For recent work on connections between science and commerce in early modern Europe, see Harold J. Cook, *Matters of Exchange: Commerce, Medicine, and Science in the Dutch Golden Age* (New Haven, CT: Yale University Press, 2007); and Pamela H. Smith and Paula Findlen, eds., *Merchants and Marvels: Commerce, Science, and Art in Early Modern Europe* (New York: Routledge, 2002).

28. Cole, "General Proposalls," in Turner, "Forgotten Naturalist," 36.

29. William Cole, "A Minute Book of Heads . . . Begun 1 Jan. 1691/2"; Wellcome Library MS 1708, fol. 160.

30. Ibid.

31. Ibid., fol. 159.

32. Ibid.

33. A compelling synthetic account is Horst Bredekamp, *The Lure of Antiquity and the Cult of the Machine: The Kunstkammer and the Evolution of Art, Nature and Technology*, trans. Allison Brown (Princeton, NJ: Markus Wiener, 1995).

34. For a survey of recent scholarship on natural theology, see John Henry, *The Scientific Revolution and the Origins of Modern Science* (New York: Palgrave, 2002), 95–97.

35. Cole, "Minute Book," fol. 159.

36. Spurr, *England in the 1670s*, xi.

37. On cultures of print in Bristol, see Jonathan Barry, "The Press and the Politics of Culture in Bristol, 1660–1775," in *Culture, Politics and Society in Britain, 1660–1800*, ed. Jeremy Black and Jeremy Gregory (Manchester: Manchester University Press, 1991), 49–81. On the effects of the Press Act, see Johns, *Nature of the Book*, 72.

38. William Cole to Edward and Robert Southwell, July 14, 1693; RSEL C2:18.

39. William Cole to Robert Southwell, September 16, 1693; RSEL C2:17.

40. William Cole to Richard Waller, December 12, 1693; RSEL C2:19.

41. See Turner, "Forgotten Naturalist," 33; see also Antony Griffiths, "Burghers, Michael (1647/8–1727)," in Matthew and Harrison, *Oxford Dictionary of National Biography*, 8:798.

42. William Cole to Robert Southwell, November 16, 1695; RSEL C2:20.

43. See Carol Gibson-Wood, "Classification and Value in a Seventeenth-Century Museum: William Courten's Collections," *Journal of the History of Collections* 9, no. 1 (1997): 61–77, esp. 70–74.

44. Cole, "General Proposalls," in Turner, "Forgotten Naturalist," 38; Peter Parshall, "Imago Contrafacta: Images and Facts in the Northern Renaissance," *Art History* 16, no. 4 (December 1993): 554–79.

45. William Cole to Richard Waller, December 12, 1693; RSEL C2:19.

46. Cole, "General Proposalls," in Turner, "Forgotten Naturalist," 38.

47. William Cole to Robert Plot, October 31, 1684; reproduced in *Early Science at Oxford*, vol. 12, *Dr. Plot and the Correspondence of the Philosophical Society of Oxford*, ed. R. T. Gunther (Oxford: Printed for the subscribers, 1939), 234.

48. Turner, "Forgotten Naturalist," 30. On relations between naturalists and informants, see Londa Schiebinger and Claudia Swan, eds., *Colonial Botany: Science, Commerce and Politics in the Early Modern World* (Philadelphia: University of Pennsylvania Press, 2005).

49. [William Cole], "A Letter from Mr. William Cole of Bristol, to the Philosophical Society of Oxford, containing his Observations on the Purple Fish," *Philosophical Transactions* 178 (December 1685): 1285–86.

50. Ibid., 1281.

51. See Katherine Gibson, "'The Best Belov'd of Kings': The Iconography of Charles II" (PhD diss., Courtauld Institute of Art, 1997), esp. 1:97–101, 200–201.

52. On cultures of chemical performance in the 1670s, see Jan Golinski, "A Noble Spectacle: Phosphorous and the Public Cultures of Science in the Early Royal Society," *Isis* 80, no. 1 (1989): 11–39.

53. Cole, "Letter," 1282. On the difficulties of transporting live specimens involved in dyeing, see Neil Safier, "Spies, Dyes and Leaves: Agro-Intermediaries, Luso-Brazilian Couriers, and the Worlds They Sowed," in *The Brokered World: Go-Betweens and Global Intelligence, 1770–1820*, ed. Simon Schaffer et al. (Sagamore Beach, MA: Science History Publications, 2009), 239–69.

54. Cole, "Letter," 1282.

55. William Cole to the Royal Society, December 3, 1686; RSEL C2:15.

56. Exemplification, Nelson Goodman argues, "is not mere possession of a feature but requires also reference to that feature"; see Nelson Goodman, *Of Mind and Other Matters* (Cambridge, MA: Harvard University Press, 1984), 59.

57. Cole to Royal Society, RSEL C2:15.

58. Ibid.

59. See Victor Stater, "Mordaunt, Henry, second earl of Peterborough (bap. 1623, d. 1697)," in Matthew and Harrison, *Oxford Dictionary of National Biography*, 39:21–22.

60. Mario Biagioli, *Galileo Courtier: The Practice of Science in the Culture of Absolutism* (Chicago: University of Chicago Press, 1993), 30.

61. On the culture of the gift in early modern scientific patronage networks, see ibid., esp. 36–54. See also Natalie Zemon Davis, "Beyond the Market: Books as Gifts in Sixteenth-Century France," *Transactions of the Royal Historical Society* 5, no. 33 (1983): 69–88.

62. See William Cole, *Purpura Anglicana; being a Discovery of a Small Fish found on the Shores of the Severn* (London: Joseph Streater, 1689).

63. Cole to Robert Boyle, November 29, 1689; Royal Society Boyle Letters (henceforth RSBL), vol. 7, fol. 48.

64. Ibid.

65. John Ray to Edward Lhwyd, December 26, 1693; reproduced in *Further Correspondence of John Ray*, ed. R. T. Gunther (London: Ray Society, 1928), 240.

66. Cole's Wellcome notebook records the following details of the postal schedule: "Post from Chippenham [Wiltshire] to London goes thence {Sunday, Tuesday, Thurs.} morning; from thence to Bristol and to the West {Wed., Friday, Sunday} nights"; see Cole, "Minute Book," inside back binding.

67. For a contemporaneous provincial naturalist who was able to produce his own engraved images, see Robert Unwin, "A Provincial Man of Science at Work: Martin Lister, F.R.S. and His Illustrators, 1670–1683," *Notes and Records of the Royal Society of London* 49, no. 2 (1995): 209–30. For a recent survey of the struggles between scientists and illustrators endemic to early modern practice, see Lorraine Daston and Peter Galison, *Objectivity* (New York: Zone Books, 2007).

68. John Evelyn, *Numismata: A Discourse of Medals, Antient and Modern* (London: Benjamin Tooke, 1697), unpaginated [BL C.71.g.3]; all citations in this paragraph come from the inscribed front matter in that volume.

69. Ibid., 341.

70. For an image of Pepys's curated collections, see Jan van der Waals, "The Print Collection of Samuel Pepys," *Print Quarterly* 1, no. 4 (1984): 236–57.

71. Evelyn, *Numismata*, 261.

72. For this description of Hooke, see John Aubrey, *Brief Lives*, ed. J. Buchanan-Brown (New York: Penguin, 2000), 396.

73. Barbara Maria Stafford, *Voyage into Substance: Art, Science, Nature, and the Illustrated Travel Account, 1760–1840* (London: MIT Press, 1984), 48. On composition, see Thomas Puttfarken, *The Discovery of Pictorial Composition: Theories of Visual Order in Painting, 1400–1800* (London: Yale University Press, 2000).

74. On physiognomic analysis, see Evelyn, *Numismata*, 339.

75. For the sources of the images etched and engraved into Hooke's plate, see Antoni van Leeuwenhoek to Nehemiah Grew, April 25, 1679, RSEL L1:40; Anonymous, "Extrait d'une Lettre . . . d'une Nouvelle invention pour vôler en l'air," *Journal des Scavans* (December 12, 1678): 452–55; and Johannes Hevelius to Nehemiah Grew, Royal Society Letter Book (Original) (henceforth RSLBO), vol. 8, fols. 78–83.

76. For a longer history of this illustrative paradigm, see Byron Ellsworth Hamann, "Drawing Glyphs Together," in *Past Presented: Archaeological Illustration and the Ancient Americas*, ed. J. Pillsbury (Washington, DC: Dumbarton Oaks Research Library and Collection, 2012), 230–81.

77. *Philosophical Collections* first appears in Hooke's diary on Sunday, October 5, 1679; see *The Diary of Robert Hooke, 1672–1680*, ed. H. W. Robinson and W. Adams (London: Taylor and Francis, 1935), 426.

78. See ibid. For an interesting discussion of preparing cut-and-paste mock-ups of plates in nineteenth-century natural history, see Martin Rudwick, "Georges Cuvier's Paper Museum of Fossil Bones," *Archives of Natural History* 27, no. 1 (2000): 51–68, esp. 59.

79. On December 7, 1674, Hooke instructed "Harry [Hunt] about plate 3rd" and then "gave Plate to [Francis] Lamb to Letter." Once "lettered," these plates were returned to the printer, John Martyn, who produced Hooke's volume *Animadversions on the first part of the Machina Cœlestis* (1674). See Hooke, *Diary*, 134. For print proofs of these engravings, see Bloomsbury Auctions, *Manuscripts & Autograph Letters, Literature & History, Modern First Editions*, November 16, 2006 (London: Bloomsbury Auctions, 2006), Lot 52, pp. 11–12.

80. The progress of the *Collections* can be followed through October and November 1679; see Hooke, *Diary*, 426, 428–29, 430.

81. See Adrian Johns, "Martyn, John 1617/18–1680," in Matthew and Harrison, *Oxford Dictionary of National Biography*, 37:37–38.

82. For Hunt's "ad-vivum" preparatory drawings, see Royal Society MS 131 ("Waller Scrapbook"), fols. 96 a–b. On this language of depiction, see Claudia Swan, "Ad Vivum, Naer het Leven, From the Life: Defining a Mode of Representation," *Word & Image* 11 (1995): 353–72.

83. I am differentiating these modes of visualization in an unapologetically analytic manner. For a definition of depiction on which I have drawn, see Flint Schier, *Deeper into Pictures: An Essay on Pictorial Representation* (Cambridge: Cambridge University Press, 1986). For historicizing treatments of early modern diagrammatic strategies, see Krzysztof Lenk and Paul Kahn, "To Show and Explain: The Information Graphics of Stevin and Comenius," *Visible Language* 26, nos. 3–4 (1992): 272–81. On period conventions of cutaway representation, see Bert S. Hall, "The Didactic and the Elegant: Some Thoughts on Scientific and Technological Illustration in the Middle Ages and the Renaissance," in *Picturing Knowledge: Historical and Philosophical Problems concerning the Use of Art in Science*, ed. Brian S. Baigrie (Buffalo, NY: University of Toronto Press, 1996), 3–39. On early modern "paper instruments," see Catherine Eagleton and Boris Jardine, "Collections and Projections: Henry Sutton's Paper Instruments," *Journal of the History of Collections* 12, no. 1 (2005): 1–13; and Suzanne Karr Schmidt and Kimberly Nichols, *Altered and Adorned: Using Renaissance Prints in Daily Life* (Chicago: Art Institute of Chicago, 2011), esp. 73–91.

84. [Dr. William Croone], "An Hypothesis of the Structure of a Muscle, and the Reason of its Contractions," *Philosophical Collections* 2 (1681): 22.

85. Ibid.

86. Among the problems Croone tackles are why each muscle has an "Antagonist," and what produces spasmodic convulsions in freshly slaughtered animals; ibid., 23–24.

87. Ibid., 25. For this earlier work, see William Croone, *On the Reason of the Movement of the Muscles* [1667], trans. Paul Maquet, introduced by Margaret Nayler (Philadelphia: American Philosophical Society, 2000).

88. [Croone], "An Hypothesis of the Structure of a Muscle," 25. Compare G. A. Borelli, *De Motu Animalium: Pars Prima* (Roma: Angeli Bernabo, 1680), 181–91; table 9, figs. 6, 7, 8, and 10.

89. [Croone], "An Hypothesis of the Structure of a Muscle," 25.

90. Alois Riegl, *The Group Portraiture of Holland*, introduced by Wolfgang Kemp, trans. Evelyn M. Kain and David Britt (Los Angeles: Getty Research Center, 1999), 103.

91. For this experiment of December 4, 1679, see Birch 3:511.

92. For an important rereading of relations between alchemy and mechanical philosophy, see William R. Newman, *Atoms and Alchemy: Chymistry and the Experimental Origins of the Scientific Revolution* (Chicago: University of Chicago Press, 2006).

93. Robert Hooke, *Lectures de Potentia Restitutiva, or of Spring. . . .* (London: J. Martyn, 1678), 9.

94. After the initial trial of December 4, 1679, Hooke's metal-mixing experiments were shown to the Royal Society on December 11, 1679, and January 15 and 22, 1680; see Birch 3:516; 4:3, 5. For Hooke's subsequent presentation of metal-mixing trials performed at separate weekly sessions from the end of January through early April 1680, see Birch 4:6, 8–9, 11–12, 13–14, 17, 23, 25, 29, 30–31, 32.

95. For Hooke's registered work of this period, see Royal Society Register Book (Original) (henceforth RSRBO), vol. 5, fols. 216–20, 222, 224–26.

96. See BL MS Sloane 1039, fol. 114. For the experiment this lecture accompanied, see Birch 4:6.

97. BL MS Sloane 1039, fol. 120.

98. A compelling account of William Petty's contemporaneous alchemical engineering of populations in Ireland is Ted McCormick, *William Petty and the Ambitions of Political Arithmetic* (New York: Oxford University Press, 2009).

99. For a contextualizing account of these documents, see Hunter and Wood, "Towards Solomon's House," 49–108.

100. See ibid., 73. On gentlemanly conduct in the Royal Society, see especially Steven Shapin, *A Social History of Truth: Civility and Science in Seventeenth-Century England* (Chicago: University of Chicago Press, 1994).

101. RSCP 20:50 (fol. 94).

102. For an authoritative reading of the relations between the Royal Society's laudable goals and its actual performance in the 1670s, see Michael Hunter, *The Royal Society and Its Fellows 1660–1700: The Morphology of an Early Scientific Institution* (Oxford: BSHS, 1994), esp. 35–49.

103. See Inwood, *Forgotten Genius*, 208–9.

104. These documents can be found in RSCP 20:50 (fols. 85–110). For the dating of these documents, I follow Hunter and Wood, "Towards Solomon's House," 95–96.

105. RSCP 20:50 (fol. 94). Requirements placed on foreign Fellows were even more strict; see especially RSCP 20:50 (fol. 94v).

106. RSCP 20:50 (fol. 94); for rates of dues, see, for example, RSCP 20:50 (fol. 87).

107. For oaths of secrecy, see RSCP 20:50 (fols. 94, 96–97). For fee payment, see RSCP 20:50 (fols. 94v, 96, 97v); for research deadlines, see RSCP 20:50 (fols. 92v, 94v); for threats of expulsion, see RSCP 20:50 (fol. 96).

108. RSCP 20:50 (fol. 94).

109. For such proscription, see RSCP 20:50 (fols. 85–87, 94v, 96–98).

110. A classic account of the social-reformist currents in English experimental philosophy is Charles Webster, *The Great Instauration: Science, Medicine, and Reform, 1626–1660* (New York: Holmes and Meier, 1976).

111. RSCP 20:50 (fol. 99v).

112. RSCP 20:50 (fol. 86).

113. Ibid.

114. RSCP 20:50 (fol. 95).

115. Ibid.; compare with RSCP 20:50 (fol. 94v).

116. RSCP 20:50 (fol. 86).

117. BL MS Sloane 1039, fol. 171.

118. See RSCP 20:50 (fol. 92 v).

119. RSCP 20:50 (fol. 97).

120. RSCP 20:50 (fols. 94, 95).

121. RSCP 20:50 (fol. 92).

122. Ibid.

123. Ibid.

124. RSCP 20:50 (fols. 99–100).

125. RSCP 20:50 (fol. 94).

126. On Hooke's relationship with Pitt, see E. G. R. Taylor, "'The English Atlas' of Moses Pitt, 1680–83," *Geographical Journal* 95, no. 4 (April 1940): 292–99. While missing in Hooke's first number of *Philosophical Collections*, this is not to say that such editorial segues had been absent from Henry Oldenburg's *Philosophical Transactions*; see, for example, the concluding paragraph to [Robert Hooke], "A Contrivance to make the Picture of any thing appear on a Wall. . . . ," *Philosophical Transactions of the Royal Society of London* 4, no. 38 (August 17, 1668): 741–43.

127. [Robert Hooke], *Philosophical Collections* 2 (London: Moses Pitt, 1681; reprint, New York: Johnson Reprint, 1965): 1.

128. Ibid.

129. Ibid., 7.

130. See ibid., 11; and Edward Tyson, "Some Anatomical Observations of Hair found in several parts of the Body," in ibid., 2:11–17.

131. See Hugh Amory, "Chiswell, Richard, the Elder (1640–1711)," in Matthew and Harrison, *Oxford Dictionary of National Biography*, 11:494–95.

132. For an example of this cascade of graphic features, see [Hooke], *Philosophical Collections* 4 (January 10, 1682; London: Printed for Richard Chiswel): 92.

133. For *mise-en-page*, see D. F. McKenzie, "Typography and Meaning: The Case of William Congreve," in *The Book and the Book Trade in Eighteenth-Century Europe*, ed. Giles Barber and Bernhard Fabian (Hamburg: Dr. Ernst Hauswedell, 1981), 81–126.

134. [Hooke], "Account of Fontaney," *Philosophical Collections* 4 (January 10, 1682): 108.

135. See RSCP 20:50 (fol. 110). For a more expansive reading of linkages and cross-references in period scientific images, see John Bender and Michael Marrinan, *The Culture of Diagram* (Stanford, CA: Stanford University Press, 2010).

136. The standard chronology of Hooke's publications is Geoffrey Keynes, *A Bibliography of Robert Hooke* (Oxford: Clarendon Press, 1960), 43.

137. For Wren, Boyle, and Leeuwenhoek, respectively, see Robert Hooke, *Lectures and Collections* (London: J. Martyn, 1678), 41–42, 57–66, and 81–82; 84–89.

138. See, for example, ibid., 53–54; and Hooke, *Lectures de Potentia Restitutiva*, 28.

139. Robert Hooke, *Micrographia; or, Some Physiological Descriptions of Minute Bodies made by Magnifying Glasses* (London: John Martyn and James Allestry, 1665), sig. Aiv. On authorship in *Micrographia*, see Michael Aaron Dennis, "Graphic Understanding: Instruments and Interpretation in Robert Hooke's *Micrographia*," *Science in Context* 3, no. 2 (1989): 309–65, esp. 311. That said, *Micrographia* itself does contain material that Hooke credits to other researchers. See, for example, Prince Rupert's recipe for making bullets: Hooke, *Micrographia*, 22–24.

140. See Robert Hooke, *Lectiones Cutlerianæ; or, A Collection of Lectures. . . .* (London: J. Martyn, 1679). For a compelling account of how Hooke used this assembled volume in 1679 as a means of extracting payment from his parsimonious patron, Sir John Cutler, see Jardine, *Curious Life*, 235–45.

141. Hooke, *Micrographia*, sig. Biii.

142. On Hooke's relations to printers and booksellers, see especially Johns, *Nature of the Book*, esp. 521–42; and Rob Iliffe, "Material Doubts: Hooke, Artisan Culture and the Exchange of Information in 1670s London," *British Journal of the History of Science* 28 (1995): 285–318.

143. For Hooke's reaction to Halley's voyages of the 1670s, see Jardine, *Curious Life*, 259–61.

144. On the Royal Society's limited success in hiring a collector to gather specimens, see Michael Hunter, *Science and the Shape of Orthodoxy: Intellectual Change in Late Seventeenth-Century Britain* (Woodbridge: Boydell Press, 1995), esp. 141–47.

145. See Hooke to Henry Jacobs, December 2, 1679; RSEL H3:59.

146. Hooke to Henry Jacobs, October 10, 1681; RSEL H3:66.

147. Ibid.; more broadly, see E. G. R. Taylor, "Robert Hooke and the Cartographical Projects of the Late Seventeenth Century (1666–1696)," *Geographical Journal* 90, no. 6 (December 1932): 529–40.

148. I thank Dr. Richard Dunn (Curator of the History of Navigation, National Maritime Museum) for information on instruments of this type.

149. Hooke to Henry Jacobs, March 3, 1682; RSEL H3:67.

150. BL MS Sloane 1039, fol. 113.

151. Hooke to Jacobs, RSEL H3:67v.

152. Edward Tyson, *Phocaena; or, The Anatomy of a Porpess, dissected at Gresham College. . . .* (London: Benj. Tooke, 1680), 3.

153. For this phrase, see Michael Hunter, *Royal Society and Its Fellows*, 7.

154. Richard Waller to Robert Hooke, August 1, 1687; Trinity College Library, Cambridge (henceforth TCLC), o.11a.1 (fol. 26).

155. Ibid.

156. Robert Hooke to Richard Waller, [n.d.], TCLC o.11a.1 (26v).

157. Ibid.

158. Ray to Lhwyd, in *Further Correspondence of John Ray*, 240.

159. Turner, "Forgotten Naturalist," 35.

160. For a sympathetic reading of Hooke's Secretaryship, see Mulligan and Mulligan, "Reconstructing Restoration Science," 327–64. For a critique of this account that has set the tone for much recent interpretation, see Michael Hunter, *Establishing the New Science: The Experience of the Early Royal Society* (Woodbridge: Boydell Press, 1989), 339–55.

CHAPTER FIVE

1. Edward Topsell, *The historie of foure-footed beastes: Describing the true and liuely figure of euery beast* (London: William Iaggard, 1607), 296–99. On horse breeding, see Nicholas Russell, *Like Engend'ring Like: Heredity and Animal Breeding in Early Modern England* (New York: Cambridge University Press, 1986).

2. See Katherine Acheson, "Gesner, Topsell, and the Purposes of Pictures in Early Modern Natural Histories," in *Printed Images in Early Modern Britain: Essays in Interpretation*, ed. Michael Hunter (Aldershot: Ashgate, 2010), 127–44.

3. Topsell, *Historie of foure-footed beastes*, 295.

4. Robert Boyle to Henry Oldenburg, June 20, 1665, in *The Correspondence of Henry Oldenburg*, ed. and trans. A. Rupert Hall, Marie Boas Hall, and Eberhard Reichmann, vol. 2 (Madison: University of Wisconsin Press, 1966), 409.

5. See Birch 2:50.

6. Robert Boyle, *Some Considerations Touching the Usefulness of Experimental Philosophy* [1663], in *The Works of Robert Boyle*, ed. Michael Hunter and E. B. Davis (London: Pickering and Chatto, 1999), 3:310. For broader interests in monstrosity, see Lorraine Daston and Katharine Park, *Wonders and the Order of Nature 1150–1750* (New York: Zone Books, 1998).

7. For the most famous narrative of early summer 1665, see Daniel Defoe, *A Journal of the Plague Year*, ed. Paula Backschneider, Norton Critical Edition (New York: Norton, 1992), 8–10.

8. Boyle to Oldenburg, in *Correspondence of Henry Oldenburg*, 2:409. See also [Robert Hooke], "Observables upon a Monstrous Head," *Philosophical Transactions of the Royal Society of London* 1, no. 5 (Munday, July 3, 1665): 85–86.

9. For Boyle's presentation of the head, see Birch 2:50.

10. Henry Oldenburg to Robert Boyle, ca. June 16–18, 1665, in *Correspondence of Henry Oldenburg*, 2:405.

11. Birch 2:52.

12. See Samuel Colepresse to Henry Oldenburg, April 13, 1667; Royal Society Early Letters, vol. C1, fol. 14.

13. Ibid.

14. On uselessness, see Krzysztof Pomian, *Collectors and Curiosities: Paris and Venice, 1500–1800*, trans. Elizabeth Wiles-Portier (Cambridge: Polity Press, 1990), esp. 7–44.

15. See Michael Hunter, *Establishing the New Science: The Experience of the Early Royal Society* (Woodbridge: Boydell Press, 1989), 125–29.

16. See October 19, 1663; Birch 1:158.

17. While the broader literature on early modern museums is now vast, key works include Oliver Impey and Arthur MacGregor, eds., *The Origins of Museums: The Cabinet of Curiosities in Sixteenth- and Seventeenth-Century Europe* (Oxford: Clarendon Press, 1985); Paula Findlen, *Possessing Nature: Museums,*

Collecting, and Scientific Culture in Early Modern Italy (Los Angeles: University of California Press, 1994); Peter Parshall, ed., "Art and Curiosity," special issue, *Word & Image* 11, no. 4 (1995), 327–404; and Arthur MacGregor, *Curiosity and Enlightenment: Collections and Collectors from the Sixteenth to the Nineteenth Century* (New Haven, CT: Yale University Press, 2007).

18. See A. D. C. Simpson, "Newton's Telescope and the Cataloguing of the Royal Society's Repository," *Notes and Records of the Royal Society of London* 38, no. 2 (1984): 187–214, esp. 187–94.

19. Jenni Thomas, "A 'Philosophical Storehouse': The Life and Afterlife of the Royal Society Repository" (PhD diss., Queen Mary and Westfields, 2009), 31.

20. [William Petty], *Advice of W.P. to Mr. Samuel Hartlib for the Advancement of Some Particular Parts of Learning* (London, 1647), 8. Petty himself had studied at a Jesuit school in Caen, France; for a recent intellectual biography of Petty, see Ted McCormick, *William Petty and the Ambitions of Political Arithmetic* (New York: Oxford University Press, 2009).

21. For a fascinating study of the implications of Jesuit institutional consciousness on its architectural production, see Evonne Levy, *Propaganda and the Jesuit Style* (Los Angeles: University of California Press, 2004). See also Mordechai Feingold, ed., *Jesuit Science and the Republic of Letters* (Cambridge, MA: MIT Press, 2003).

22. For the "Hartlib circle," see Charles Webster, *The Great Instauration: Science, Medicine, and Reform, 1626–1660* (New York: Holmes and Meier, 1976).

23. [Petty], *Advice of W.P. to Mr. Samuel Hartlib*, 5–6.

24. Ibid., 22. For the broader economic discourse, see Steven Pincus, *Protestantism and Patriotism: Ideologies and the Making of English Foreign Policy, 1650–1668* (New York: Cambridge University Press, 1996).

25. [Petty], *Advice of W.P. to Mr. Samuel Hartlib*, 4.

26. See Tony Aspromourgos, "Political Economy, Political Arithmetic and Political Medicine in the Thought of William Petty," in *Physicians and Political Economy: Six Studies of the Work of Doctor-Economists*, ed. Peter D. Groenewegen (London: Routledge, 2001), 10–25.

27. [Petty], *Advice of W.P. to Mr. Samuel Hartlib*, 7.

28. Ibid. On Bacon's *New Atlantis* and attempts to "complete" it, see Adrian Johns, *The Nature of the Book: Print and Knowledge in the Making* (Chicago: University of Chicago Press, 1998), 478–80. For Bacon's text, see Francis Bacon, *"The Advancement of Learning" and "New Atlantis,"* ed. Arthur Johnston (Oxford: Clarendon Press, 1974), 239–45.

29. [Petty], *Advice of W.P. to Mr. Samuel Hartlib*, 8.

30. See Arthur MacGregor, ed., *Tradescant's Rarities: Essays on the Foundation of the Ashmolean Museum* (Oxford: Clarendon Press, 1983).

31. See Jonathan Brown, *Kings and Connoisseurs: Collecting Art in Seventeenth-Century Europe* (Princeton, NJ: Princeton University Press, 1995), esp. 66–93; and David Howarth, *Lord Arundel and His Circle* (New Haven, CT: Yale University Press, 1985).

32. See Jonathan Brown and J. H. Elliott, eds., *The Sale of the Century: Artistic Relations between Spain and Great Britain, 1604–1655* (London: Yale University Press, 2002); and Jerry Brotton, *The Sale of the Late King's Goods: Charles I and His Art Collection* (London: Macmillan, 2006).

33. [Thomas Tenison], *Baconiana; or, Certain Genuine Remains of Sr. Francis Bacon* (London: J. D. for Richard Chiswell, 1679), 57. On the collection and display of the Arundel marbles, see Elizabeth

Angelicoussis, "The Collection of Classical Sculptures of the Earl of Arundel, 'Father of Vertu in England,'" *Journal of the History of Collections* 16, no. 2 (2004): 143–59.

34. On connections between medicine and collecting, see Craig Hanson, *The English Virtuoso: Art, Medicine, and Antiquarianism in the Age of Empiricism* (Chicago: University of Chicago Press, 2009).

35. [Petty], *Advice of W.P. to Mr. Samuel Hartlib*, 10.

36. McCormick, *William Petty and the Ambitions of Political Arithmetic*, 71.

37. See, for example, "Alphabetical Account of the Publick Statues in and about the City," in Edward Hatton, *A New View of London* (London: J. Nicholson, 1708), 2:799–802. On London's growing natural-historical collections, see Carol Gibson-Wood, "Classification and Value in a Seventeenth-Century Museum: William Courten's Collections," *Journal of the History of Collections* 9, no. 1 (1997): 61–77. See chapter 6 for literature on London's expanding art markets and consumer culture.

38. [Ned Ward], *The London-Spy Compleat, in Eighteen-Parts* (London: J. How, 1703), pt. 3, p. 56. Richard Altick, *The Shows of London* (London: Belknap Press of Harvard University Press, 1978), esp. 1–4. For a contextualization of Ward, see Miles Ogborn, *Spaces of Modernity: London's Geographies 1680–1780* (New York: Guilford Press, 1998), esp. 104–14.

39. [Ward], *The London-Spy Compleat*, pt. 3, p. 57.

40. Ibid., pt. 3, pp. 57–58.

41. Ibid., pt. 3, pp. 58–59. See also Susan Stewart, "Death and Life, in That Order, in the Works of Charles Wilson Peale," in *Visual Display: Culture beyond Appearances*, ed. Lynne Cooke and Peter Wollen (Seattle: Bay Press, 1995), 30–53.

42. For a scintillating reading of reanimation in the reception of Michelangelo, see Carolina Mangone, "Becoming Bernini: Imitation, Motion and Identity Formation" (paper circulated at the "Repetition, Emulation and Innovation" workshop at UC Riverside, January 2011).

43. [Ward], *The London-Spy Compleat*, pt. 3, p. 59.

44. Ibid.

45. Altick, *Shows of London*, 2.

46. On "visual quackery," see Barbara Maria Stafford, *Body Criticism: Imaging the Unseen in Enlightenment Art and Medicine* (London: MIT Press, 1991), 362–78.

47. Thomas Sprat, *The History of the Royal-Society of London, For the Improving of Natural Knowledge* (London: T. R. for J. Martyn, 1667), 253.

48. For more on this relocation, see Steven Shapin, "The House of Experiment in Seventeenth-Century England," in *The Science Studies Reader*, ed. Mario Biagioli (New York: Routledge, 1999), 479–504.

49. See A. O. Lovejoy, "Milton and the Paradox of the Fortunate Fall," *English Literary History* 4, no. 3 (September 1937): 161–79.

50. Sprat, *History of the Royal-Society*, 253.

51. For an account of Wren's lost anamorphoses, see Nehemiah Grew, *Musaeum Regalis Societatis; or, A Catalogue and Description of the Natural and Artificial Rarities Belonging to the Royal Society and Preserved at Gresham College* (London: W. Rawlins, 1681), 375–76.

52. Compare Thomas, "'Philosophical Storehouse,'" esp. 7–117; and Lisa Jardine, "Paper Monuments and Learned Societies: Hooke's Royal Society Repository," in *Enlightening the British: Knowledge, Discovery and the Museum in the Eighteenth Century*, ed. R. G. W. Anderson et al. (London: British Museum, 2003), 49–55.

53. Michael Hunter, *Establishing the New Science*, 150.

54. On Grew and his catalogue, see ibid., 261–78. For interesting discussion and further citations on relations between museums and their catalogues, see A. MacGregor, *Curiosity and Enlightenment*, 60–64.

55. Titles from which Grew drew include Michael Rupertus Besler, *Gazophylacium rerum naturalium e regno vegetabili, animali et minerali depromptarum.* . . . (Leipzig, 1642); Benedictus Cerutus, *Musæum Franc. Calceolarii.* . . . (Verona: A. Tamum, 1622); and Olaus Worm, *Museum Wormianum seu historia rerum rariorum* (Lugduni, 1655).

56. Grew, *Musaeum*, 1–10.

57. This inference was certainly encouraged by Sprat, who claims, "This *Repository* he [Robert Hooke] has begun to reduce under its several heads, according to the exact method of the ranks of all the *Species* of *Nature*, which has been compos'd by Doctor *Wilkins*, and will shortly be publish'd in his *Universal Language*"; Sprat, *History of the Royal-Society*, 251. As Hunter notes, however, all evidence of these taxonomies reportedly drafted for the Repository by Fellows such as Hooke and John Aubrey is now lost; see Hunter, *Establishing the New Science*, 140–42.

58. For complaints about Grew's system, see Hunter, *Establishing the New Science*, 145–48.

59. Zacharias Conrad von Uffenbach, *London in 1710: From the Travels of Conrad von Uffenbach*, trans. and ed. W. H. Quarrell and M. Mare (London: Faber and Faber, 1936), 98.

60. For other cases of discrepancy between presentation of artifacts in early modern museums and their catalogues, see Pomian, *Collectors and Curiosities*, 102.

61. Uffenbach, *London in 1710*, 98.

62. On connections between museums and spaces of safekeeping, see Lorraine Daston, "The Factual Sensibility," *Isis* 79, no. 3 (1988): 452–67.

63. For this table, see Huntington Library RB 752249, front matter.

64. Boyle's pump is textually described but not depicted in Grew's catalogue. In personal communication, Michael Hunter has cogently proposed that this drawing is likely based on Boyle's second phase of instrumental design for the air-pump. For a standard account of the evolving form of the air-pump, see George Wilson, "On the Early History of the Air-Pump in England," *Edinburgh New Philosophical Journal* 46 (1848–49): 330–54.

65. Royal Society MS (General) 415/2, fol. 1.

66. See ibid., fol. 17. In his section on human artifacts, Grew mentions only one drawing: "the Draught of another Tooth, taken also out of the Ovary of a Woman," which was given by Dr. Edward Tyson; Grew, *Musaeum*, 8.

67. See MS (General) 415/2, fol. 21.

68. See Brian O'Doherty, *Inside the White Cube: The Ideology of the Gallery Space*, expanded ed. (Los Angeles: University of California Press, 1999).

69. Fabio Rambelli and Eric Reinders, "What Does Iconoclasm Create? What Does Preservation Destroy? Reflections on Iconoclasm in East Asia," in *Iconoclasm: Contested Images, Contested Terms*, ed. Richard Clay and Stacy Boldrick (Aldershot: Ashgate, 2007), 20.

70. A highly impressionistic biography of Papin is Charles-Armand Klein, *Denis Papin illustre savant blaisois* (Chambray: CLD, 1987).

71. Birch 3:486.

72. Ibid.

73. Ibid., 3:501.

74. See Denis Papin, *A New Digester or Engine for Softning Bones* (London: J. M. for Henry Bonwicke, 1681), 6.

75. Birch 3:487.

76. See, respectively, Birch 3:487, 489, 492. For a brilliant account of a Restoration aristocrat's palate, see Jane Stevenson and Peter Davidson, eds., *The Closet of the Eminently Learned Sir Kenelme Digbie Kt. Opened* (Totnes, Devon: Prospect Books, 1997).

77. It is true that the Digester does not appear in Grew's *Musaeum*, which was published in 1681. However, as Michael Hunter has demonstrated, Grew's text had actually been complete in the spring of 1679, nearly two years before Papin entrusted his instrument to the Royal Society; see Hunter, *Establishing the New Science*, 142.

78. Birch 4:72.

79. See ibid., 4:74, 77, 87, 88.

80. February 23, 1681; ibid., 4:71.

81. See Horst Bredekamp, *The Lure of Antiquity and the Cult of the Machine: The Kunstkammer and the Evolution of Nature, Art and the Machine*, trans. Allison Brown (Princeton, NJ: Markus Wiener, 1995).

82. Birch 3:73.

83. Ibid. For an image of this device, see Jacob Leupold, *Theatrum Arithmetico-Geometricum, Das ist: Schau-Platz der Rechen- und Meß-Kunst* (Leipzig: Christoph Zunkel, 1727), tab. 8. More broadly, see Ernst Martin, *The Calculating Machines (Die Rechenmaschinen): Their History and Development*, trans. and ed. Peggy Aldrich Kidwell and Michael R. Williams (Cambridge, MA: MIT Press, 1992), esp. 38–42.

84. G. W. Leibniz to Henry Oldenburg, March 8, 1673, in *Correspondence of Henry Oldenburg*, 9:493.

85. Ibid.

86. For a defense of reverse engineering's legitimate place in science, see Kathryn A. Ingle, *Reverse Engineering* (New York: McGraw-Hill, 1994).

87. For an excellent exposition of these points, see Rob Iliffe, "'In the Warehouse': Privacy, Property and Priority in the Early Royal Society," *History of Science* 30 (1992): 29–68, esp. 37–39.

88. *Correspondence of Henry Oldenburg*, 9:493.

89. Birch 3:75.

90. *Correspondence of Henry Oldenburg*, 9:493. Leibniz himself was subsequently embroiled in accusations of theft by Newton and his followers; see Mordechai Feingold, *The Newtonian Moment: Isaac Newton and the Making of Modern Culture* (New York: Oxford University Press, 2004), esp. 33–51.

91. For this provenance, compare Robert Hooke, *Micrographia; or, Some Physiological Descriptions of Minute Bodies made by Magnifying Glass* (London: John Martyn and James Allestry, 1665), 105–6; and Hooke, *Lectures and Discourse of Earthquakes*, in *The Posthumous Works of Robert Hooke*, ed. Richard Waller (London: Sam. Smith and Benj. Walford, 1705), 339.

92. See Andrew C. Scott and David Freedberg, *Fossil Woods and Other Geological Specimens: The Paper Museum of Cassiano dal Pozzo*, Series B: Natural History, pt. 3 (Turnhout, Belgium: Harvey Miller, 2000); and David Freedberg, *The Eye of the Lynx: Galileo, His Friends, and the Beginnings of Modern Natural History* (Chicago: University of Chicago Press, 2002).

93. For more on fossil specimens in the Repository, see Grew, *Musaeum*, 269–70.

94. Ibid., 253. On fossilization theories more broadly, see William R. Newman, *Promethan Ambitions: Alchemy and the Quest to Perfect Nature* (Chicago: University of Chicago Press, 2004), esp. 145–63.

95. See Hooke, *Micrographia*, 106.

96. Royal Society Classified Papers 20:50, fol. 102.

97. Ibid.

98. Ibid.

99. On the connections between Hooke's engagements with fossils and his controversial theories of the earth's history, see Ellen Tan Drake, *Restless Genius: Robert Hooke and His Earthly Thoughts* (New York: Oxford University Press, 1996); and Paolo Rossi, *The Dark Abyss of Time: The History of the Earth and the History of Nations from Hooke to Vico*, trans. L. Cochrane (Chicago: University of Chicago Press, 1984), esp. 12–17.

100. August 2, 1678; Birch 3:430.

101. Grew, *Musaeum*, 380.

102. Birch 3:430.

103. Ibid., 4:278. On Lister and glass-making, see Anna Marie Roos, "A Speculum of Chymical Practice: Isaac Newton, Martin Lister (1639–1712), and the Making of Telescopic Mirrors," *Notes and Records of the Royal Society of London* 64, no. 2 (June 2010): 105–20.

104. Birch 4:279.

105. Ibid.

106. Ibid.

107. See Michele and Robert Root-Bernstein, "Thinkering," *Psychology Today* (August 21, 2008): http://www.psychologytoday.com/blog/imagine/200808/thinkering.

108. Zacharias Conrad von Uffenbach, *Oxford in 1710 from the Travels of Zacharias Conrad von Uffenbach*, ed. and trans. W. H. Quarrell and W. J. C. Quarrell (Oxford: Basil Blackwell, 1928), 31.

109. For an argument effectively along these lines, see Ronald Paulson, *Breaking and Remaking: Aesthetic Practice in England, 1700–1820* (New Brunswick, NJ: Rutgers University Press, 1989).

110. Uffenbach, *Oxford in 1710*, 31.

111. Ibid., 101.

112. On these techniques, see Harold J. Cook, "Time's Bodies: Crafting the Preparation and Preservation of Naturalia," in *Merchants and Marvels: Commerce, Science and Art in Early Modern Europe*, ed. Pamela H. Smith and Paula Findlen (New York: Routledge, 2002), 223–47.

113. Boyle, *Some Considerations*, in *Works of Robert Boyle*, 3:308.

114. Ibid., 307.

115. Other recent studies that have connected Hooke's 1682 lecture to the Repository at Gresham College include Nick Wilding, "Graphic Technologies," in *Robert Hooke: Tercentennial Studies*, ed. Michael Cooper and Michael Hunter (Aldershot: Ashgate, 2006), 123–34; Ken Arnold, *Cabinets for the Curious: Looking Back at Early English Museums* (Aldershot: Ashgate, 2006), esp. 195–98; and Richard Yeo, "Before Memex: Robert Hooke, John Locke, and Vannevar Bush on External Memory," *Science in Context* 20, no. 1 (2007): 21–47.

116. For a transcription (which I have followed here) and fuller contextualization of this lecture, see D. R. Oldroyd, "Some 'Philosophical Scribbles' Attributed to Robert Hooke," *Notes and Records of the Royal Society of London* 35 (1980): 17–32. On the early modern imagination more broadly, see Ioan P.

Couliano, *Eros and Magic in the Renaissance*, trans. Margaret Cook (Chicago: University of Chicago Press, 1987); Stuart Clark, *Vanities of the Eye: Vision in Early Modern European Culture* (New York: Oxford University Press, 2007); Claudia Swan, "Eyes Wide Shut: Early Modern Imagination, Demonology, and the Visual Arts," *Zeitsprünge: Forschungen zur Frühen Neuzeit* 7 (2003): 156–81; and Krzysztof Pomian, "Vision and Cognition," in *Picturing Science, Producing Art*, ed. Peter Galison and Caroline Jones (New York: Routledge, 1998), 211–31.

117. Oldroyd, "Some 'Philosophical Scribbles,'" 17.

118. Ibid.

119. Ibid., 18.

120. Ibid., 17.

121. Ibid., 18. For more on these arguments, see Matthew C. Hunter, "The Theory of the Impression according to Robert Hooke," in Michael Hunter, *Printed Images*, 167–90.

122. Classic studies of this theme include Richard Rorty, *Philosophy and the Mirror of Nature* (Princeton, NJ: Princeton University Press, 1979); and Jonathan Crary, *Techniques of the Observer: On Vision and Modernity in the Nineteenth Century* (Cambridge, MA: MIT Press, 1990).

123. Although Locke's *Essay* was not published until 1690, it had circulated since the early 1670s. Relations between Locke and Hooke (who both simultaneously attended Westminster School and Christ's Church College, Oxford) deserve much fuller examination than I can give them here. For a brief note in that direction, see Brian Singer, "Robert Hooke on Memory, Association and Time Perception," *Notes and Records of the Royal Society of London* 31, no. 1 (1976): 115–31.

124. John Locke, *An Essay concerning Human Understanding*, ed. P. H. Nidditch (Oxford: Clarendon Press, 1975), bk. 2, chap. 11, §17, p. 163.

125. Ibid.

126. Hooke's "long discourse, being the substance of three lectures" was first delivered on June 21, 1682, and then read again on June 28, 1682; see Birch 4:153–54.

127. Robert Hooke, "Sect. VII," *Lectures of Light*, in Waller, *Posthumous Works of Robert Hooke*, 146.

128. Ibid., 142. An average fifty-year-old will form nearly two billion ideas; ibid., 143.

129. Ibid., 140.

130. Ibid., 141.

131. Recent readings include Arnold, *Cabinets for the Curious*; Wilding, "Graphic Technologies"; Lotte Mulligan, "Robert Hooke's 'Memoranda': Memory and Natural History," *Annals of Science* 49 (1992): 47–61; and Christine Stevenson, "Robert Hooke, Monuments and Memory," *Art History* 28, no. 1 (February 2005): 43–73.

132. Creating a "repository" was standard practice in the early modern technique of artificial memory, which certainly persisted into the Restoration. For a period example including a depiction of this repository, see John Willis, *Mnemonica; or, The Art of Memory* (London: L. Sowersby, 1661), esp. 52–134. More broadly, see Frances Yates, *The Art of Memory* (Chicago: University of Chicago Press, 1966).

133. Compare Douwe Draaisma, "Hooke on Memory and the Memory of Hooke," in Cooper and Hunter, *Robert Hooke: Tercentennial Studies*, 111–21; John Sutton, *Philosophy and Memory Traces: Descartes to Connectionism* (New York : Cambridge University Press, 1998), esp. 129–48; and Jamie C. Kassler, *Inner Music: Hobbes, Hooke and North on Internal Character* (London: Athlone, 1995), esp. 129–59.

134. Hooke, "Sect. VII," 139–40.

135. Ibid., 141. Famously and controversially, Descartes had seated the soul in the pineal gland; see Réne Decartes, *The Passions of the Soul*, in *Descartes: Selected Philosophical Writings*, ed. and trans. John Cottingham and Dugald Murdoch (New York: Cambridge University Press, 1988), 218–38.

136. Hooke, "Sect. VII," 141.

137. Ibid.

138. See Jan Golinski, "A Noble Spectacle: Phosphorous and the Public Cultures of Science in the Early Royal Society," *Isis* 80, no. 1 (1989): 11–39. For phosphorous in the Repository, see Grew, *Musaeum*, 353–57. For Hooke's recipes for making phosphorous, see Robert Hooke, *Philosophical Experiments and Observations of the Late Eminent Dr. Robert Hooke*, ed. William Derham (London: W. and J. Innys, 1726), 174–83. For a tale of rubbing phosphorous on the face of a leading experimentalist published by Hooke, see [Frederick Slare], "An Account of Several Experiments made with the Shining Substance of the Liquid and of the Solid Phosphorous. . . . ," *Philosophical Collections* 3 (December 10, 1681): 48–50.

139. Hooke, "Sect. VII," 141.

140. Thomas Willis, *Two Discourses concerning the Soul of Brutes* [1683], trans. S. Pordage and introduced by S. Diamond (Gainesville, FL: Scholars' Facsimiles and Reprints, 1971), 24–25. On the personal importance of Willis's *Two Discourses* to Hooke, see Lisa Jardine, *The Curious Life of Robert Hooke: The Man Who Measured London* (London: Harper Perennial, 2003), 73–74.

141. Willis, *Two Discourses*, 25. On connections between Willis and Locke, see G. S. Rousseau, "Nerves, Spirits, and Fibres: Towards Defining the Origins of Sensibility," *Studies in Eighteenth Century Culture* 3 (1976): 137–57.

142. Willis, *Two Discourses*, 25.

143. Hooke's lecture thus utilized a distinction effectively effaced by Locke. Writing fifty years later, David Hume explained that by insisting on a differentiation between impressions and ideas, he aimed to "restore the word, idea, to its original sense, from which Mr. *Locke* had perverted it, in making it stand for all our perceptions"; David Hume, *A Treatise of Human Nature* [1739], ed. L. A. Shelby-Bigge (Oxford: Clarendon Press, 1896), 1, 2. On Locke's notoriously confusing conception of ideas, see Keith Allen, "Locke and the Nature of Ideas," *Archiv für Geschichte der Philosophie* 92, no. 3 (2010): 236–55.

144. Here and elsewhere in the lecture, Hooke demonstrates his familiarity with a broad range of alchemical ideas. For an interesting examination of the influence of the alchemical tradition on later seventeenth-century theories of mind, see Antonio Clericuzio, "The Internal Laboratory: The Chemical Reinterpretation of Medical Spirits in England (1650–1680)," in *Alchemy and Chemistry in the 16th and 17th Centuries*, ed. Piyo Rattansi and Antonio Clericuzio (Boston: Kluwer, 2002), 51–83.

145. Hooke, "Sect. VII," 146.

146. At the lecture's second reading at the Royal Society in June 1682, Hooke had to answer the objection "that this discourse seemed to tend to prove the soul mechanical"; Birch 4:154. See Oldroyd, "Some 'Philosophical Scribbles,'" 21–23; and, more broadly, Simon Schaffer, "Godly Men and Mechanical Philosophers: Souls and Spirits in Restoration Natural Philosophy," *Science in Context* 1, no. 1 (1987): 55–85.

147. On Willis's sources, see John P. Wright, "Locke, Willis, and the Seventeenth-Century Epicurean Soul," in *Atoms, Pneuma, and Tranquility: Epicurean and Stoic Themes in European Thought*, ed. M. J. Osler (New York: Cambridge University Press, 1991), 239–58.

148. Willis, *Two Discourses*, 41.

149. Ibid.

150. Carl Zimmer, *Soul Made Flesh: The Discovery of the Brain—and How It Changed the World* (New York: Free Press, 2004), 228.

151. Robert G. Frank Jr., "Thomas Willis and His Circle: Brain and Mind in Seventeenth-Century Medicine," in *The Languages of Psyche: Mind and Body in Enlightenment Thought*, ed. G. S. Rousseau (Berkeley: University of California Press, 1990), 146.

152. Hooke, *Micrographia*, sig. Biii.

153. Hooke, "Sect. VII," 147.

154. Ibid.

155. Hooke, "Sect. VII," 145.

156. See *The Diary of Robert Hooke, 1672–1680*, ed. H. W. Robinson and W. Adams (London: Taylor and Francis, 1935), 147. On the dealer's identity, see Felicity Henderson, "Unpublished Material from the Memorandum Book of Robert Hooke, Guildhall Library MS 1758," *Notes and Records of the Royal Society of London* 61 (2007): 129–75, esp. 164. For more on this commission, see Allison Stoesser, "Robert Hooke's Montagu House," in Cooper and Hunter, *Robert Hooke: Tercentennial Studies*, 165–80.

157. Hooke, *Diary*, 149.

158. On Dee, see William H. Sherman, *John Dee: The Politics of Reading and Writing in the English Renaissance* (Amherst: University of Massachusetts Press, 1995); and Deborah E. Harkness, "Managing an Experimental Household: The Dees of Mortlake and the Practice of Natural Philosophy," *Isis* 88, no. 2 (1997): 247–62.

159. Robert Hooke, "Of Dr. Dee's Book of Spirits," in Waller, *Posthumous Works of Robert Hooke*, 204.

160. Peter J. French, *John Dee: The World of an Elizabethan Magus* (London: Routledge and Kegan Paul, 1972), 167.

161. John Dee, "Mathematicall Praeface," in *The Elements of Geometrie of the Most Auncient Philosopher Euclide of Megara*, trans. Henry Billingsley (London: John Daye, 1570), sig. IJ.

162. Ibid., * verso.

163. Nicholas H. Clulee, *John Dee's Natural Philosophy: Between Science and Religion* (New York: Routledge and Kegan Paul, 1988), 160.

164. Compare James S. Ackerman, "Architectural Practice in the Italian Renaissance," *Journal of the Society of Architectural Historians* 13, no. 3 (October 1954): 3–11; Martha Hollingsworth, "The Architect in Fifteenth-Century Florence," *Art History* 7 (December 1984): 385–410; and Robert Williams, *Art, Theory, and Culture in Sixteenth-Century Italy: From Techne to Metatechne* (New York: Cambridge University Press, 1997).

165. Dee, "Mathematicall Praeface," sig. D.iij. verso–D.iiij; compare Vitruvius, *Ten Books on Architecture*, trans. Ingrid Rowland (New York: Cambridge University Press, 1999), bk. 1, chap. 1, pp. 21–24.

166. Dee, "Mathematicall Praeface," sig. D.iiij; compare Leon Battista Alberti, "Prologue," in *On the Art of Building*, trans. Joseph Rykwert et al. (Cambridge, MA: MIT Press, 1991), 3.

167. Dee, "Mathematicall Praeface," sig. D.iiij.–D.iiij.v; compare Plato, *The Statesman*, trans. H. N. Fowler (New York: Putnam, 1925), 259 E.

168. Christopher Wren, "Letter to a Friend from Paris (late September/October 1665)," in *Wren's "Tracts" on Architecture and Other Writings*, ed. Lydia Soo (New York: Cambridge University Press, 1998), 105.

169. See Anthony Geraghty, "Robert Hooke's Collection of Architectural Books and Prints," *Architectural History* 47 (2004): 113–25.

170. On the broader ambit of this model of the architect, see Simon Varey, *Space and the Eighteenth-Century English Novel* (New York: Cambridge University Press, 1990), esp. 42–61.

171. Moxon not only advertised his globes in Hooke's *Philosophical Collections* 1 (1679) but then petitioned him for the post of printer to the Royal Society less than a week after the death of John Martyn; see Hooke, *Diary*, 448.

172. Joseph Moxon, *Mechanick Exercises on the Whole Art of Printing* [1683–84], ed. H. Davis and H. Carter (London: Oxford University Press, 1962), 10–11.

173. Ibid., 11.

174. Ibid., 11–12. On Moxon's use of Dee, see Johns, *Nature of the Book*, 79–82. More broadly, see Hentie Louw, "The 'Mechanick Artist' in Late Seventeenth Century English and French Architecture: The Work of Robert Hooke, Christopher Wren and Claude Perrault Compared as Products of an Interactive Science/Architecture Relationship," in Cooper and Hunter, *Robert Hooke: Tercentennial Studies*, 181–99.

175. Hooke, "Sect. VII," 147.

176. Grew, *Musaeum*, unpaginated.

177. See Edwin Hutchins, "The Social Organization of Distributed Cognition," in *Perspectives on Socially Shared Cognition*, ed. L. B. Resnick et al. (Washington, DC: American Psychological Association, 1991), 283–307.

178. Hooke, *Micrographia*, sig. Biv.

179. Hooke, "Sect. VII," 146.

CHAPTER SIX

1. See Norman H. Robinson, *The Royal Society Catalogue of Portraits* (London: Royal Society, 1980), 332. For the donation of this portrait by Wren's grandson, see Royal Society Journal Book (Original), vol. 21 (1748–51), fols. 310–11.

2. For this attribution, see Malcolm Rogers, *John Closterman: Master of the English Baroque 1660–1711* (London: National Portrait Gallery, 1981), 11.

3. For relevant drawings, see Anthony Geraghty, *The Architectural Drawings of Sir Christopher Wren at All Souls College, Oxford: A Complete Catalogue* (Burlington, VT: Lund Humphries, 2007), 68–69.

4. A defense of the west portico and its classical precedents is outlined in *Parentalia*, the volume assembled by Christopher Wren Jr. and published by Stephen Wren, the architect's grandson. See Christopher Wren Jr., *Parentalia; or, Memoirs of the Family of the Wrens* (London: T. Osborn and R. Dodsley, 1750), 287–89.

5. Robert Hooke, *Micrographia; or, Some Physiological Descriptions of Minute Bodies made by Magnifying Glass* (London: John Martyn and James Allestry, 1665), sig. Giv.

6. Thomas Willis, *The Anatomy of the Brain*, in *The Remaining Medical Works of that Famous and Renowned Physician Dr. Thomas Willis* (London: T. Dring, 1681), 78. See also Martin Kemp and Nathan Flis, "Mapping the Cerebral Globe," *Nature* 456, no. 7219 (November 2008): 174.

7. David Abercromby, MD, *A Discourse of Wit* (London: John Weld, 1685), 69. On Abercromby's

relations with Boyle, see Edward B. Davis, "The Anonymous Works of Robert Boyle and the *Reasons Why a Protestant Should not Turn Papist* (1687)," *Journal of the History of Ideas* 55, no. 4 (1994): 611–29.

8. Abercromby, *Discourse of Wit*, 73.

9. A rich exposition of this point is Simon Varey, *Space and the Eighteenth-Century English Novel* (New York: Cambridge University Press, 1990), esp. 42–61.

10. Note, for example, John Locke's claim that in an age of "Master-Builders" such as Robert Boyle and Isaac Newton, he hoped but "to be employed as an Under-Labourer in clearing Ground a little, and removing some of the Rubbish, that lies in the way to Knowledge"; John Locke, "The Epistle to the Reader," in *An Essay concerning Human Understanding*, ed. P. H. Nidditch (Oxford: Clarendon Press, 1975), 9–10. On art academies and markets, compare Ilaria Bignamini, "George Vertue, Art Historian, and Art Institutions in London, 1689–1768," *Walpole Society* 54 (1988): 1–148; and Brian Cowan, "Art and Connoisseurship in the Auction Market of Later Seventeenth-Century London," in *Mapping Markets for Paintings in Europe, 1450–1750*, ed. Neil de Marchi and Hans J. Van Miegroet (Turnhout, Belgium: Brepols, 2006), 263–84.

11. Anonymous, *Abridgment of the Architecture of Vitruvius* (London: Abel Swall and T. Child, 1692), 5. On broader discourses of architecture and knowledge, compare Claudia Brodsky Lacour, *Lines of Thought: Discourse, Architectonics and the Origin of Modern Philosophy* (Durham, NC: Duke University Press, 1996); and J. A. Bennett and Scott Mandelbrote, *The Garden, the Ark, the Tower, the Temple: Biblical Metaphors of Knowledge in Early Modern Europe* (Oxford: Museum of the History of Science, 1998).

12. Anonymous, *Abridgment*, 5.

13. For these styles, see John Summerson, *Architecture in Britain 1530–1830* (New Haven, CT: Yale University Press, 1953), 142–56; and Timothy Mowl and Brian Earnshaw, *Architecture without Kings: The Rise of Puritan Classicism under Cromwell* (New York: Manchester University Press, 1995).

14. See Antoine Picon, *French Architects and Engineers in the Age of Enlightenment*, trans. M. Thom (New York: Cambridge University Press, 1992).

15. For the king's satire, see the entry for February 1, 1664, in *The Diary of Samuel Pepys*, ed. Robert Latham and William Matthews (Los Angeles: University of California Press, 1971), 5:32–33.

16. Thomas Brown, *Amusements Serious and Comical, calculated for the Meridian of London* (London, 1700), 86–87.

17. John Summerson, "The Tyranny of the Intellect," *Journal of the Royal Institute of British Architects* (February 20, 1937): 390.

18. For this assessment, see Hentie Louw, "The 'Mechanick Artist' in Late Seventeenth Century English and French Architecture: The Work of Robert Hooke, Christopher Wren and Claude Perrault Compared as Products of an Interactive Science/Architecture Relationship," in *Robert Hooke: Tercentennial Studies*, ed. Michael Cooper and Michael Hunter (Aldershot: Ashgate, 2006), 186.

19. For example, see Joseph Rykwert, *The First Moderns: The Architects of the Eighteenth Century* (Cambridge, MA: MIT Press, 1980), esp. 134–51; and Vaughn Hart, *St. Paul's Cathedral: Sir Christopher Wren* (London: Phaidon, 1995).

20. The most prominent exponent of this position has been J. A. Bennett; see especially J. A. Bennett, "Christopher Wren: The Natural Causes of Beauty," *Architectural History* 15 (1972): 5–22; Bennett, "Christopher Wren: Astronomy, Architecture, and the Mathematical Sciences," *Journal for the History of Astronomy* 6 (1975): 149–84; and Bennett, *The Mathematical Science of Christopher Wren* (New York: Cambridge University Press, 1982). Most recently, see Anthony Gerbino and Stephen Johnston, *Compass*

and Rule: Architecture as Mathematical Practice in England 1600–1750 (New Haven, CT: Yale University Press, 2009); and Matthew Walker, "*Architectus Ingenio*: Robert Hooke, the Early Royal Society, and the Practices of Architecture" (PhD diss., York University, 2009).

21. See, for example, Lisa Jardine, "Monuments and Microscopes: Scientific Thinking on a Grand Scale in the Early Royal Society," *Notes and Records of the Royal Society of London* 55, no. 2 (2001): 289–308; and James W. P. Campbell, *Building St. Paul's* (New York: Thames and Hudson, 2007). For a dissenting view of Hooke and Wren as collaborative partners, see Matthew Walker, "The Limits of Collaboration: Robert Hooke, Christopher Wren and the Designing of the Monument to the Great Fire of London," *Notes and Records of the Royal Society* 65, no. 2 (2011): 121–43.

22. Many archival documents and images relating to Wren's architecture were edited and published by A. T. Bolten and H. D. Hendry between 1923 and 1943 in volumes henceforth referred to as *Wren Society*. See Christopher Wren, "Report on Old St. Paul's After the Fire," *Wren Society* 13 (1936): 21.

23. Christopher Wren to William Sancroft, April 28, 1668; *Wren Society* 13:46.

24. Such traffic between buildings and brains was not missed by Robert Boyle's friend David Abercromby. Tall men, he argues, are often less clever than those of average heights, just as "in High Houses, the uppermost Room is commonly the worst furnished"; Abercromby, *Discourse of Wit*, 82.

25. On the persistence of this tendency in the early modern British tradition, see Susan Scott Parrish, *American Curiosity: Cultures of Natural History in the Colonial British Atlantic World* (Chapel Hill, NC: University of North Carolina Press, 2006).

26. See Germaine Aujac, "The Foundations of Theoretical Cartography in Archaic and Classical Greece," in *The History of Cartography*, ed. J. B. Harley and D. Woodward, vol. 1, *Cartography in Prehistoric, Ancient, and Medieval Europe and the Mediterranean* (Chicago: University of Chicago Press, 1987), 130–48.

27. Pliny the Elder, *Natural History*, trans. H. Rackham, Loeb Classical Library (Cambridge, MA: Harvard University Press, 1938), vol. 1, bk. 2, chap. 80, pp. 321–23. Ideas like these were completely conventional in Restoration London. The Lord Mayor's Day parade of 1677 included a pageant featuring "two young Persons of different Climats, the one an *European*, the other an *Indian*; one white, the other black"; see Thomas Jordan, *Londons Triumphs: Illustrated with many Magnificent Structures & Pageants* (London: Printed for John Playford, 1677), 7.

28. For a magisterial treatment of this tradition of climatic geography, see Nicolás Wey Gómez, *The Tropics of Empire: Why Columbus Sailed South to the Indies* (Cambridge, MA: MIT Press, 2008).

29. Hooke, *Micrographia*, 206.

30. See Thomas Sprat, *The History of the Royal-Society of London, For the Improving of Natural Knowledge* (London: T. R. for J. Martyn, 1667), 5–7.

31. For a scintillating reading of the Zeuxis legend in the early modern period, see Leonard Barkan, "The Heritage of Zeuxis: Painting, Rhetoric, and History," in *Antiquity and Its Interpreters*, ed. Alina Payne et al. (New York: Cambridge University Press, 2000), 99–109; Sprat, *History of the Royal-Society*, 64.

32. See Vitruvius, *Ten Books on Architecture*, trans. Ingrid Rowland (New York: Cambridge University Press, 1999), bk. 6, chap. 1, pp. 76–78.

33. Sprat, *History of the Royal-Society*, 114.

34. On this confluence of Dutch science and commerce, see Harold C. Cook, *Matters of Exchange: Commerce, Medicine and Science in the Dutch Golden Age* (New Haven, CT: Yale University Press, 2007).

On the Anglo-Dutch Wars, see Steven Pincus, *Protestantism and Patriotism: Ideologies and the Making of English Foreign Policy, 1650–1668* (New York: Cambridge University Press, 1996).

35. Vanessa Harding, "City, Capital and Metropolis: The Changing Shape of Seventeenth-Century London," in *Imagining Early Modern London: Perceptions and Portrayals of the City from Stow to Strype, 1598–1720*, ed. J. F. Merritt (New York: Cambridge University Press, 2001), 128.

36. See Peter Borsay, *The English Urban Renaissance: Culture and Society in the Provincial Town, 1660–1770* (New York: Oxford University Press, 1989).

37. T. Brown, *Amusements*, 23.

38. On this point, see Steven Zwicker, *Lines of Authority: Politics and English Literary Culture, 1649–1689* (Ithaca, NY: Cornell University Press, 1993), esp. 17–36.

39. Abraham Cowley, "Of Wit," in *The Works of Mr. Abraham Cowley* (London: J. M. for Henry Herringman, 1668), 4.

40. See Thomas Hobbes, *Leviathan*, ed. R. Tuck (New York: Cambridge University Press, 1991), pt. 1, chap. 8, pp. 50–51.

41. Anonymous, *The Polite Gentleman; or, Reflections Upon the Several Kinds of Wit* (London: R. Basset, 1700), 7.

42. For this assessment, see John H. Wilson, *The Court Wits of the Restoration: An Introduction* (New York: Octagon Books, 1967), 108.

43. Thomas Sprat to Christopher Wren, "Recital of a mutual Discourse on the Subject of the Wit of Conversation. 1663," in Wren Jr., *Parentalia*, 258.

44. Ibid., 259.

45. More broadly, see J. Douglas Canfield, "Women's Wit: Subversive Women Tricksters in Restoration Comedy," in *The Restoration Mind*, ed. W. G. Marshall (Newark: University of Delaware Press, 1997), 147–76.

46. On the persistence of occupation-based neighborhoods, see M. J. Power, "The Social Topography of Restoration London," in *London 1500–1700: The Making of the Metropolis*, ed. A. C. Beier and R. Findlay (New York: Longman, 1986), 199–223.

47. Sprat to Wren, "Recital of a mutual Discourse," 258.

48. See Michelle O'Callaghan, *The English Wits: Literature and Sociability in Early Modern England* (New York: Cambridge University Press, 2007).

49. [Joseph Addison], *The Spectator* 403 (Thursday, June 12, 1712), in *The Spectator*, ed. G. A. Aitken (London: John C. Nimmo, 1898), 6:37 (my emphasis).

50. A definitive study is Brian Cowan, *The Social Life of Coffee: The Emergence of the British Coffeehouse* (New Haven, CT: Yale University Press, 2005). On the importance of the coffee shop in Hooke's circles, see especially Rob Iliffe, "Material Doubts: Hooke, Artisan Culture and the Exchange of Information in 1670s London," *British Journal of the History of Science* 28 (1995): 285–318.

51. John Aubrey, *Brief Lives*, ed. J. Buchanan-Brown (New York: Penguin, 2000), 3.

52. See Steven Pincus, "'Coffee Politicians Does Create': Coffeehouses and Restoration Political Culture," *Journal of Modern History* 67 (December 1995): 807–34. On spatial layout, see Cowan, *Social Life of Coffee*, 79–88.

53. Anonymous, *Coffee-houses Vindicated in Answer to the late Published Character of a Coffee-house* (London: J. Lock, 1675), 4.

54. Ibid.

55. See Cowan, *Social Life of Coffee*, esp. 89–99.

56. See, for example, J. D., *A Word Without Doors Concerning the Bill for Succession* (London, 1679); and *A Coffee-House Dialogue; or, A Discourse between Captain Y—— and a Young Barrester of the Middle-Temple, with some Reflections upon the Bill against the D. of Y.* (London, 1679).

57. Anonymous, *Knavery in All Trades; or, The Coffee-House, A Comedy* (London: J. B. for W. Gilbertson, 1664), sig. D2v. For more on this play and coffeehouse culture in general, see Joseph Monteyne, *The Printed Image in Early Modern London: Urban Space, Visual Representation, and Social Exchange* (Burlington, VT: Ashgate, 2007), esp. 31–72.

58. Anonymous, *The Character of a Coffee-House, with the Symptomes of a Town-Wit* (London: Jonathan Edwin, 1673), 4.

59. Ibid., 3.

60. See *The Diary of Robert Hooke, 1672–1680*, ed. H. W. Robinson and W. Adams (London: Taylor and Francis, 1935), 199.

61. Sprat, *History of the Royal-Society*, 88.

62. Ibid., 19.

63. Ibid., 88. On the discourse of gentility implicit in Sprat's claims, see John Barrell, "The Public Prospect and the Private View: The Politics of Taste in Eighteenth-Century Britain," in *The Birth of Pandora and the Division of Knowledge* (Philadelphia: University of Pennsylvania Press, 1992), 41–61.

64. Hooke, *Micrographia*, sig. Gii.

65. Sprat, *History of the Royal-Society*, 89.

66. John Evelyn, *Fumifugium; or, The Inconveniencie of the Aer and Smoak of London Dissipated* (London: W. Godbid, 1661), sig. A2.

67. See Jane Lang, *Rebuilding St. Paul's after the Great Fire of London* (New York: Oxford University Press, 1956), 5–6.

68. Anonymous, *A Character of England, As it was lately presented to a Noble Man of France* (London: Jo. Crooke, 1659), 11–12.

69. Ibid., 12.

70. See Valentine Knight, *Proposals of a New Model for Rebuilding the City of London, with Houses, Streets and Wharfs, to be forthwith set out by His Majesty's and the City Surveyors* (London, 1666).

71. See Geraghty, *Architectural Drawings*, 254–56. On opticality in Wren's architecture more broadly, see Lydia Soo, ed., *Wren's "Tracts" on Architecture and Other Writings* (New York: Cambridge University Press, 1998), esp. 199–201.

72. Wren Jr., *Parentalia*, 268.

73. For these competing sensorial urbanisms, see Marvin Trachtenberg, *Dominion of the Eye: Urbanism, Art and Power in Early Modern Florence* (New York: Cambridge University Press, 1997); and Alain Corbin, *The Foul and the Fragrant: Odor and the French Social Imagination*, trans. M. L. Kochan et al. (Cambridge, MA: Harvard University Press, 1986).

74. Evelyn, *Fumifugium*, sig. A1v.

75. Ibid., fol. 6.

76. John Evelyn, *London Revived: Considerations for Its Rebuilding in 1666*, ed. E. S. de Beer (Oxford: Clarendon Press, 1938), 35–36.

77. Ibid., 36.

78. Ibid., 40.

79. Michael Cooper convincingly argues against the traditional claim that Hooke's city plan is represented in a Dutch engraving of London from 1666 by Marcel Willemz Doornick; see Michael Cooper, *"A More Beautiful City": Robert Hooke and the Rebuilding of London after the Great Fire* (Thrupp: Sutton, 2003), 111–14.

80. Richard Waller, "The Life of Dr. Robert Hooke," in *The Posthumous Works of Robert Hooke*, ed. Richard Waller (London: Sam. Smith and Benj. Walford, 1705), xiii.

81. On the older tradition, see Maurice Howard, "The Ideal House and Healthy Life: The Origins of Architectural Theory in England," in *Les Traités d'architecture de la Renaissance*, ed. Jean Guillaume (Paris: Picard, 1988), 425–33.

82. For classical precedents, see Vitruvius, *Ten Books on Architecture*, bk. 6, chap. 1, pp. 26–31, 101.

83. Christine Stevenson, "Robert Hooke's Bethlem," *Journal of the Society of Architectural Historians* 55, no. 3 (September 1996): 256. See also M. I. Batten, "The Architecture of Dr. Robert Hooke FRS," *Walpole Society* 25 (1936–37): 83–113.

84. For more on Lodwick, see William Poole, *Francis Lodwick (1619–1694), "A Country Not Named" (MS. Sloane 913, Fols. 1R-33R): An Edition with an Annotated Primary Bibliography and an Introductory Essay on Lodwick and His Intellectual Context* (Tempe, AZ: ACMRS, 2007). More broadly, see Mireille Galinou, ed., *City Merchants and the Arts, 1670–1720* (Wetherby: Oblong for the Corporation of London, 2004).

85. Citations of this plan come from "Mr. Lodowick's Proposals for the Rebuilding of the City," Royal Society Classified Papers (henceforth RSCP), vol. 17, fols. 7–8.

86. Peter Borsay, "Early Modern Urban Landscapes, 1540–1800," in *The English Urban Landscape*, ed. P. J. Waller (New York: Oxford University Press, 2000), 99–124.

87. On the surveying of the City, see Cooper, *"A More Beautiful City."*

88. A brilliant account of these developments is Elizabeth McKellar, *The Birth of Modern London: The Development and Design of the City 1660–1720* (New York: Manchester University Press, 1999), esp. 18–37.

89. Ibid., 184.

90. Nicholas Hawksmoor to Dr. George Clarke, February 17, 1715; reproduced in Kerry Downes, *Hawksmoor* (London: A. Zwemmer, 1959), 241.

91. Ibid., 242.

92. Nicholas Barbon, *An Apology for the Builder; or, A Discourse Shewing the Causes and Effects of the Increase of Building* (London: Cave Pullen, 1685), 2.

93. Ibid., 22.

94. Ibid., 23–24. For a useful overview of literature on anxieties about surplus populations in early modern England, see Ronald Hutton, *The Rise and Fall of Merry England: The Ritual Year 1400–1700* (New York: Oxford University Press, 1996).

95. For the causal relationship between housing expansion and population growth that Barbon traces via the techniques of statistical demographics pioneered by Captain John Graunt and William Petty, see Barbon, *An Apology*, 26–35.

96. A programmatic statement with citation of Barbon is Neil McKendrick, "The Consumer Revo-

lution of Eighteenth-Century England," in *The Birth of a Consumer Society: The Commercialization of Eighteenth-Century England*, ed. Neil McKendrick et al. (Bloomington: Indiana University Press, 1982), 9–33. On this point, see McKellar, *Birth of Modern London*, esp. 30–34.

97. Barbon, *An Apology*, 5.

98. Christopher Wren to Charles II, ca. 1670/1; *Wren Society* 13:18.

99. Howard M. Colvin, *The History of the King's Works*, vol. 5, *1660–1782* (London: H.M. Stationery Office, 1976), 25–26.

100. Barbon, *An Apology*, 2.

101. Ibid., 30 (my emphasis). On Barbon's medical training and its place in his economic ideas, see Marina Bianchi, "The Infinity of Human Desires and the Advantages of Trade: Nicholas Barbon and the Wants of the Mind," in *Physicians and Political Economy: Six Studies of the Work of Doctor-Economists*, ed. Peter D. Groenewegen (London: Routledge, 2001), 48–66.

102. John Greaves, *Pyramidographia; or, A Description of the Pyramids in Aegypt* (London: George Badger, 1646), 86. See also Zur Shalev, "Measurer of All Things: John Greaves (1602–1652), the Great Pyramid, and Early Modern Metrology," *Journal of the History of Ideas* 63, no. 4 (October 2002): 555–75.

103. See John Greaves, "A Fragment of a Tract of Mr. John Greaves concerning the Mountains of the Earth, according to the Arabians," in *Miscellaneous Works of Mr. John Greaves, Professor of Astronomy in the University of Oxford*, ed. Thomas Birch (London: J. Hughs, 1737), 2:473–74.

104. For this argument, see Greaves, *Pyramidographia*, 118.

105. In general, see James Stevens Curl, *Egyptomania: The Egyptian Revival, a Recurring Theme in the History of Taste* (New York: Manchester University Press, 1994).

106. For example, Robert Hooke, *Lectures and Discourses of Earthquakes*, in Waller, *Posthumous Works of Robert Hooke*, 353.

107. Compare Pliny the Elder, *Natural History*, trans. D. E. Eichholz (Cambridge, MA: Harvard University Press, 1962), vol. 10, bk. 36, chap. 15, p. 59. For more on experimentalists' approach to ancient monuments, see Li Shiqiao, *Power and Virtue: Architecture and Intellectual Change in England 1660–1730* (New York: Routledge, 2007).

108. Architectural historian Lydia M. Soo has compiled many of Wren's writings, including his "Tracts"; I use her transcriptions here. See Christopher Wren, "Tract V, 'Discourse on Architecture' (beginning in the mid-1670s)," in Soo, *Wren's "Tracts,"* 190.

109. Ibid. For modern analysis of relations between ancient Egyptian agriculture and state-building, see Robert J. Wenke, "Egypt: Origins of Complex Societies," *Annual Review of Anthropology* 18 (1989): 129–55.

110. William Petty, *A Treatise of Taxes and Contributions* (London: N. Brooke, 1662), 13. For more on Petty, see Ted McCormick, *William Petty and the Ambitions of Political Arithmetic* (New York: Oxford University Press, 2009).

111. Wren, "Tract V," in Soo, *Wren's "Tracts,"* 190.

112. Ibid.

113. Ibid.

114. I draw the sense of "fable" used here from Nancy Cartwright, "Models: Parables v. Fables," in *Beyond Mimesis and Convention: Representation in Art and Science*, ed. Roman Frigg and Matthew C. Hunter (New York: Springer, 2010), 19–31.

115. Executed under royal warrant, the construction was funded from 1670 by Parliament's tax of 4½ pence per chaldron on coal unloaded in London and, later, within the greater Thames estuary; see Campbell, *Building St. Paul's*, 60–69.

116. Christopher Wren, "Letter to a Friend from Paris (late September/October 1665)," in Soo, *Wren's "Tracts,"* 103.

117. Ibid.

118. Christopher Wren, quoted in Edward Browne to Sir Thomas Browne, September 30, 1665; *Wren Society* 18:180.

119. The unofficial pay for the Surveyor-General could be far higher. Wren reported that his predecessor, Sir John Denham, received 7,000 pounds per annum; see Aubrey, *Brief Lives*, 108. In the 1660s, Master Mason Edward Marshall was paid nearly 200 pounds, while Sergeant Painter Sir Robert Howard received the paltry official salary of 10 pounds per year; see Colvin, *King's Works*, 7–9.

120. Colvin, *King's Works*, 10.

121. On Hulsbergh, see Timothy Clayton, "Hulsberg, Henry (d. 1729)," in *Oxford Dictionary of National Biography*, ed. H. C. G. Matthew and Brian Harrison (New York: Oxford University Press, 2004), 28:720–21.

122. On administration of the fiscal-military state, see John Brewer, *The Sinews of Power: War, Money and the English State, 1688–1783* (New York: Knopf, 1989).

123. Campbell, *Building St. Paul's*, 170. A similar view is John Wilton-Ely, "The Rise of the Professional Architect in England," in *The Architect: Chapters in the History of the Profession*, ed. Spiro Kostof (New York: Oxford University Press, 1977), 180–208, esp. 185–86.

124. Cooper, *"A More Beautiful City,"* 192.

125. For classic accounts of bureaucracy, see Max Weber, "Bureaucracy," in *From Max Weber: Essays in Sociology*, trans., ed., and introduced by H. H. Gerth and C. W. Mills (London: Routledge and Kegan Paul, 1948), 198–244; and Robert K. Merton, "Bureaucratic Structure and Personality," in *Social Theory and Social Structure* (New York: Free Press, 1968), 249–60.

126. *Wren Society* 13:81.

127. Ibid., 107; compare ibid., 84.

128. Ibid., 74, 101.

129. See Lang, *Rebuilding St. Paul's*, 87.

130. Judith Hook, *The Baroque Age in England* (London: Thames and Hudson, 1976), 52.

131. On Woodroffe, see Paul Jeffrey, *The City Churches of Sir Christopher Wren* (London: Hambledon Press, 1996), esp. 37–38; and Lang, *Rebuilding St. Paul's*, 88.

132. See *Wren Society* 13:75.

133. Ibid., 69–71.

134. On Longland and Knowles, see Campbell, *Building St. Paul's*, 84–86.

135. See *Wren Society* 16:7–10. For a revealing comparison of Strong and Marshall, see Lang, *Rebuilding St. Paul's*, 80–85.

136. Wren's Clerk of the Works at Greenwich Palace, John Scarborough, was paid £14 10 s. in December 1688 for measuring masons' work; *Wren Society* 14:60.

137. Ibid., 84. Hawksmoor was paid 20 d. per day—a very modest sum considering that a decade earlier, plasterer Sam Clarke was being paid nearly twice that for making models for the choir's architrave frieze and cornice; see *Wren Society* 13:175. For a recent treatment of Wren's relations to his draftsmen, see

Geraghty, *Architectural Drawings*, esp. 8–14. A deliciously different reading of architectural drawings in this period is McKellar, *Birth of Modern London*, 116–35.

138. See Campbell, *Building St. Paul's*, 66–69.

139. Christopher Wren to William Sancroft, May 24, 1668; *Wren Society* 13:48.

140. For various calculations of these stages, compare Kerry Downes, *Sir Christopher Wren: The Design of St. Paul's Cathedral* (London: Trefoil, 1988); and Downes, "Wren and the New Cathedral," in *St. Paul's: The Cathedral Church of London 604–2004*, ed. Derek Keene et al. (New Haven, CT: Yale University Press, 2004), 190–206.

141. For this split, see Geraghty, *Architectural Drawings*, 11–13; and Gerbino and Johnston, *Compass and Rule*, esp. 104–10.

142. See especially Gordon Higgott, "The Revised Design for St. Paul's Cathedral 1685–1690: Wren, Hawksmoor and Les Invalides," *Burlington Magazine* 146, no. 1217 (August 2004): 534–47.

143. For more on these traditions, see Henry A. Millon et al., eds., *The Renaissance from Brunelleschi to Michelangelo: The Representation of Architecture* (New York: Rizzoli, 1997), esp. 19–74.

144. Christopher Wren, "Proposals to the Right Honorable the Commissioners for the Reparation of St Paul's Cathedral," May 1, 1666; *Wren Society* 13:17.

145. See John Wilton-Ely, "Wren, Hawksmoor and the Architectural Model," in *English Architecture Public and Private: Essays for Kerry Downes*, ed. John Bold and E. Chaney (London: Hambledon, 1993), 147–58.

146. See February 21, 1674; Hooke, *Diary*, 87.

147. Wren Jr., *Parentalia*, 283.

148. *Wren Society* 13:101–2.

149. For an example of the diagrams and figures sent by Christopher Wren to an agent in the quarries at Portland detailing dimensions of the "Scantling Blockstones for the Quires of St. Pauls from the Plinth to the Architrave of the Isles," see Guildhall Library MS 25, 579/1.

150. Henry Compton, *You perceive by His Majesties letters patents for rebuilding the Cathedral Church of St. Pauls how zealously His Majesty is concerned. . . .* (London, 1678), sig. Ai.

151. *Wren Society* 13:105, 125, 168.

152. Over three thousand of these returns survive in the Guildhall Library; see http://www.history.ac.uk/gh/briefs.htm.

153. BL MS Sloane 1039, fol. 171. On this project, see Lang, *Rebuilding St. Paul's*, 93–97; a less sanguine view of this campaign is Campbell, *Building St. Paul's*, esp. 64–65.

154. Robert Hooke, "Sect. VII," *Lectures of Light*, in Waller, *Posthumous Works of Robert Hooke*, 147.

155. Wren, "Tract V," 191.

156. RSCP 20:50 (fol. 94). For Barbon's syndicate, see McKellar, *Birth of Modern London*, 52.

157. On Kempster's prominent role in the west portico, see Lang, *Rebuilding St. Paul's*, 231–33. On this division of space, see John Newman, "Fittings and Liturgy in Post-Fire St Paul's," in Keene et al, *St. Paul's*, 220–32.

158. A. Welby Pugin, *The True Principles of Pointed or Christian Architecture* (London: John Weale, 1841), 5. An excellent account of these points is Campbell, *Building St. Paul's*, 109–14.

159. Higgott, "Revised Design," 547.

160. Lang, *Rebuilding St. Paul's*, 240.

161. See Carol Gibson-Wood, "The Political Background to Thornhill's Paintings in St. Paul's Cathedral," *Journal of the Warburg and Courtauld Institutes* 56 (1993): 229–37.

162. On the paving of the south choir aisle, see Lang, *Rebuilding St. Paul's*, 170–71.

163. Wren Jr., *Parentalia*, 292.

164. Christopher Wren, "Letter to a Friend on the Commission for Building Fifty New City Churches" (1711), in Soo, *Wren's "Tracts,"* 115.

165. For an authoritative account of the library's design and construction, see Howard Colvin, "The Building," in *The Making of the Wren Library*, ed. David McKitterick (New York: Cambridge University Press, 1995), 28–49.

166. Christopher Wren to Isaac Barrow, in *Wren Society* V (1928): 33. For surviving drawings for the library's furniture prepared by Wren's studio, see *Wren Society* 5:44, plate 24.

167. See Lawrence I. Lipking, *The Ordering of the Arts in Eighteenth-Century England* (Princeton, NJ: Princeton University Press, 1970). For the sociological sense of *art world* as I mean it here, compare Howard S. Becker, *Art Worlds* (Los Angeles: University of California Press, 2008); and Tod Volpe, *Framed: America's Art Dealer to the Stars Tells All* (Toronto: ECW Press, 2003).

168. David Esterly, *Grinling Gibbons and the Art of Carving* (New York: Harry N. Abrams, 1998), 180.

169. See Jean Tijou, *A New Booke of Drawings Invented and Desined by John Tijou* (London: J. Tijou, 1693). This title page was designed by decorative painter Louis Laguerre, Tijou's son-in-law; see Walter A. Dyer, "The Art of Jean Tijou," *Art World* 2, no. 4 (July 1917): 355–59.

170. At nearly the same time, Edward Pearce (sculptor and master mason at St. Paul's) also published his designs; see E. Pearce, *Twelve Engravings of Friezes* (London: I. Overton, n.d., ca. 1680).

171. For the *paragone* in various early modern contexts, compare Jonathan Brown, *Images and Ideas in Seventeenth-Century Spanish Painting* (Princeton, NJ: Princeton University Press, 1978), esp. 44–62; and Andrea Bolland, "*Desiderio* and *Diletto*: Vision, Touch and the Poetics of Bernini's *Apollo and Daphne*," *Art Bulletin* 82, no. 2 (2000): 309–30.

172. On this definition of the baroque, see Heinrich Wölfflin, *Principles of Art History: The Problem of the Development of Style in Later Art*, trans. M. D. Hottinger (New York: Dover, 1950), esp. 155–95. The extent to which Wren's Office of Works controlled design has been an object of interesting disagreement in recent scholarship. Whereas Caroline van Eck emphasizes the relative freedom of masons from the Office of Works in the design and execution of architectural details, Anthony Gerbino and Stephen Johnston stress how such ornament "was clearly an area of autonomy that Wren sought to wrest from them"; see Caroline van Eck, ed., *British Architectural Theory 1540–1730* (Burlington, VT: Ashgate, 2003), 53–55; and Gerbino and Johnston, *Compass and Rule*, 105.

173. James Ralph, *A Critical Review of the Publick Buildings, Statues and Ornaments in, and about London and Westminster* (London: C. Ackers, 1734), 22.

174. Ibid.

175. Summerson, "Tyranny of the Intellect," 386.

176. Christopher Wren, "Tract I," in Soo, *Wren's "Tracts,"* 153.

177. Robert Hooke, *A General Scheme, or Idea of the Present State of Natural Philosophy*, in Waller, *Posthumous Works of Robert Hooke*, 18. Compare, for example, the privilege of economy in Vitruvius, *Ten Books on Architecture*, bk. 1; and Roger Pratt's 1665 "Rules for the Guidance of Architects," cited in van Eck, *British Architectural Theroy*, 73–75.

178. Hooke, *A General Scheme*, 18.

179. Hooke, *Lectures and Discourses of Earthquakes*, 329.

180. On this biblical pairing of building and planting in seventeenth-century English devotional writing, see John Coolidge, *The Pauline Renaissance in England: Puritanism and the Bible* (Oxford: Clarendon Press, 1970), esp. 27–36.

181. [Ned Ward], *The London-Spy Compleat, in Eighteen Parts*, 4th ed. (London: J. How, 1709), 104.

182. Ibid., 105.

183. Ibid., 101.

184. Peter Parshall, "Prints as Objects of Consumption in Early Modern Europe," *Journal of Medieval and Early Modern Studies* 28, no. 1 (Winter 1998): 19–36.

185. [Nicholas Barbon], *A Discourse of Trade* (London: Thomas Milbourn, 1690), 15.

186. [Ward], *The London-Spy Compleat*, 101.

187. Ibid.

188. Ibid.

CONCLUSION

1. Anonymous, *Frauds and Abuses at St. Paul's* (London, 1712), 4.

2. For these accusations, see ibid., 6, 19–20, 21–23, 26–27.

3. Ibid., 31.

4. John Summerson, *Architecture in Britain, 1530–1830* (New Haven, CT: Yale University Press, 1953), 295.

5. On the contemporaneous emergence of such a public in France, see Thomas Crow, *Painters and Public Life in Eighteenth-Century Paris* (New Haven, CT: Yale University Press, 1985).

6. Anthony Ashley Cooper, Lord Shaftesbury, "A Letter Concerning Design," in *Characteristicks of Men, Manners, Opinions, Times*, 2nd ed. (London, 1714), 3:401, 403.

7. See Howard Colvin, "Lord Burlington and the Office of Works," in *Essays in English Architectural History* (New Haven, CT: Yale University Press, 1999), 263–67.

8. For various treatments of this rupture, see Margaret 'Espinasse, *Robert Hooke* (Los Angeles: University of California Press, 1956), esp. 1–41; and Mordechai Feingold, *The Newtonian Moment: Isaac Newton and the Making of Modern Culture* (New York: Oxford University Press, 2004).

9. Rob Iliffe, "Butter for Parsnips: Authorship, Audience, and the Incomprehensibility of the *Principia*," in *Scientific Authorship: Credit and Intellectual Property in Science*, ed. Mario Biagioli and Peter Galison (New York: Routledge, 2003), 44.

10. For this legend, see Lisa Jardine, *The Curious Life of Robert Hooke: The Man who Measured London* (London: Harper Perennial, 2003), 15.

11. Anonymous, "Hooke, Robert," in *Biographia Britannica* (London: W. Innys et al., 1750), 3:2662–63.

12. See Arthur MacGregor, ed., *Sir Hans Sloane: Collector, Scientist, Antiquary, Founding Father of the British Museum* (London: British Museum, 1994).

13. Christopher Wren, "Tract I," in *Wren's "Tracts" on Architecture and Other Writings*, ed. Lydia Soo (New York: Cambridge University Press, 1998), 153.

14. This point is most strongly stated in the recent claim that Hooke largely published to extract

payment from his reluctant patrons; see Michael Hunter, "Hooke the Natural Philosopher," in *London's Leonardo: The Life and Work of Robert Hooke*, ed. J. A. Bennett et al. (New York: Oxford University Press, 2003), 105–62.

15. See Larry Stewart, *The Rise of Public Science: Rhetoric, Technology, and Natural Philosophy in Newtonian Britain, 1660–1750* (New York: Cambridge University Press, 1992).

16. Compare D. C. Martin, "Former Homes of the Royal Society," *Notes and Records of the Royal Society of London* 22, nos. 1–2 (1967): 12–19; and Elizabeth McKellar, *The Birth of Modern London: The Development and Design of the City 1660–1720* (New York: Manchester University Press, 1999), 43.

BIBLIOGRAPHY

ARCHIVAL SOURCES

Bodleian Library, Oxford University: MS Aubrey 12

British Library, London: Add. MS 22911, Add. MS 22950, Add. MS 57495, Add. MS 6194, EVE.A.149, MS Ewr D 1076, MS 31.g.8, MS Sloane 1039, Sloane 3340, Sloane 4065

Guildhall Library, London: MS 25, MS 579

London Metropolitan Archives: CLC/495/MSO 1758

Royal Society of London: MS 131 ("Waller Scrapbook"); MS (General) 415/2

Royal Society Boyle Letters

Royal Society Early Letters

Royal Society Classified Papers, vols. 2, 13, 17, 20, 24

Royal Society Library Letter Books (Original)

Royal Society Register Books (Original)

Trinity College Library, Cambridge: MS o.11a.1

Wellcome Library, London: MS 1708

PRINTED SOURCES

Abercromby, David. *A Discourse of Wit*. London: John Weld, 1685.

Acheson, Katherine. "Gesner, Topsell, and the Purposes of Pictures in Early Modern Natural Histories." In Michael Hunter, *Printed Images in Early Modern Britain*, 127–44.

Ackerman, James S. "Architectural Practice in the Italian Renaissance." *Journal of the Society of Architectural Historians* 13, no. 3 (October 1954): 3–11.

Addison, Joseph, and Richard Steele. *The Spectator*. Edited by G. A. Aitken. 6 vols. London: John C Nimmo, 1898.

Alberti, Leon Battista. *On the Art of Building*. Translated by Joseph Rykwert et al. Cambridge, MA: MIT Press, 1991.

———. *"On Painting" and "On Sculpture": The Latin Texts of "De pictura" and "De statua."* Edited and translated by Cecil Grayson. London: Phaidon, 1972.

Alexander, Julia Manciari. "Beauties, Bawds and Bravura: The Critical History of Restoration Portraits of Women." In MacLeod and Manciari Alexander, *Painted Ladies*, 62–71.

Alexander, Julia Manciari, and Catharine MacLeod, eds. *Politics, Transgression, and Representation at the Court of Charles II*. New Haven, CT: Yale University Press, 2007.

Allen, Keith. "Locke and the Nature of Ideas." *Archiv für Geschichte der Philosophie* 92, no. 3 (2010): 236–55.

Alpers, Svetlana. *The Art of Describing: Dutch Art in the Seventeenth Century*. Chicago: University of Chicago Press, 1983.

———. *Rembrandt's Enterprise: The Studio and the Market*. Chicago: University of Chicago Press, 1988.

Alpers, Svetlana, and Michael Baxandall. *Tiepolo and the Pictorial Intelligence*. New Haven, CT: Yale University Press, 1994.

Altena, I. Q. Van Regteren. *Jacques de Gheyn: Three Generations*. 3 vols. London: Martinus Nijhoff, 1983.

Altick, Richard. *The Shows of London*. London: Belknap Press of Harvard University Press, 1978.

Amory, Hugh. "Chiswell, Richard, the Elder (1640–1711)." In Matthew and Harrison, *Oxford Dictionary of National Biography*, 11:494–95.

Andersen, Kirsti. *The Geometry of an Art: The History of Mathematical Theory of Perspective from Alberti to Monge*. New York: Springer, 2007.

Andrade, E. N. da C. "Samuel Pepys and the Royal Society." *Notes and Records of the Royal Society of London* 18, no. 2 (1963): 82–93.

Angelicoussis, Elizabeth. "The Collection of Classical Sculptures of the Earl of Arundel, 'Father of Vertu in England.'" *Journal of the History of Collections* 16, no. 2 (2004): 143–59.

Anonymous. *Abridgment of the Architecture of Vitruvius*. London: Abel Swall and T. Child, 1692.

———. *The Character of a Coffee-House, with the Symptomes of a Town-Wit*. London: Jonathan Edwin, 1673.

———. *A Character of England, As it was lately presented to a Noble Man of France*. London: Jo. Crooke, 1659.

———. *A Coffee-House Dialogue; or, A Discourse between Captain Y—— and a Young Barrester of the Middle-Temple, with some Reflections upon the Bill against the D. of Y*. London, 1679.

———. *Coffee-houses Vindicated in Answer to the late Published Character of a Coffee-house*. London: J. Lock, 1675.

———. "Extrait d'une Lettre . . . d'une Nouvelle invention pour vôler en l'air." *Journal des Scavans* (December 12, 1678): 452–55.

———. *Frauds and Abuses at St. Paul's*. London, 1712.

———. "Hooke, Robert." In *Biographia Britannica*, 3:2662–63. London: W. Innys et al., 1750.

———. *Knavery in All Trades; or, The Coffee-House, A Comedy*. London: J. B. for W. Gilbertson, 1664.

———. *The Polite Gentleman; or, Reflections Upon the Several Kinds of Wit*. London: R. Basset, 1700.

Anselment, Raymond A. "'Clouded Majesty': Richard Lovelace, Sir Peter Lely, and the Royalist Spirit." *Studies in Philology* 86, no. 3 (Summer 1989): 367–87.

Armenini, Giovanni Battista. *On the True Precepts of the Art of Painting*. 1586. Edited and translated by E. J. Olszewski, New York: Burt Franklin, 1977.

Arnold, Ken. *Cabinets for the Curious: Looking Back at Early English Museums*. Aldershot: Ashgate, 2006.

Ashley Cooper, Anthony, Lord Shaftesbury. *Characteristicks of Men, Manners, Opinions, Times*. 2nd ed. 3 vols. London, 1714.

Aspromourgos, Tony. "Political Economy, Political Arithmetic and Political Medicine in the Thought of William Petty." In Groenewegen, *Physicians and Political Economy*, 10–25.

Aubrey, John. *Brief Lives*. Edited by J. Buchanan-Brown. New York: Penguin, 2000.

Aujac, Germaine. "The Foundations of Theoretical Cartography in Archaic and Classical Greece." In *The History of Cartography*, edited by J. B. Harley and David Woodward. Vol. 1, *Cartography in Prehistoric, Ancient, and Medieval Europe and the Mediterranean*. Chicago: University of Chicago Press, 1987.

Auzout, Adrien. "An Extract of a Letter Written Decemb. 28. 1666. By M. Auzout to the Publisher, Concerning a Way of His, for Taking the Diameters of the Planets. . . ." *Philosophical Transactions* 21, no. 1 (1667): 373–75.

Aylett, Robert. *Svsanna; or, The Arraignment of the Two Vnjvst Elders*. London: John Teage, 1620.

Bacon, Francis. *The Advancement of Learning*. 1605. Reprint, Chicago: Encyclopaedia Britannica, 1952.

———. *"The Advancement of Learning" and "New Atlantis."* Edited by Arthur Johnston. Oxford: Clarendon Press, 1974.

———. *Magna Instauratio*. In *Francis Bacon: Essays*, edited by R. F. Jones. New York: Odyssey Press, 1937.

Barber, Tabitha. *Mary Beale: Portrait of a Seventeenth-Century Painter, Her Family and Her Studio*. London: Geffrye Museum, 1999.

Barbon, Nicholas. *An Apology for the Builder; or, A Discourse Shewing the Causes and Effects of the Increase of Building*. London: Cave Pullen, 1685.

———. *A Discourse of Trade*. London: Thomas Milbourn, 1690.

Barkan, Leonard. "The Heritage of Zeuxis: Painting, Rhetoric, and History." In *Antiquity and Its Interpreters*, edited by Alina Payne et al., 99–109. New York: Cambridge University Press, 2000.

Barnes, Susan, et al. *Van Dyck: A Complete Catalogue of the Paintings*. New Haven, CT: Yale University Press, 2004.

Barrell, John. *The Birth of Pandora and the Division of Knowledge*. Philadelphia: University of Pennsylvania Press, 1992.

Barrera-Osorio, Antonio. *Experiencing Nature: The Spanish American Empire and the Early Scientific Revolution*. Austin: University of Texas Press, 2006.

Barry, Jonathan. "The Press and the Politics of Culture in Bristol, 1660–1775." In Black and Gregory, *Culture, Politics and Society in Britain, 1660–1800*, 49–81.

Barthes, Roland. "The World as Object." In *Calligram: Essays in New Art History from France*, edited by Norman Bryson, 106–15. New York: Cambridge University Press, 1988.

Batten, M. I. "The Architecture of Dr. Robert Hooke FRS." *Walpole Society* 25 (1936–37): 83–113.

Becker, Howard S. *Art Worlds*. Los Angeles: University of California Press, 2008.

Becker, Ulrich, ed. *Alte Galerie: Masterpieces*. Graz, Austria: Landesmuseum Joanneum, 2005.

Beckett, R. B. *Lely*. London: Routledge and Keegan Paul, 1951.

Beier, Linda McCray. "Experience and Experiment: Robert Hooke, Illness and Medicine." In Michael Hunter and Simon Schaffer, *Robert Hooke: New Studies*, 235–51.

Bendall, A. Sarah. *Maps, Land and Society: A History, with a Carto-Bibliography of Cambridgeshire Estate Maps, c. 1600–1836*. New York: Cambridge University Press, 1992.

Bender, John, and Michael Marrinan. *The Culture of Diagram*. Stanford, CA: Stanford University Press, 2010.

Bennett, J. A. "Christopher Wren: Astronomy, Architecture, and the Mathematical Sciences." *Journal for the History of Astronomy* 6 (1975): 149–84.

———. "Christopher Wren: The Natural Causes of Beauty." *Architectural History* 15 (1972): 5–22.

———. "Hooke and Wren and the System of the World: Some Points toward an Historical Account." *British Journal for the History of Science* 8, no. 1 (March 1975): 32–61.

———. "Hooke's Instruments." In Bennett et al., *London's Leonardo*, 63–104.

———. "Hooke's Instruments for Astronomy and Navigation." In Michael Hunter and Simon Schaffer, *Robert Hooke: New Studies*, 21–32.

———. *The Mathematical Science of Christopher Wren*. New York: Cambridge University Press, 1982.

———. "The Mechanics' Philosophy and the Mechanical Philosophy." *History of Science* 24, no. 1 (March 1986): 1–28.

Bennett, J. A., et al., eds. *London's Leonardo: The Life and Work of Robert Hooke*. New York: Oxford University Press, 2003.

Bennett, J. A., and Scott Mandelbrote. *The Garden, the Ark, the Tower, the Temple: Biblical Metaphors of Knowledge in Early Modern Europe*. Oxford: Museum of the History of Science, 1998.

Berger, John. *Ways of Seeing*. New York: Viking Press, 1972.

Bermingham, Ann. *Landscape and Ideology: The English Rustic Tradition, 1740–1860*. Los Angeles: University of California Press, 1986.

———. *Learning to Draw: Studies in the Cultural History of a Polite and Useful Art*. New Haven, CT: Yale University Press, 2000.

Biagioli, Mario. *Galileo Courtier: The Practice of Science in the Culture of Absolutism*. Chicago: University of Chicago Press, 1993.

Biagioli, Mario, and Peter Galison, eds. *Scientific Authorship: Credit and Intellectual Property in Science*. New York: Routledge, 2003.

Bianchi, Marina. "The Infinity of Human Desires and the Advantages of Trade: Nicholas Barbon and the Wants of the Mind." In Groenewegen, *Physicians and Political Economy*, 48–66.

Bignamini, Ilaria. "George Vertue, Art Historian, and Art Institutions in London, 1689–1768." *Walpole Society* 54 (1991): 1–148.

Bignamini, Ilaria, and Martin Postle, eds. *The Artist's Model: Its Role in British Art from Lely to Etty*. Nottingham: Nottingham University Art Gallery, 1991.

Biles, Jeremy. "I, Insect, or Bataille and the Crush Freaks." *Janus Head* 7, no. 1 (2004): 115–31.

Birch, Thomas, ed. *The History of the Royal Society of London for Improving of Natural Knowledge*. 4 vols. London: Printed for A. Millar, 1756–57.

Black, Jeremy, and Jeremy Gregory, eds. *Culture, Politics and Society in Britain, 1660–1800*. Manchester: Manchester University Press, 1991.

Blair, Ann. "Reading Strategies for Coping with Information Overload ca. 1550–1700." *Journal of the History of Ideas* 64, no. 1 (2003): 11–28.

Boccaccio, Giovanni. *The Decameron: Containing An Hundred Pleasant Novels*. London: James Iaggard, 1620.

Bolland, Andrea. "*Desiderio* and *Diletto*: Vision, Touch and the Poetics of Bernini's *Apollo and Daphne*." *Art Bulletin* 82, no. 2 (2000): 309–30.

Bolten, A. T., and H. D. Hendry, eds. *Wren Society*. Oxford: Oxford University Press, 1923–43.

Boogaart, Ernest van den. *Civil and Corrupt Asia: Image and Text in the "Itinerario" and "Icones" of Jan Hugyen van Linschoten*. Chicago: University of Chicago Press, 2003.

Borelli, G. A. *De Motu Animalium: Pars Prima*. Rome: Angeli Bernabo, 1680.

Borsay, Peter. "Early Modern Urban Landscapes, 1540–1800." In *The English Urban Landscape*, edited by P. J. Waller, 99–124. New York: Oxford University Press, 2000.

———. *The English Urban Renaissance: Culture and Society in the Provincial Town, 1660–1770*. New York: Oxford University Press, 1989.

Boyle, Robert. *New Experiments Physico-Mechanicall, Touching the Spring of the Air*. 1660. In Michael Hunter and E. B. Davis, *The Works of Robert Boyle*, vol. 1.

———. *Some Considerations Touching the Usefulness of Experimental Philosophy*. 1663. In Michael Hunter and E. B. Davis, *The Works of Robert Boyle*, vol. 3.

———. "Mr. Boyle's Account of the above-mentioned *Invention*, and the *Experiments* thereon." In *Parentalia; or, Memoirs of the Family of the Wrens*, edited by Christopher Wren Jr. London: T. Osborn and R. Dodsley, 1750.

———. *The Works of Robert Boyle*. Edited by Michael Hunter and E. B. Davis. 14 vols. London: Pickering and Chatto, 1999.

Bredekamp, Horst. *The Lure of Antiquity and the Cult of the Machine: The Kunstkammer and the Evolution of Art, Nature and Technology*. Translated by Allison Brown. Princeton, NJ: Markus Wiener, 1995.

Brewer, John. *The Sinews of Power: War, Money and the English State, 1688–1783*. New York: Knopf, 1989.

Brooks, Randall C. "The Development of Micrometers in the Seventeenth, Eighteenth and Nineteenth Centuries." *Journal for the History of Astronomy* 22 (1991): 127–73.

Brotton, Jerry. *The Sale of the Late King's Goods: Charles I and His Art Collection*. London: Macmillan, 2006.

Brown, Jonathan. *Images and Ideas in Seventeenth-Century Spanish Painting*. Princeton, NJ: Princeton University Press, 1978.

———. *Kings and Connoisseurs: Collecting Art in Seventeenth-Century Europe*. Princeton, NJ: Princeton University Press, 1995.

———. *Velázquez, Painter and Courtier*. New Haven, CT: Yale University Press, 1986.

Brown, Jonathan, and J. H. Elliott, eds. *The Sale of the Century: Artistic Relations between Spain and Great Britain, 1604–1655*. London: Yale University Press, 2002.

Brown, Thomas. *Amusements Serious and Comical, calculated for the Meridian of London*. London, 1700.

Browne, Alexander. *Ars pictoria; or, An academy Treating of Drawing, Painting, Limning, and Etching*. London: J. Redmayne, 1669.

Brusati, Celeste. *Artifice and Illusion: The Art and Writing of Samuel van Hoogstraten*. Chicago: University of Chicago Press, 1995.

Buchwald, J. Z. "Discrepant Measurements and Experimental Knowledge in the Early Modern Era." *Archive for History of Exact Sciences* 60, no. 6 (2006): 565–649.

Buckeridge, Bainbrigg. *An Essay Towards an English School*. In *The Art of Painting . . . : Third edition*. 1754. Reprint, London: Cornmarket Press, 1969.

Bynum, William F. "The Anatomical Method, Natural Theology, and the Functions of the Brain." *Isis* 64, no. 4 (December 1973): 445–68.

Campbell, James W. P. *Building St. Paul's*. New York: Thames and Hudson, 2007.

Canfield, J. Douglas. "Women's Wit: Subversive Women Tricksters in Restoration Comedy." In *The Restoration Mind*, edited by W. G. Marshall, 147–76. Newark: University of Delaware Press, 1997.

Carlino, Andrea. *Paper Bodies: A Catalogue of Anatomical Fugitive Sheets 1538–1687*. Translated by Noga Arikha. London: Wellcome Institute for the History of Medicine, 1999.

Cartwright, Nancy. "Models: Parables v. Fables." In Frigg and Hunter, *Beyond Mimesis and Convention*, 19–31.

Casaubon, Meric. *A True & Faithful Relation of what Passed for Many Yeers between Dr. John Dee and Some Spirits*. London: D. Maxwell, 1659.

Cavazzini, Patrizia. *Painting as Business in Early Seventeenth-Century Rome*. University Park: Pennsylvania State University Press, 2008.

Cennini, Cennino. *The Craftsman's Handbook*. Circa 1437. Translated by Daniel V. Thompson Jr., New York: Dover, 1960.

Chapman, Allan. "Jeremiah Horrocks, the Transit of Venus, and the 'New Astronomy' in Early Seventeenth-Century England." *Quarterly Journal of the Royal Astronomical Society* 31 (1990): 333–57.

Clark, Stuart. *Vanities of the Eye: Vision in Early Modern European Culture*. New York: Oxford University Press, 2007.

Clarke, Arthur C. *2001: A Space Odyssey*. New York: Roc, 1999.

Clayton, Timothy. "Hulsberg, Henry (d. 1729)." In Matthew and Harrison, *Oxford Dictionary of National Biography*, 28:720–21.

Clericuzio, Antonio. "The Internal Laboratory: The Chemical Reinterpretation of Medical Spirits in England (1650–1680)." In *Alchemy and Chemistry in the 16th and 17th Centuries*, edited by Piyo Rattansi and Antonio Clericuzio, 51–83. Boston: Kluwer, 2002.

Clulee, Nicholas H. *John Dee's Natural Philosophy: Between Science and Religion*. New York: Routledge and Kegan Paul, 1988.

Cole, Michael. *Cellini and the Principles of Sculpture*. New York: Cambridge University Press, 2002.

———. "Cellini's Blood." *Art Bulletin* 81, no. 2 (June 1999): 215–35.

Cole, William. "General Proposalls of the Termes and Conditions for Surrender of my Collections." MS Ashmole 1820 a, fols. 65–69. Reproduced in A. J. Turner, "A Forgotten Naturalist of the Seventeenth Century: William Cole of Bristol and His Collections." *Archives of Natural History* 11, no. 1 (1982): 27–41.

———. "A Letter from Mr. William Cole of Bristol, to the Philosophical Society of Oxford, containing his Observations on the Purple Fish." *Philosophical Transactions* 178 (December 1685): 1285–86.

———. *Purpura Anglicana; being a Discovery of a Small Fish found on the Shores of the Severn*. London: Joseph Streater, 1689.

Colvin, Howard M. "The Building." In *The Making of the Wren Library*, edited by David McKitterick, 28–49. New York: Cambridge University Press, 1995.

———. *Essays in English Architectural History*. New Haven, CT: Yale University Press, 1999.

———. *The History of the King's Works*. Vol. 5, *1660–1782*. London: H.M. Stationery Office, 1976.

Compton, Henry. *You perceive by His Majesties letters patents for rebuilding the Cathedral Church of St. Pauls how zealously His Majesty is concerned. . . .* London, 1678.

Cook, Harold J. "The Cutting Edge of a Revolution? Medicine and Natural History near the Shores of the North Sea." In *Renaissance and Revolution: Humanists, Scholars, Craftsmen and Natural Philosophers in Early Modern Europe*, edited by J. V. Field and Frank A. J. L. James, 45–61. Cambridge: Cambridge University Press, 1993.

———. *Matters of Exchange: Commerce, Medicine and Science in the Dutch Golden Age*. New Haven, CT: Yale University Press, 2007.

———. "Time's Bodies: Crafting the Preparation and Preservation of Naturalia." In P. Smith and Findlen, *Merchants and Marvels*, 223–47.

Coolidge, John. *The Pauline Renaissance in England: Puritanism and the Bible*. Oxford: Clarendon Press, 1970.

Cooper, Michael. *"A More Beautiful City": Robert Hooke and the Rebuilding of London after the Great Fire*. Thrupp: Sutton, 2003.

Cooper, Michael, and Michael Hunter, eds. *Robert Hooke: Tercentennial Studies*. Aldershot: Ashgate, 2006.

Corbin, Alain. *The Foul and the Fragrant: Odor and the French Social Imagination*. Translated by M. L. Kochan et al. Cambridge, MA: Harvard University Press, 1986.

Couliano, Ioan P. *Eros and Magic in the Renaissance*. Translated by Margaret Cook. Chicago: University of Chicago Press, 1987.

Cowan, Brian. "Art and Connoisseurship in the Auction Market of Later Seventeenth-Century London." In *Mapping Markets for Paintings in Europe, 1450–1750*, edited by Neil de Marchi and Hans J. Van Miegroet, 263–84. Turnhout, Belgium: Brepols, 2006.

———. *The Social Life of Coffee: The Emergence of the British Coffeehouse*. New Haven, CT: Yale University Press, 2005.

Cowley, Abraham. *The Works of Mr. Abraham Cowley*. London: J. M. for Henry Herringman, 1668.

Crary, Jonathan. *Techniques of the Observer: On Vision and Modernity in the Nineteenth Century*. Cambridge, MA: MIT Press, 1990.

[Croone, Dr. William]. "An Hypothesis of the Structure of a Muscle, and the Reason of its Contractions." *Philosophical Collections* 2 (1681): 23–25.

———. *On the Reason of the Movement of the Muscles*. 1667. Translated by Paul Maquet and introduced by Margaret Nayler, Philadelphia: American Philosophical Society, 2000.

Crow, Thomas. *Painters and Public Life in Eighteenth-Century Paris*. New Haven, CT: Yale University Press, 1985.

Curd, Mary Bryan H. *Flemish and Dutch Artists in Early Modern England: Collaboration and Competition, 1460–1680*. Burlington, VT: Ashgate, 2010.

Curl, James Stevens. *Egyptomania: The Egyptian Revival, a Recurring Theme in the History of Taste*. New York: Manchester University Press, 1994.

D., J. *A Word Without Doors Concerning the Bill for Succession*. London, 1679.

Dackerman, Susan. *Prints and the Pursuit of Knowledge in Early Modern Europe*. Cambridge, MA: Harvard Art Museums, 2011.

Daston, Lorraine. "Curiosity in Early Modern Science." *Word & Image* 11, no. 4 (1995): 391–404.

———. "The Factual Sensibility." *Isis* 79, no. 3 (1988): 452–67.

———. "The Moral Economy of Science." *Osiris* 10 (1995): 2–24.

———. "On Scientific Observation." *Isis* 99, no. 1 (March 2008): 97–110.

Daston, Lorraine, and Peter Galison. "The Image of Objectivity." *Representations* 40 (Autumn 1992): 81–120.

———. *Objectivity*. New York: Zone Books, 2007.

Daston, Lorraine, and Katharine Park. *Wonders and the Order of Nature 1150–1750*. New York: Zone Books, 1998.

Davenport-Hines, Richard. *The Pursuit of Oblivion: A Social History of Drugs*. London: Phoenix Press, 2001.

David Winton Bell Gallery. *Children of Mercury: The Education of Artists in the Sixteenth and Seventeenth Century*. Providence, RI: Brown University Press, 1984.

Davies, David. "Learning through Fictional Narratives in Art and Science." In Frigg and Hunter, *Beyond Mimesis and Convention*, 51–70.

Davis, Edward B. "The Anonymous Works of Robert Boyle and the *Reasons Why a Protestant Should not Turn Papist* (1687)." *Journal of the History of Ideas* 55, no. 4 (1994): 611–29.

Davis, Natalie Zemon. "Beyond the Market: Books as Gifts in Sixteenth-Century France." *Transactions of the Royal Historical Society* 5, no. 33 (1983): 69–88.

Davis, Ralph. *Aleppo and Devonshire Square: English Traders in the Levant in the Eighteenth Century*. Toronto: Macmillan, 1967.

Debus, Allen G. *Science and Education in the Seventeenth Century: The Webster-Ward Debate*. New York: American Elsevier, 1970.

Dee, John. "Mathematicall Praeface." In *The Elements of Geometrie of the Most Auncient Philosopher Euclide of Megara*, translated by Henry Billingsley. London: John Daye, 1570.

Defoe, Daniel. *A Journal of the Plague Year*. Edited by Paula Backschneider. Norton Critical Edition. New York: Norton, 1992.

Dennis, Michael Aaron. "Graphic Understanding: Instruments and Interpretation in Robert Hooke's *Micrographia*." *Science in Context* 3, no. 2 (1989): 309–65.

Descartes, René. *The Passions of the Soul*. In *Descartes: Selected Philosophical Writings*, edited and translated by John Cottingham and Dugald Murdoch. New York: Cambridge University Press, 1988.

———. *The Philosophical Writings of Descartes*. Translated by John Cottingham et al. Vol. 1. New York: Cambridge University Press, 1985.

———. *Principles of Philosophy*. 1647. Translated by Valentine Rodger Miller and Reese P. Miller, Boston: D. Reidel, 1984.

Dethloff, Diana. "The Executors' Account Book and the Dispersal of Sir Peter Lely's Collection." *Journal of the History of Collections* 8, no. 1 (1996): 15–51.

———. "Lely, Sir Peter (1618–1680)." In Matthew and Harrison, *Oxford Dictionary of National Biography*, 33:303–9.

———. "Patterns of Drawing Collecting in Late Seventeenth and Early Eighteenth Century England." In *Drawing: Masters and Methods, Raphael to Rodin*, edited by Diana Dethloff, 197–207. London: Royal Academy of Art, 1992.

Detienne, Marcel, and Jean-Pierre Vernant. *Cunning Intelligence in Greek Culture and Society*. Translated by Janet Lloyd. Atlantic Highlands, NJ: Humanities Press, 1978.

Dickie, Simon. "Hilarity and Pitilessness in the Mid-Eighteenth Century: English Jestbook Humor." *Eighteenth-Century Studies* 37, no. 1 (Fall 2003): 1–22.

Dikötter, Frank, Lars Laamann, and Zhou Xun. *Narcotic Culture: A History of Drugs in China*. London: Hurst, 2004.

Downes, Kerry. *Hawksmoor*. London: A. Zwemmer, 1959.

———. *Sir Christopher Wren: The Design of St. Paul's Cathedral*. London: Trefoil, 1988.

———, ed. *Sir Christopher Wren: An Exhibition Selected by Kerry Downes at the Whitechapel Art Gallery*. London: Trefoil Books, 1982.

———. "Wren and the New Cathedral." In Keene et al., *St. Paul's*, 190–206.

Draaisma, Douwe. "Hooke on Memory and the Memory of Hooke." In Cooper and Hunter, *Robert Hooke: Tercentennial Studies*, 111–22.

Drake, Ellen Tan. *Restless Genius: Robert Hooke and His Earthly Thoughts*. New York: Oxford University Press, 1996.

Dreyer, J. L. E. "Flamsteed's Letters to Richard Towneley." *Observatory* 580 (September 1922): 280–94.

Duro, Paul. "'The Surest Measure of Perfection': Approaches to Imitation in Seventeenth-Century French Art and Theory." *Word & Image* 25, no. 4 (October–December 2009): 363–83.

Duve, Thierry de. *Pictorial Nominalism: On Marcel Duchamp's Passage from Painting to the Readymade*. Translated by Dana Polan. Minneapolis: University of Minnesota Press, 1991.

———. "The Readymade and the Tube of Paint." *Artforum* 24 (1986): 110–21.

Dyer, Walter A. "The Art of Jean Tijou." *Art World* 2, no. 4 (July 1917): 355–59.

Eagleton, Catherine, and Boris Jardine. "Collections and Projections: Henry Sutton's Paper Instruments." *Journal of the History of Collections* 12, no. 1 (2005): 1–13.

Eck, Caroline van, ed. *British Architectural Theory 1540–1730*. Burlington, VT: Ashgate, 2003.

Edgerton, Samuel Y., Jr. "Galileo, Florentine 'Disegno,' and the 'Strange Spottednesse of the Moon.'" *Art Journal* 44, no. 3 (Fall 1984): 225–32.

"Editorial: Sir Peter Lely's Collection." *Burlington Magazine* 83, no. 485 (1943): 188.

Ehrlich, Mark E. "Mechanism and Activity in the Scientific Revolution: The Case of Robert Hooke." *Annals of Science* 52 (1992): 127–51.

Ekserdjian, David. "A Portrait by Lely and a Drawing after Correggio: An Artist's Use of His Collection of Drawings." *Apollo* (May 1998): 28–29.

Elkins, James. *The Domain of Images*. Ithaca, NY: Cornell University Press, 1999.

———. *Six Stories from the End of Representation: Images in Painting, Photography, Astronomy, Microscopy, Particle Physics, and Quantum Mechanics, 1980–2000*. Stanford, CA: Stanford University Press, 2008.

———. "On Visual Desperation and the Bodies of Protozoa." *Representations* 40 (Autumn 1992): 33–56.

———. "Visual Practices across the University." In Frigg and Hunter, *Beyond Mimesis and Convention*, 169–92.

———. *Visual Practices across the University*. Munich: Wilhelm Fink Verlag, 2007.

———. *Why Are Our Pictures Puzzles? On the Modern Origins of Pictorial Complexity*. New York: Routledge, 1999.

Emmons, Paul. "Size Matters: Virtual Scale and Bodily Imagination in Architectural Drawing." *Architectural Research Quarterly* 9, nos. 3–4 (2005): 227–35.

'Espinasse, Margaret. *Robert Hooke*. Los Angeles: University of California Press, 1956.

Esterly, David. *Grinling Gibbons and the Art of Carving*. New York: Harry N. Abrams, 1998.

Evelyn, John. *The Diary of John Evelyn*. Edited by A. Dobson. 3 vols. London: Macmillan, 1906.

———. *Fumifugium; or, The Inconveniencie of the Aer and Smoak of London Dissipated*. London: W. Godbid, 1661.

———. *London Revived: Considerations for Its Rebuilding in 1666*. Edited by E. S. de Beer. Oxford: Clarendon Press, 1938.

———. *Numismata: A Discourse of Medals, Antient and Modern*. London: Benjamin Tooke, 1697.

———. *Sculptura; or, The History, and Art of Chalcography and Engraving in Copper*. London: J. C. for G. Beedle and T. Collins, 1662.

Ezell, Mary. "Richard Waller S.R.S: 'In Pursuit of Nature.'" *Notes and Records of the Royal Society of London* 38, no. 2 (1984): 215–33.

Feingold, Mordechai, ed. *Jesuit Science and the Republic of Letters*. Cambridge, MA: MIT Press, 2003.

———. *The Mathematician's Apprentice: Science, Universities and Society in England, 1560–1640*. London: Cambridge University Press, 1984.

———. *The Newtonian Moment: Isaac Newton and the Making of Modern Culture*. New York: Oxford University Press, 2004.

———. "Robert Hooke: Gentleman of Science." In Cooper and Hunter, *Robert Hooke: Tercentennial Studies*, 203–17.

Feyerabend, Paul. *Against Method*. London: Verso, 1978.

Findlen, Paula. *Possessing Nature: Museums, Collecting, and Scientific Culture in Early Modern Italy*. Los Angeles: University of California Press, 1994.

Fischel, Angela. "Sehen, Darstellen, Beschreiben: Mikroskopische Beobachtung in den Kupferstichen der *Micrographia*." *Kunsttexte.de* 1 (2002): 1–10.

Fish, Stanley. *Surprised by Sin: The Reader in "Paradise Lost."* Cambridge, MA: Harvard University Press, 1997.

Flecknoe, Richard. *A Treatise of the Sports of Wit*. London: Printed for Flecknoe, 1675.

Forbes, Eric G. "The Comet of 1680–1681." In Thrower, *Standing on the Shoulders of Giants*, 312–23.

Foster, Hal, ed. *Vision and Visuality*. Seattle: Bay Press, 1988.

Foucart, Jacques. "Peter Lely, Dutch History Painter." *Hoogsteder-Naumann Mercury* 2 (1985): 17–26.

Foucault, Michel. *The Order of Things: An Archaeology of the Human Sciences*. London: Routledge Classics, 2004.

Fox, Celina. *The Arts of Industry in the Age of Enlightenment*. New Haven, CT: Yale University Press, 2009.

Fox, Robert, ed. *Thomas Harriot: An Elizabethan Man of Science*. Burlington, VT: Ashgate, 2000.

Frank, Robert G., Jr. *Harvey and the Oxford Physiologists: A Study of Scientific Ideas*. Berkeley: University of California Press, 1980.

———. "Thomas Willis and His Circle: Brain and Mind in Seventeenth-Century Medicine." In *The Languages of Psyche: Mind and Body in Enlightenment Thought*, edited by G. S. Rousseau, 107–46. Berkeley: University of California Press, 1990.

Freedberg, David. *The Eye of the Lynx: Galileo, His Friends, and the Beginnings of Modern Natural History*. Chicago: University of Chicago Press, 2002.

French, Peter J. *John Dee: The World of an Elizabethan Magus*. London: Routledge and Kegan Paul, 1972.

French, Roger. *Dissection and Vivisection in the European Renaissance*. Brookfield: Ashgate, 1999.

Fried, Michael. *Absorption and Theatricality: Painting and Beholder in the Age of Diderot*. Los Angeles: University of California Press, 1980.

———. *Realism, Writing, Disfiguration: On Thomas Eakins and Stephen Crane*. Chicago: University of Chicago Press, 1987.

Frigg, Roman. "Models and Fiction." *Synthese* 172 (2010): 251–68.

Frigg, Roman, and Matthew C. Hunter, eds. *Beyond Mimesis and Convention: Representation in Art and Science*. New York: Springer, 2010.

Fyfe, Gordon, and John Law, eds. *Picturing Power: Visual Depiction and Social Relations*. London: Routledge, 1988.

Gal, Ofer. *Meanest Foundations and Nobler Superstructures: Hooke, Newton, and the "Compounding of the Celestiall Motions of the Planetts."* London: Kluwer, 2002.

Galinou, Mireille, ed. *City Merchants and the Arts, 1670–1720*. Wetherby: Oblong for the Corporation of London, 2004.

Galison, Peter. *Image and Logic: A Material Culture of Microphysics*. Chicago: University of Chicago Press, 1997.

———. "Theory Bound and Unbound: Superstrings and Experiment." In *Laws of Nature: Essays on the Philosophical, Scientific and Historical Dimensions*, edited by Fridel Weinert, 269–408. New York: Walter de Gruyter, 1995.

Gardner, Howard. *Frames of Reference: The Theory of Multiple Intelligences*. New York: Basic Books, 1983.

Garrard, Mary D. "Artemisia and Susanna." In *Feminism and Art History: Questioning the Litany*, edited by Norma Broude and Mary D. Garrard. New York: Harper and Row, 1982.

Gaskell, Philip. *A New Introduction to Bibliography*. Oxford: Clarendon Press, 1974.

Gately, Iain. *Tobacco: A Cultural History of How an Exotic Plant Seduced Civilization*. New York: Grove Press, 2001.

Gell, Alfred. *The Art of Anthropology: Essays and Diagrams*. Edited by Eric Hirsch. London: Athlone Press, 1999.

Gent, Lucy, ed. *Albion's Classicism: The Visual Arts in Britain, 1550–1660*. New Haven, CT: Yale University Press, 1995.

Genuth, Sara Schechner. *Comets, Popular Culture and the Birth of Modern Cosmology*. Princeton, NJ: Princeton University Press, 1997.

Geraghty, Anthony. *The Architectural Drawings of Sir Christopher Wren at All Souls College, Oxford: A Complete Catalogue*. Burlington, VT: Lund Humphries, 2007.

———. "Robert Hooke's Collection of Architectural Books and Prints." *Architectural History* 47 (2004): 113–25.

Gerbino, Anthony, and Stephen Johnston. *Compass and Rule: Architecture as Mathematical Practice in England 1600–1750*. New Haven, CT: Yale University Press, 2009.

Gibson, Katherine. "'The Best Belov'd of Kings': The Iconography of Charles II." PhD diss., Courtauld Institute of Art, 1997.

Gibson, W. C. "The Bio-Medical Pursuits of Christopher Wren." *Medical History* 14, no. 4 (October 1970): 331–41.

Gibson-Wood, Carol. "Classification and Value in a Seventeenth-Century Museum: William Courten's Collections." *Journal of the History of Collections* 9, no. 1 (1997): 61–77.

———. "Picture Consumption in London at the End of the Seventeenth Century." *Art Bulletin* 84, no. 3 (September 2002): 491–500.

———. "The Political Background to Thornhill's Paintings in St. Paul's Cathedral." *Journal of the Warburg and Courtauld Institutes* 56 (1993): 229–37.

Goldgar, Anne. *Impolite Learning: Conduct and Community in the Republic of Letters, 1680–1750*. New Haven, CT: Yale University Press, 1995.

Golinski, Jan. "A Noble Spectacle: Phosphorous and the Public Cultures of Science in the Early Royal Society." *Isis* 80, no. 1 (1989): 11–39.

Goodman, Nelson. *Languages of Art: An Approach to a Theory of Symbols*. New York: Bobbs-Merrill, 1968.

———. *Of Mind and Other Matters*. Cambridge, MA: Harvard University Press, 1984.

Gorges, Sir Ferdinando. *America painted to the life: The history of the Spaniards proceeding in America*. London: T. J. for N. Brook, 1659.

Gouk, Penelope. *Music, Science and Natural Magic in Seventeenth-Century England*. New Haven, CT: Yale University Press, 1999.

———. "The Role of Acoustics and Music Theory in the Scientific Work of Robert Hooke." *Annals of Science* 37 (1980): 573–605.

Graham, Richard. *A Short Account of the most Eminent Painters, both Ancient and Modern. . . .* In *De Arte Graphica: The Art of Painting by C.A. Du Fresnoy with Remarks*, translated by John Dryden. London: J. Heptinstall for W. Rogers, 1695.

Grant, James William. "Remarks on Hooke's Observations of the Comets of 1680 and 1682." *Monthly Notices of the Royal Astronomical Society* 14 (January 1854): 77.

Greaves, John. *Miscellaneous Works of Mr. John Greaves, Professor of Astronomy in the University of Oxford*. Edited by Thomas Birch. 2 vols. London: J. Hughs, 1737.

———. *Pyramidographia; or, A Description of the Pyramids in Aegypt*. London: George Badger, 1646.

Grew, Nehemiah. *Musaeum Regalis Societatis; or, A Catalogue and Description of the Natural and Artificial Rarities Belonging to the Royal Society and Preserved at Gresham College*. London: W. Rawlins, 1681.

Griffiths, Antony. "Burghers, Michael (1647/8–1727)." In Matthew and Harrison, *Oxford Dictionary of National Biography*, 8:798.

———. "The Etchings of John Evelyn." In *Art and Patronage in the Caroline Courts: Essays in Honour of Sir Oliver Millar*, edited by David Howarth, 51–67. Cambridge: Cambridge University Press, 1993.

Groenewegen, Peter D., ed. *Physicians and Political Economy: Six Studies of the Work of Doctor-Economists*. London: Routledge, 2001.

Grotius, Hugo. *Syntagma Arateorum*. Lyon: Ex Officina Plantiniana, 1600.

Guerrini, Anita. "The Ethics of Animal Experimentation in Seventeenth-Century England." *Journal of the History of Ideas* 50, no. 3 (1999): 391–407.

Gunther, R. T. *Early Science in Oxford*. Vol. 6, *The Life and Work of Robert Hooke*. Oxford: Oxford University Press, 1930.

———. *Early Science in Oxford*. Vol. 12, *Dr. Plot and the Correspondence of the Philosophical Society of Oxford*. Oxford: Printed for the subscribers, 1939.

Hall, Bert S. "The Didactic and the Elegant: Some Thoughts on Scientific and Technological Illustration in the Middle Ages and the Renaissance." In *Picturing Knowledge: Historical and Philosophical Problems concerning the Use of Art in Science*, edited by Brian S. Baigrie, 13–39. Buffalo, NY: University of Toronto Press, 1996.

Hall, Marie Boas. *Henry Oldenburg: Shaping the Royal Society*. New York: Oxford University Press, 2002.

Halley, Edmond. *Astronomical Tables with Precepts in both English and Latin*. London: Printed for William Innys, 1752.

Hamann, Byron Ellsworth. "Drawing Glyphs Together." In *Past Presented: Archaeological Illustration and the Ancient Americas*, edited by Joanne Pillsbury, 230–81. Washington, DC: Dunbarton Oaks Research Library and Collection, 2012.

Hamilton, Anthony. *Memoirs of the Comte de Gramont*. Translated by Peter Quennell. London: Routledge, 1932.

Hammond, Paul. "The King's Two Bodies: Representations of Charles II." In Black and Gregory, *Culture, Politics and Society in Britain, 1660–1800*, 13–48.

Hanson, Craig Ashley. *The English Virtuoso: Art, Medicine, and Antiquarianism in the Age of Empiricism*. Chicago: University of Chicago Press, 2009.

Harding, Vanessa. "City, Capital and Metropolis: The Changing Shape of Seventeenth-Century London." In *Imagining Early Modern London: Perceptions and Portrayals of the City from Stow to Strype, 1598–1720*, edited by J. F. Merritt. New York: Cambridge University Press, 2001.

Harkness, Deborah E. *The Jewel House: Elizabethan London and the Scientific Revolution*. New Haven, CT: Yale University Press, 2007.

———. "Managing an Experimental Household: The Dees of Mortlake and the Practice of Natural Philosophy." *Isis* 88, no. 2 (1997): 247–62.

Hart, Vaughn. *St. Paul's Cathedral: Sir Christopher Wren*. London: Phaidon, 1995.

Harwood, John T. "Rhetoric and Graphics in *Micrographia*." In Michael Hunter and Simon Schaffer, *Robert Hooke: New Studies*, 119–47.

Hatton, Edward. *A New View of London*. London: J. Nicholson, 1708.

Heering, Peter. "The Enlightened Microscope: Re-enactment and Analysis of Projections with Eighteenth-Century Solar Microscopes." *British Journal for the History of Science* 41, no. 3 September 2008): 345–67.

Heidarzadeh, Tofigh. *A History of Physical Theories of Comets, from Aristotle to Whipple*. New York: Springer, 2008.

Henderson, Felicity. "Unpublished Material from the Memorandum Book of Robert Hooke, Guildhall Library MS 1758." *Notes and Records of the Royal Society of London* 61 (2007): 129–75.

Hendriks, Ella, and Karen Groen. "Lely's Studio Practice." *Bulletin of the Hamilton Kerr Institute* 2 (1994): 21–38.

Henry, John. "Robert Hooke: The Incongruous Mechanist." In Michael Hunter and Simon Schaffer, *Robert Hooke: New Studies*, 149–80.

———. *The Scientific Revolution and the Origins of Modern Science*. New York: Palgrave, 2002.

Hesse, Mary B. "Hooke's Philosophical Algebra." *Isis* 57, no. 1 (Spring 1966): 67–83.

Hevelius, Johannes. *Cometographia*. Danzig: Simon Reiniger, 1668.

Higgott, Gordon. "The Revised Design for St. Paul's Cathedral 1685–1690: Wren, Hawksmoor and Les Invalides." *Burlington Magazine* 146, no. 1217 (August 2004): 534–47.

Hobbes, Thomas. *Leviathan*. Edited by Richard Tuck. New York: Cambridge University Press, 1991.

Hodgson, Barbara. *In the Arms of Morpheus: The Tragic History of Laudanum, Morphine and Patent Medicines*. Vancouver: Greystone, 2001.

Holger-Maehle, Andreas. "Pharmacological Experimentation with Opium in the Eighteenth Century." In *Drugs and Narcotics in History*, edited by Roy Porter and Mikulás Teich, 52–76. New York: Cambridge University Press, 1995.

Holger-Maehle, Andreas, and Ulrich Tröhler. "Animal Experimentation from Antiquity to the End of the Eighteenth Century: Attitudes and Arguments. " In *Vivisection in Historical Perspective*, edited by Nicolaas Rupke, 14–47. New York: Routledge, 1987.

Hollingsworth, Martha. "The Architect in Fifteenth-Century Florence." *Art History* 7 (December 1984): 385–410.

Hook, Judith. *The Baroque Age in England*. London: Thames and Hudson, 1976.

Hooke, Robert. *Animadversions on the first part of the Machina Coelestis of the Honourable, Learned and deservedly Famous Astronomer Johannes Hevelius, &c.* London: T. R. for John Martyn, 1674.

———. *An Attempt to Prove the Motion of the Earth by Observations*. London: J. Martyn, 1674.

———. "A Contrivance to make the Picture of any thing appear on a Wall. . . ." *Philosophical Transactions of the Royal Society of London* 4, no. 38 (August 17, 1668): 741–43.

———. *The Diary of Robert Hooke, 1672–1680*. Edited by H. W. Robinson and W. Adams. London: Taylor and Francis, 1935.

———. *Lectiones Cutlerianae*. London: J. Martyn, 1679.

———. *Lectures and Collections*. London: J. Martyn, 1678.

———. *Lectures de Potentia Restitutiva, or of Spring. . . .* London: J. Martyn, 1678.

———. *Micrographia; or, Some Physiological Descriptions of Minute Bodies made by Magnifying Glass*. London: John Martyn and James Allestry, 1665.

———. "More Wayes For the same Purpose, Intimated by M. Hook." *Philosophical Transactions of the Royal Society of London* 25, no. 2 (May 6, 1667): 459.

———. "Observables upon a Monstrous Head." *Philosophical Transactions of the Royal Society of London* 1, no. 5 (Munday, July 3, 1665): 85–86.

———. *Philosophical Collections*. London: Various publishers, 1679–82. New York: Johnson Reprint, 1965.

———. *Philosophical Experiments and Observations of the Late Eminent Dr. Robert Hooke*. Edited by William Derham. London: W. and J. Innys, 1726.

———. *The Posthumous Works of Robert Hooke*. Edited by Richard Waller. London: Sam. Smith and Benj. Walford, 1705.

———. "Some Observations, and Conjectures Concerning the *Chinese* Characters." *Philosophical Transactions* 16 (1686): 63–78.

Howard, Maurice. "The Ideal House and Healthy Life: The Origins of Architectural Theory in England." In *Les Traités d'architecture de la Renaissance*, edited by Jean Guillaume, 425–33. Paris: Picard, 1988.

Howarth, David. *Lord Arundel and His Circle*. New Haven, CT: Yale University Press, 1985.

Howse, Derek. *Francis Place and the History of the Greenwich Observatory*. New York: Science History Publications, 1975.

———. *Greenwich Observatory: The Royal Observatory at Greenwich and Herstmonceux, 1675–1975*. London : Taylor and Francis, 1975.

Hughes, David W. "Edmond Halley: His Interest in Comets." In Thrower, *Standing on the Shoulders of Giants*, 324–72.

Hume, David. *A Treatise of Human Nature*. 1739. Edited by L. A. Shelby-Bigge, Oxford: Clarendon Press, 1896.

Hunter, Matthew C. "Experiment, Theory, Representation: Robert Hooke's Material Models." In Frigg and Hunter, *Beyond Mimesis and Convention*, 193–219.

———. "Hooke's Figurations: A Figural Drawing Attributed to Robert Hooke." *Notes and Records of the Royal Society of London* 64 (2010): 251–60.

———. "Robert Hooke Fecit: Making and Knowing in Restoration London." PhD diss., University of Chicago, 2007.

———. "The Theory of the Impression according to Robert Hooke." In Michael Hunter, *Printed Images in Early Modern Britain*, 164–93.

Hunter, Michael. *Establishing the New Science: The Experience of the Early Royal Society*. Woodbridge: Boydell Press, 1989.

———. "Hooke the Natural Philosopher." In Bennett et al., *London's Leonardo*, 105–62.

———, ed. *Printed Images in Early Modern Britain: Essays in Interpretation*. Aldershot: Ashgate, 2010.

———. *The Royal Society and Its Fellows 1660–1700: The Morphology of an Early Scientific Institution*. Oxford: BSHS, 1994.

———. *Science and the Shape of Orthodoxy: Intellectual Change in Late Seventeenth-Century Britain*. Woodbridge: Boydell Press, 1995.

———. *Science and Society in Restoration England*. London: Cambridge University Press, 1981.

Hunter, Michael, and Paul B. Wood. "Towards Solomon's House: Rival Strategies for Reforming the Early Royal Society." *History of Science* 24 (1986): 49–108.

Hunter, Michael, and Simon Schaffer, eds. *Robert Hooke: New Studies*. Woodbridge: Boydell Press, 1989.

Hutchins, Edwin. "The Social Organization of Distributed Cognition." In *Perspectives on Socially Shared Cognition*, edited by L. B. Resnick et al., 283–307. Washington, DC: American Psychological Association, 1991.

Hutton, Ronald. *The Rise and Fall of Merry England: The Ritual Year 1400–1700*. New York: Oxford University Press, 1996.

Hyde, Lewis. *Trickster Makes This World: Mischief, Myth and Art*. New York: North Point, 1998.

Iliffe, Rob. "Butter for Parsnips: Authorship, Audience, and the Incomprehensibility of the *Principia*." In *Scientific Authorship: Credit and Intellectual Property in Science*, edited by Mario Biagioli and Peter Galison, 33–65. New York: Routledge, 2003.

———. "'In the Warehouse': Privacy, Property and Priority in the Early Royal Society." *History of Science* 30 (1992): 29–68.

———. "Material Doubts: Hooke, Artisan Culture and the Exchange of Information in 1670s London." *British Journal of the History of Science* 28 (1995): 285–318.

Impey, Oliver, and Arthur MacGregor, eds. *The Origins of Museums: The Cabinet of Curiosities in Sixteenth- and Seventeenth-Century Europe*. Oxford: Clarendon Press, 1985.

Ingle, Kathryn A. *Reverse Engineering*. New York: McGraw-Hill, 1994.

Inwood, Stephen. *The Forgotten Genius: The Biography of Robert Hooke 1635–1703*. San Francisco: Macadam/Cage, 2003.

———. *The Man Who Knew Too Much: The Strange and Inventive Life of Robert Hooke 1635–1703*. London: Macmillan, 2002.

Ivins, William M., Jr. *On the Rationalization of Sight*. New York: Metropolitan Museum of Art, 1938.

Jardine, Lisa. *The Curious Life of Robert Hooke: The Man Who Measured London*. London: Harper Perennial, 2003.

———. "Hooke the Man: His Diary and His Health." In Bennett et al., *London's Leonardo*, 214–47.

———. "Monuments and Microscopes: Scientific Thinking on a Grand Scale in the Early Royal Society." *Notes and Records of the Royal Society of London* 55, no. 2 (2001): 289–308.

———. *On a Grander Scale: The Outstanding Career of Sir Christopher Wren*. London: HarperCollins, 2003.

———. "Paper Monuments and Learned Societies: Hooke's Royal Society Repository." In *Enlightening the British: Knowledge, Discovery and the Museum in the Eighteenth Century*, edited by R. G. W. Anderson et al., 49–55. London: British Museum, 2003.

Jaucourt, Louis de. "Wight L'ile de." In *Encyclopédie, ou Dictionnaire Raisonné*, vol. 26, *Venerien–Z*, pp. 613–15. Neufchastel: Samuel Faulche, 1765.

Jeffrey, Paul. *The City Churches of Sir Christopher Wren*. London: Hambledon Press, 1996.

Johns, Adrian. "Martyn, John (1617/18–1680)." In Matthews and Harrison, *Oxford Dictionary of National Biography*, 37:37–38.

———. *The Nature of the Book: Print and Knowledge in the Making*. Chicago: University of Chicago Press, 1998.

———. "Prudence and Pedantry in Early Modern Cosmology: The Trade of Al Ross." *History of Science* 25 (1997): 23–59.

Jones, J. R. *Country and Court: England 1658–1714*. Cambridge, MA: Harvard University Press, 1978.

Jones, Richard Foster. "Science and English Prose Style in the Third Quarter of the Seventeenth Century." *Proceedings of the Modern Language Association* 45 (1930): 977–1009.

Jongh, Eddy de. "The Art of Describing: Dutch Art in the Seventeenth Century." *Simiolus* 14, no. 1 (1984): 51–59.

Jordan, Thomas. *Londons Triumphs: Illustrated with many Magnificent Structures & Pageants*. London: Printed for John Playford, 1677.

Juricek, John T. "English Territorial Claims in North America under Elizabeth and the Early Stuarts." *Terrae Incognitae* 7–8 (1976): 7–22.

Kant, Immanuel. *Critique of Judgment*. 1790. Translated by Werner S. Pluhar, Cambridge: Hackett, 1987.

Kassler, Jamie C. *Inner Music: Hobbes, Hooke and North on Internal Character*. London: Athlone, 1995.

Keene, Derek, et al., eds. *St. Paul's: The Cathedral Church of London 604–2004*. New Haven, CT: Yale University Press, 2004.

Kemp, Martin. *The Science of Art: Optical Themes in Western Art from Brunelleschi to Seurat*. New Haven, CT: Yale University Press, 1990.

Kemp, Martin, and Nathan Flis. "Mapping the Cerebral Globe." *Nature* 456, no. 7219 (November 2008): 174.

Keynes, Geoffrey. *A Bibliography of Robert Hooke*. Oxford: Clarendon Press, 1960.

Klein, Charles-Armand. *Denis Papin illustre savant blaisois*. Chambray: CLD, 1987.

Klein, Ursula. "Techniques of Modelling and Paper-Tools in Classical Chemistry." In Morgan and Morrison, *Models as Mediators*, 146–67.

Knight, Valentine. *Proposals of a New Model for Rebuilding the City of London, with Houses, Streets and Wharfs, to be forthwith set out by His Majesty's and the City Surveyors*. London, 1666.

Knoppers, Laura Lunger. "The Politics of Portraiture: Oliver Cromwell and the Plain Style." *Renaissance Quarterly* 12, no. 22 (1998): 1281–1319.

Knox, Robert. *An Historical Relation of the Island of Ceylon in the East Indies*. London: Richard Chiswell, 1681.

Knuuttila, Tarja, and Mieke Boon. "How Do Models Give Us Knowledge? The Case of Carnot's Ideal Heat Engine." *European Journal for Philosophy of Science* 1, no. 3 (2011): 309–34.

Knuuttila, Tarja, and Atro Voutilainen. "A Parser as an Epistemic Artifact: A Material View on Models." *Philosophy of Science* 70 (December 2003): 1484–95.

Koerner, Joseph Leo. *The Moment of Self-Portraiture in German Renaissance Art*. Chicago: University of Chicago Press, 1993.

Koyré, Alexandre. "La gravitation universelle de Kepler à Newton." *Archives Internationales d'Histoire des Sciences* 4, no. 14 (January 1951): 638–53.

———. "An Unpublished Letter of Robert Hooke to Isaac Newton." *Isis* 43, no. 4 (December 1952): 312–37.

Kris, Ernst, and Otto Kurz. *Legend, Myth, and Magic in the Image of the Artist*. New Haven, CT: Yale University Press, 1979.

Kronick, David A. K. "Notes on the Printing History of the Early *Philosophical Transactions*." *Libraries & Culture* 25, no. 2 (1990): 243–68.

Kronk, Gary W. *Cometography: A Catalog of Comets*. Vol. 1, *Ancient–1799*. New York: Cambridge University Press, 1999.

Kuhn, Thomas. *The Essential Tension: Selected Studies in Scientific Tradition and Change*. Chicago: University of Chicago Press, 1977.

———. *The Structure of Scientific Revolutions*. Chicago: University of Chicago Press, 1965.

Kultermann, Udo. "Woman Asleep and the Artist." *Artibus et Historiae* 11, no. 22 (1990): 129–61.

Kusukawa, Sachiko. "The *Historia Piscium* (1686)." *Notes and Records of the Royal Society of London* 54, no. 2 (2000): 179–97.

———. "Leonhart Fuchs on the Importance of Pictures." *Journal of the History of Ideas* 58, no. 3 (1997): 403–28.

———. "Picturing Knowledge in the Early Royal Society: The Examples of Richard Waller and Henry Hunt." *Notes and Records of the Royal Society* 65, no. 3 (2011): 273–94.

Lacour, Claudia Brodsky. *Lines of Thought: Discourse, Architectonics and the Origin of Modern Philosophy*. Durham, NC: Duke University Press, 1996.

Landau, David, and Peter Parshall. *The Renaissance Print, 1470–1550*. New Haven, CT: Yale University Press, 1996.

Lang, Jane. *Rebuilding St. Paul's after the Great Fire of London*. New York: Oxford University Press, 1956.

Latour, Bruno. "Drawing Things Together." In *Representation in Scientific Practice*, edited by Michael Lynch and Steve Woolgar, 19–68. London: MIT Press, 1990.

———. *We Have Never Been Modern*. Translated by Catherine Porter. Cambridge, MA: Harvard University Press, 1993.

Le Brun, Charles. *Conférence de monsieur le Brun, premier peintre du roi . . . sus l'expression générale et particulière*. Amsterdam: J. L. De Lorme, 1698.

———. *Conference of Monsieur Le Brun . . . Upon Expression, General and Particular*. London: John Smith, 1701.

Leddy, Thomas. "Is the Creative Process a Form of Puzzle Solving?" *Journal of Aesthetic Education* 24, no. 3 (Autumn 1990): 83–97.

Leedham-Green, Elisabeth. "Covel [Colvill], John (1638–1722)." In Matthew and Harrison, *Oxford Dictionary of National Biography*, 13:726–27.

[Leeuwenhoek, Antoni van]. "Microscopical Observations of Mr. Leewenhoeck, Concerning the Optic Nerve. . . ." *Philosophical Transactions* 10 (1675): 378–80.

———. "A Specimen of Some Observations Made by a Microscope, Contrived by M. Leewenhoeck in Holland, Lately Communicated by Dr. Regnerus de Graaf." *Philosophical Transactions* 8 (1673): 6037–38.

Leja, Michael. "Peirce, Visuality and Art." *Representations* 72 (Autumn 2000): 97–122.

Lenk, Krzysztof, and Paul Kahn. "To Show and Explain: The Information Graphics of Stevin and Comenius." *Visible Language* 26, nos. 3–4 (1992): 272–81.

Leonardo da Vinci. *Leonardo on Painting*. Edited by Martin Kemp, selected and translated by Martin Kemp and Margaret Walker. New Haven, CT: Yale University Press, 1989.

Leupold, Jacob. *Theatrum Arithmetico-Geometricum, Das ist: Schau-Platz der Rechen- und Meß-Kunst*. Leipzig: Christoph Zunkel, 1727.

Levi-Strauss, Claude. *The Savage Mind*. Chicago: University of Chicago Press, 1968.

Levine, Joseph M. *Dr. Woodward's Shield: History, Science, and Satire in Augustan England*. Berkeley: University of California Press, 1977.

Levy, Evonne. *Propaganda and the Jesuit Style*. Los Angeles: University of California Press, 2004.

Liedtke, Walter. "Pepys and the Pictorial Arts: 'Great Plenty of Good Pictures.'" *Apollo* 133 (1991): 227–37.

Lindberg, David. *Theories of Vision from Al-Kindi to Kepler*. Chicago: University of Chicago Press, 1976.

Linschoten, Jan Huygen van. *Iohn Huighen van Linschoten his Discourse of Voyages into the Easte & West Indies*. Translated by John Wolfe. London: John Wolfe, 1598.

Lipking, Lawrence I. *The Ordering of the Arts in Eighteenth-Century England*. Princeton, NJ: Princeton University Press, 1970.

Locke, John. "An Account of One Who Had Horny Excrescencies or Extraordinary Large Nails on His Fingers and Toes, by Mr. Locke." *Philosophical Transactions* 19, no. 230 (July 1697): 694–96.

———. *An Essay concerning Human Understanding*. Edited by P. H. Nidditch. Oxford: Clarendon Press, 1975.

Loh, Maria H. "New and Improved: Repetition as Originality in Italian Baroque Practice and Theory." *Art Bulletin* 86, no. 3 (September 2004): 477–504.

Lohne, Johannes. "Hooke *versus* Newton." *Centaurus* 7, no. 1 (1960): 6–52.

Louw, Hentie. "The 'Mechanick Artist' in Late Seventeenth Century English and French Architecture: The Work of Robert Hooke, Christopher Wren and Claude Perrault Compared as Products of an Interactive Science/Architecture Relationship." In Cooper and Hunter, *Robert Hooke: Tercentennial Studies*, 181–99.

Lovejoy, A. O. "Milton and the Paradox of the Fortunate Fall." *English Literary History* 4, no. 3 (September 1937): 161–79.

Lovelace, Richard. *Lucasta: Epodes, Odes, Sonnets, Songs, &c.* London: Thomas Harper, 1649.

———. *Lucasta: Posthume Poems of Richard Lovelace Esq.* London: William Godbid for Clement Darby, 1659.

Lynch, Michael. "Discipline and the Material Form of Images: An Analysis of Scientific Visibility." *Social Studies of Science* 15, no. 1 (February 1985): 37–66.

Lynch, William. *Solomon's Child: Method in the Early Royal Society of London*. Stanford, CA: Stanford University Press, 2001.

MacGregor, Arthur. *Curiosity and Enlightenment: Collections and Collectors from the Sixteenth to the Nineteenth Century*. New Haven, CT: Yale University Press, 2007.

———. *Sir Hans Sloane: Collector, Scientist, Antiquary, Founding Father of the British Museum*. London: British Museum, 1994.

———, ed. *Tradescant's Rarities: Essays on the Foundation of the Ashmolean Museum*. Oxford: Clarendon Press, 1983.

MacGregor, William B. "The Authority of Prints: An Early Modern Perspective." *Art History* 22, no. 3 (September 1999): 389–420.

MacLeod, Catharine. "'Good but Not Like': Peter Lely, Portrait Practice and the Creation of a Court Look." In MacLeod and Marciari Alexander, *Painted Ladies*.

MacLeod, Catharine, and Julia Marciari Alexander, eds. *Painted Ladies: Women at the Court of Charles II*. London: National Portrait Gallery, 2001.

———. "The 'Windsor Beauties' and the Beauties Series in Restoration London." In Alexander and MacLeod, *Politics, Transgression, and Representation at the Court of Charles II*. New Haven, CT: Yale University Press, 2007.

Mander, Karel van. *The Lives of the Illustrious Netherlandish and German Painters*. 1609. 6 vols. Edited and translated by H. Miedema, Doornspijk, Netherlands: Davaco, 1999.

Mangone, Carolina. "Becoming Bernini: Imitation, Motion and Identity Formation." Unpublished paper circulated at the "Repetition, Emulation and Innovation" workshop at UC Riverside, January 2011.

Mannoni, Laurent. *The Great Art of Light and Shadow: Archaeology of the Cinema*. Translated by Richard Crangle. Exeter: Exeter University Press, 2000.

Marin, Louis. *To Destroy Painting*. 1977. Translated by Mette Hjort, Chicago: University of Chicago Press, 1995.

———. "In Praise of Appearance." *October* 37 (Summer 1986): 98–112.

Markeley, Richard. *Fallen Languages: Crises of Representation in Newtonian England, 1660–1740*. Ithaca, NY: Cornell University Press, 1993.

Marly, Diana de. "Undress in the Oeuvre of Lely." *Burlington Magazine* 120, no. 908 (November 1978): 749–50.

Martin, D. C. "Former Homes of the Royal Society." *Notes and Records of the Royal Society of London* 22, nos. 1–2 (1967): 12–19.

Martin, Ernst. *The Calculating Machines (Die Rechenmaschinen): Their History and Development*. Translated and edited by Peggy Aldrich Kidwell and Michael R. Williams. Cambridge, MA: MIT Press, 1992.

Matthew, H. C. G., and Brian Harrison, eds. *Oxford Dictionary of National Biography*. 60 vols. New York: Oxford University Press, 2004.

McCormick, Ted. *William Petty and the Ambitions of Political Arithmetic*. New York: Oxford University Press, 2009.

McKellar, Elizabeth. *The Birth of Modern London: The Development and Design of the City 1660–1720*. New York: Manchester University Press, 1999.

McKendrick, Neil. "The Consumer Revolution of Eighteenth-Century England." In *The Birth of a Con-*

sumer Society: The Commercialization of Eighteenth-Century England, edited by Neil McKendrick et al., 9–33. Bloomington: Indiana University Press, 1982.

McKenzie, D. F. "Typography and Meaning: The Case of William Congreve." In *The Book and the Book Trade in Eighteenth-Century Europe*, edited by Giles Barber and Bernhard Fabian, 81–125. Hamburg: Dr. Ernst Hauswedell, 1981.

McKeon, Robert M. "Établissement de l'astronomie de précision et œuvre d'Adrien Auzout." PhD diss., École des hautes études en sciences sociales, Paris, 1965.

McTighe, Sheila. "Abraham Bosse and the Language of Artisans: Genre and Perspective in the Académie royale de peinture et de sculpture, 1640–1670." *Oxford Art Journal* 21, no. 1 (1998): 1–26.

Meli, Domenico Bertoloni. "The Representation of Insects in the Seventeenth Century: A Comparative Approach." *Annals of Science* 67, no. 3 (2010): 405–29.

———. *Thinking with Objects: The Transformation of Mechanics in the Seventeenth Century*. Baltimore: Johns Hopkins University Press, 2006.

———. "Who Is Afraid of Centrifugal Force?" *Early Science and Medicine* 10, no. 4 (2005): 535–43.

Merton, Robert K. "Bureaucratic Structure and Personality." In *Social Theory and Social Structure*, 249–60. New York: Free Press, 1968.

Millar, Oliver, ed. "The Inventories and Valuations of the King's Goods, 1649–51." *Walpole Society* 43 (1972): 1–443.

Millon, Henry A., et al., eds. *The Renaissance from Brunelleschi to Michelangelo: The Representation of Architecture*. New York: Rizzoli, 1997.

Mintz, Sidney W. *Sweetness and Power: The Place of Sugar in Modern History*. New York: Viking, 1985.

Monteyne, Joseph. *The Printed Image in Early Modern London: Urban Space, Visual Representation, and Social Exchange*. Burlington, VT: Ashgate, 2007.

Morgan, Mary S., and Margaret C. Morrison, eds. *Models as Mediators: Perspectives on Natural and Social Science*. Cambridge: Cambridge University Press, 1999.

Mowl, Timothy, and Brian Earnshaw. *Architecture without Kings: The Rise of Puritan Classicism under Cromwell*. New York: Manchester University Press, 1995.

Moxon, Joseph. *Mechanick Exercises on the Whole Art of Printing*. 1683–84. Edited by H. Davis and H. Carter, London: Oxford University Press, 1962.

———. *Practical Perspective; or, Perspective made Easie*. London: J. Moxon, 1670.

Mulligan, Lotte. "Robert Hooke's 'Memoranda': Memory and Natural History." *Annals of Science* 49 (1992): 47–61.

Mulligan, Lotte, and Glenn Mulligan. "Reconstructing Restoration Science: Styles of Leadership and Social Composition of the Early Royal Society." *Social Studies of Science* 11 (1981): 327–64.

Nauenberg, Michael. "Robert Hooke's Seminal Contribution to Orbital Dynamics." In Cooper and Hunter, *Robert Hooke: Tercentennial Studies*, 3–32.

Neri, Janice. "Between Observation and Image: Representation of Insects in Robert Hooke's *Micrographia*." In *The Art of Natural History: Illustrated Treatises and Botanical Paintings, 1400–1850*, edited by Teresa O'Malley and Amy R. W. Meyers, 83–108. New Haven, CT: Yale University Press, 2008.

———. *The Insect and the Image: Visualizing Nature in Early Modern Europe, 1500–1700*. Minneapolis: University of Minnesota Press, 2011.

———. "Some Early Drawings by Robert Hooke." *Archives of Natural History* 32, no. 1 (2005): 41–47.

Newman, John. "Fittings and Liturgy in Post-Fire St Paul's." In Keene et al., *St. Paul's*, 220–32.

Newman, William R. *Atoms and Alchemy: Chymistry and the Experimental Origins of the Scientific Revolution*. Chicago: University of Chicago Press, 2006.

———. *Promethean Ambitions: Alchemy and the Quest to Perfect Nature*. Chicago: University of Chicago Press, 2004.

Newton, Isaac. *The Correspondence of Isaac Newton*. Edited by H. W. Turnbull. Vol. 2, *1676–1687*. Cambridge: Cambridge University Press for the Royal Society, 1960.

Nickles, Thomas. "Kuhnian Puzzle Solving and Schema Theory." *Philosophy of Science* 67, pt. 2 (September 2000): S242–55.

Nicolson, Marjorie Hope. *Pepys' Diary and the New Science*. Charlottesville: University of Virginia Press, 1965.

Nuti, Lucia. "'To Make the Whole Progress a Lineall Visible Demonstration': The Journal of Edward Dummer." *Word & Image* 15, no. 3 (1999): 292–305.

Nuttall, R. H. "That Curious Curiosity: The Scotoscope." *Notes and Records of the Royal Society of London* 42 (1988): 133–38.

O'Callaghan, Michelle. *The English Wits: Literature and Sociability in Early Modern England*. New York: Cambridge University Press, 2007.

O'Connor, Erin. "Embodied Knowledge in Glassblowing: The Experience of Meaning and the Struggle towards Proficiency." *Sociological Review* 55, Suppl. 1 (May 2007): 126–41.

O'Doherty, Brian. *Inside the White Cube: The Ideology of the Gallery Space*. Expanded ed. Los Angeles: University of California Press, 1999.

Ogborn, Miles. *Spaces of Modernity: London's Geographies 1680–1780*. New York: Guilford Press, 1998.

Ogilvie, Brian W. *The Science of Describing: Natural History in Renaissance Europe*. Chicago: University of Chicago Press, 2006.

Oldenburg, Henry. *The Correspondence of Henry Oldenburg*. Edited by A. Rupert Hall, Marie Boas Hall, and Eberhard Reichmann. 13 vols. Madison: University of Wisconsin Press, 1965–75.

———. "An Experiment of a Way of Preparing a Liquor, That Shall Sink into, and Colour the Whole Body of Marble, Causing a Picture, Drawn on a Surface, to Appear Also in the Inmost Parts of the Stone." *Philosophical Transactions* 1 (1665–66): 125–27.

Oldroyd, D. R. "Robert Hooke's Methodology of Science as Exemplified in His 'Discourse of Earthquakes.'" *British Journal for the History of Science* 6, no. 22 (1972): 109–30.

———. "Some 'Philosophical Scribbles' Attributed to Robert Hooke." *Notes and Records of the Royal Society of London* 35 (1980): 17–32.

Olson, Roberta J. M. ". . . And They Saw Stars: Renaissance Representations of Comets and Pretelescopic Astronomy." *Art Journal* 44, no. 3 (1984): 216–24.

Oster, Malcolm R. "The 'Beame of Diuinity': Animal Suffering in the Early Thought of Robert Boyle." *British Journal for the History of Science* 22, no. 2 (July 1989): 151–79.

Pace, Claire. "'Delineated Lives': Themes and Variations in Seventeenth-Century Poems about Portraits." *Word & Image* 2, no. 1 (1986): 1–17.

Padelford, Frederick M. "Robert Aylett." *Huntington Library Bulletin* 10 (October 1936): 1–48.

———. "Robert Aylett: A Supplement." *Huntington Library Quarterly* 2, no. 4 (July 1939): 471–78.

Pagden, Anthony. *Lords of All the World: Ideologies of Empire in Spain, Britain, and France c.1500–c.1800.* New Haven, CT: Yale University Press, 1995.

Pallucchini, Rodolfo, and Paola Rossi. *Tintoretto: Le Opere sacre e profane.* 2 vols. Milan: Electa, 1994.

Panofsky, Erwin. "Artist, Scientist, Genius: Notes on the 'Renaissance-Dämmerung.'" In *The Renaissance: Six Essays*, edited by W. K. Ferguson et al., 121–82. New York: Harper and Row, 1962.

———. *Idea: A Concept in Art Theory.* Translated by J. S. Peake. New York: Harper and Row, 1968.

Papin, Denis. *A New Digester of Engine for Softning Bones.* London: J. M. for Henry Bonwicke, 1681.

Parrish, Susan Scott. *American Curiosity: Cultures of Natural History in the Colonial British Atlantic World.* Chapel Hill: University of North Carolina Press, 2006.

Parshall, Peter, ed. "Art and Curiosity," Special issue, *Word & Image* 11, no. 4 (1995): 327–404.

———. "Imago Contrafacta: Images and Facts in the Northern Renaissance." *Art History* 16, no. 4 (December 1993): 554–79.

———. "Prints as Objects of Consumption in Early Modern Europe." *Journal of Medieval and Early Modern Studies* 28, no. 1 (Winter 1998): 19–36.

Patterson, Louise Diehl. "Hooke's Gravitation Theory and Its Influence on Newton. I: Hooke's Gravitation Theory." *Isis* 40, no. 4 (November 1949): 327–41.

Paulson, Ronald. *Breaking and Remaking: Aesthetic Practice in England, 1700–1820.* New Brunswick, NJ: Rutgers University Press, 1989.

Pearce, E. [Edward]. *Twelve Engravings of Friezes.* London: I. Overton, n.d., ca. 1680.

Pears, Iain. *The Discovery of Painting: The Growth of Interest in the Arts in England, 1680–1768.* New Haven, CT: Yale University Press, 1988.

Peck, Linda Levy. "Uncovering the Arundel Library at the Royal Society: Changing Meanings of Science and the Fate of the Norfolk Donation." *Notes and Records of the Royal Society of London* 52, no. 1 (1998): 3–24.

Pepys, Samuel. *The Diary of Samuel Pepys.* Edited by Robert Latham and William Matthews. 10 vols. Los Angeles: University of California Press, 1970–83.

Perez-Ramos, Antonio. "Bacon's Forms and the Maker's Knowledge Tradition." In *The Cambridge Companion to Bacon*, edited by M. Peltonen, 99–120. New York: Cambridge University Press, 1996.

[Petty, William]. *Advice of W.P. to Mr. Samuel Hartlib for the Advancement of Some Particular Parts of Learning.* London, 1647.

———. *A Treatise of Taxes and Contributions.* London: N. Brooke, 1662.

Picon, Antoine. *French Architects and Engineers in the Age of Enlightenment.* Translated by M. Thom. New York: Cambridge University Press, 1992.

Pietz, William. "The Problem of the Fetish, I." *RES* 9 (1995): 5–17.

———. "The Problem of the Fetish, II: The Origins of the Fetish." *RES* 12 (1987): 23–45.

Pincus, Steven. "'Coffee Politicians Does Create': Coffeehouses and Restoration Political Culture." *Journal of Modern History* 67 (December 1995): 807–34.

———. *Protestantism and Patriotism: Ideologies and the Making of English Foreign Policy, 1650–1668.* New York: Cambridge University Press, 1996.

———. *1688: The First Modern Revolution.* New Haven, CT: Yale University Press, 2009.

Pitt, Moses, ed. *The English Atlas*. 5 vols. Oxford: M. Pitt, 1680–83.

Plato. *The Statesman*. Translated by H. N. Fowler. New York: Putnam, 1925.

Pliny the Elder. *Natural History*. Translated by H. Rackham. Loeb Classical Library. Cambridge, MA: Harvard University Press, 1938.

Pomian, Krzysztof. *Collectors and Curiosities: Paris and Venice, 1500–1800*. Translated by Elizabeth Wiles-Portier. Cambridge: Polity Press, 1990.

———. "Vision and Cognition." In *Picturing Science, Producing Art*, edited by Peter Galison and Caroline Jones, 211–31. New York: Routledge, 1998.

Poole, William. *Francis Lodwick (1619–1694), "A Country Not Named" (MS. Sloane 913, Fols. 1R-33R): An Edition with an Annotated Primary Bibliography and an Introductory Essay on Lodwick and His Intellectual Context*. Tempe, AZ: ACMRS, 2007.

Pope, Alexander. *The First Epistle of the Second Book of Horace Imitated*. In *Poems of Alexander Pope*, vol. 4, *Imitations of Horace*, edited by J. Butt. London: Methuen, 1939.

Powell, Jocelyn. *Restoration Theatre Production*. London: Routledge and Kegan Paul, 1984.

Power, Henry. *Experimental Philosophy, in Three Books*. London: T. Roycroft, 1664.

Power, M. J. "The Social Topography of Restoration London." In *London 1500–1700: The Making of the Metropolis*, edited by A. C. Beier and R. Findlay, 199–223. New York: Longman, 1986.

Povey, Thomas. "The Method, Manner and Order of the Transmutation of Copper into Brass, etc." *Philosophical Transactions of the Royal Society of London* 22 (1700–1701): 474–75.

Pugin, A. Welby. *The True Principles of Pointed or Christian Architecture*. London: John Weale, 1841.

Pugliese, Patri J. "Robert Hooke and the Dynamics of Motion in a Curved Path." In Michael Hunter and Simon Schaffer, *Robert Hooke: New Studies*, 181–205.

Pumfrey, Stephen. "Ideas above His Station: A Social Study of Hooke's Curatorship of Experiments." *History of Science* 29 (1991): 1–44.

Puttfarken, Thomas. *The Discovery of Pictorial Composition: Theories of Visual Order in Painting, 1400–1800*. London: Yale University Press, 2000.

Rabb, Theodore K. *The Struggle for Stability in Early Modern Europe*. New York: Oxford University Press, 1975.

Ralph, James. *A Critical Review of the Publick Buildings, Statues and Ornaments in, and about London and Westminster*. London: C. Ackers, 1734.

Rambelli, Fabio, and Eric Reinders. "What Does Iconoclasm Create? What Does Preservation Destroy? Reflections on Iconoclasm in East Asia." In *Iconoclasm: Contested Images, Contested Terms*, edited by Richard Clay and Stacy Boldrick, 15–33. Aldershot: Ashgate, 2007.

Ray, John. *Further Correspondence of John Ray*. Edited by R. T. Gunther. London: Ray Society, 1928.

———. "Observations made at Rome, by the late Reverend Mr. John Ray, of the Comet which appeared Anno 1664. . . ." *Philosophical Transactions* 25 (1707): 2350–52.

Rheinberger, Hans-Jörg. *Toward a History of Epistemic Things: Synthesizing Proteins in the Test Tube*. Stanford, CA: Stanford University Press, 1997.

Riegl, Alois. *The Group Portraiture of Holland*. Introduced by Wolfgang Kemp, translated by Evelyn M. Kain and David Britt. Los Angeles: Getty Research Center, 1999.

Rivington, Charles A. "Early Printers to the Royal Society, 1663–1708." *Notes and Records of the Royal Society of London* 39, no. 1 (Spetember 1984): 1–27.

Robinson, H. W., and W. Adams, eds. *The Diary of Robert Hooke, 1672–1680*. London: Taylor and Francis, 1935.

Robinson, Norman H. *The Royal Society Catalogue of Portraits*. London: Royal Society, 1980.

Rogers, Malcolm. *John Closterman: Master of the English Baroque 1660–1711*. London: National Portrait Gallery, 1981.

Roos, Anna Marie. "A Speculum of Chymical Practice: Isaac Newton, Martin Lister (1639–1712), and the Making of Telescopic Mirrors." *Notes and Records of the Royal Society of London* 64, no. 2 (June 2010): 105–20.

Root-Bernstein, Michele, and Robert Root-Bernstein. "Thinkering." *Psychology Today* (August 21, 2008), http://www.psychologytoday.com/blog/imagine/200808/thinkering.

Rorty, Richard. *Philosophy and the Mirror of Nature*. Princeton, NJ: Princeton University Press, 1979.

Rossi, Paolo. *The Dark Abyss of Time: The History of the Earth and the History of Nations from Hooke to Vico*. Translated by L. Cochrane. Chicago: University of Chicago Press, 1984.

Rostenburg, Leona. *The Library of Robert Hooke: The Scientific Book Trade of Restoration England*. Santa Monica, CA: MODOC Press, 1989.

Rousseau, G. S. "Nerves, Spirits, and Fibres: Towards Defining the Origins of Sensibility." *Studies in Eighteenth Century Culture* 3 (1976): 137–57.

Rudwick, Martin. "George Cuvier's Paper Museum of Fossil Bones." *Archives of Natural History* 27, no. 1 (2000): 51–68.

Ruestow, Edward G. *The Microscope in the Dutch Republic: The Shaping of Discovery*. New York: Cambridge University Press, 1996.

Russell, Nicholas. *Like Engend'ring Like: Heredity and Animal Breeding in Early Modern England*. New York: Cambridge University Press, 1986.

Rykwert, Joseph. *The First Moderns: The Architects of the Eighteenth Century*. Cambridge, MA: MIT Press, 1980.

Safier, Neil. "Spies, Dyes and Leaves: Agro-Intermediaries, Luso-Brazilian Couriers, and the Worlds They Sowed." In *The Brokered World: Go-Betweens and Global Intelligence, 1770–1820*, edited by Simon Schaffer et al., 239–69. Sagamore Beach, MA: Science History Publications, 2009.

Sahlins, Marshall. "The Sadness of Sweetness: The Native Anthropology of Western Cosmology." *Current Anthropology* 37, no. 3 (June 1996): 395–415.

Sawday, Jonathan. *The Body Emblazoned: Dissection and the Human Body in Renaissance Culture*. New York: Routledge, 1996.

Schaffer, Simon. "Godly Men and Mechanical Philosophers: Souls and Spirits in Restoration Natural Philosophy." *Science in Context* 1, no. 1 (1987): 55–85.

———. "Halley's Atheism and the End of the World." *Notes and Records of the Royal Society of London* 32, no. 1 (July 1977): 17–40.

———. "Regeneration: The Body of Natural Philosophers in Restoration England." In *Science Incarnate: Historical Embodiments of Natural Knowledge*, edited by Christopher Lawrence and Steven Shapin, 83–120. Chicago: University of Chicago Press, 1998.

Schiebinger, Londa, and Claudia Swan, eds. *Colonial Botany: Science, Commerce and Politics in the Early Modern World*. Philadelphia: University of Pennsylvania Press, 2005.

Schier, Flint. *Deeper into Pictures: An Essay on Pictorial Representation*. Cambridge: Cambridge University Press, 1986.

Schmidt, Suzanne Karr, and Kimberly Nichols. *Altered and Adorned: Using Renaissance Prints in Daily Life*. Chicago: Art Institute of Chicago, 2011.

Scott, Andrew C., and David Freedberg. *Fossil Woods and Other Geological Specimens: The Paper Museum of Cassiano dal Pozzo*. Series B: Natural History, pt. 3. Turnhout, Belgium: Harvey Miller, 2000.

Seltzer, Mark. *Serial Killers: Death and Life in America's Wound Culture*. New York: Routledge, 1998.

Seneca. *Ad Lucilium Epistulae Morales*. Translated by R. H. Gummere. 3 vols. London: Heinemann, 1920–25.

Shalev, Zur. "Measurer of All Things: John Greaves (1602–1652), the Great Pyramid, and Early Modern Metrology." *Journal of the History of Ideas* 63, no. 4 (October 2002): 555–75.

Shapin, Steven. "The House of Experiment in Seventeenth-Century England." In *The Science Studies Reader*, edited by Mario Biagioli. New York: Routledge, 1999.

———. "'A Scholar and a Gentleman': The Problematic Identity of the Scientific Practitioner in Early Modern England." *History of Science* 29 (1991): 279–327.

———. *A Social History of Truth: Civility and Science in Seventeenth-Century England*. Chicago: University of Chicago Press, 1994.

———. "Who Was Robert Hooke?" In Michael Hunter and Simon Schaffer, *Robert Hooke: New Studies*, 255–85.

Shapin, Steven, and Simon Schaffer. *Leviathan and the Air-Pump: Hobbes, Boyle and the Experimental Life*. Princeton, NJ: Princeton University Press, 1985.

Sharpe, Kevin. "'Thy Longing Country's Darling and Desire': Aesthetics, Sex, and Politics in the England of Charles II." In Alexander and MacLeod, *Politics, Transgression, and Representation at the Court of Charles II*, 1–32.

Sherman, William H. *John Dee: The Politics of Reading and Writing in the English Renaissance*. Amherst: University of Massachusetts Press, 1995.

Shiqiao, Li. *Power and Virtue: Architecture and Intellectual Change in England 1660–1730*. New York: Routledge, 2007.

Simpson, A. D. C. "Newton's Telescope and the Cataloguing of the Royal Society's Repository." *Notes and Records of the Royal Society of London* 38, no. 2 (1984): 187–214.

———. "Robert Hooke and Practical Optics: Technical Support at a Scientific Frontier." In Michael Hunter and Simon Schaffer, *Robert Hooke: New Studies*, 33–62.

Singer, Brian. "Robert Hooke on Memory, Association and Time Perception." *Notes and Records of the Royal Society of London* 31, no. 1 (1976): 115–31.

[Slare, Frederick]. "An Account of Several Experiments made with the Shining Substance of the Liquid and of the Solid Phosphorous. . . ." *Philosophical Collections* 3 (December 10, 1681): 48–50.

Slaughter, M. M. *Universal Languages and Scientific Taxonomy in the Seventeenth Century*. London: Cambridge University Press, 1982.

Sloane, Kim. *"A Noble Art": Amateur Arts and Drawing Masters c. 1600–1800*. London: British Museum Press, 2001.

Smentek, Kristel. "The Collector's Cut: Why Pierre-Jean Mariette Tore Up His Drawings and Put Them Back Together Again." *Master Drawings* 46, no. 1 (Spring 2008): 36–60.

Smith, Pamela H. *The Body of the Artisan: Art and Experience in the Scientific Revolution*. Chicago: University of Chicago Press, 2004.

Smith, Pamela H., and Paula Findlen, eds. *Merchants and Marvels: Commerce, Science and Art in Early Modern Europe*. New York: Routledge, 2002.

Smyth, Adam. "'Rend and Teare in Peeces': Textual Fragmentation in Seventeenth-Century England." *Seventeenth Century* 19, no. 1 (2004): 36–52.

Snyder, George Sergeant. *Maps of the Heavens*. London: André Deutsch, 1984.

Solkin, David H. *Painting for Money: The Visual Arts and the Public Sphere in Eighteenth-Century England*. New Haven, CT: Yale University Press, 1992.

Soo, Lydia, ed. *Wren's "Tracts" on Architecture and Other Writings*. New York: Cambridge University Press, 1998.

Sprat, Thomas. *History of the Royal-Society of London, For the Improving of Natural Knowledge*. London: T. R. for J. Martyn, 1667.

Spurr, John. *England in the 1670s: "This Masquerading Age."* Malden, MA: Blackwell, 2000.

Stafford, Barbara Maria. *Body Criticism: Imaging the Unseen in Enlightenment Art and Medicine*. Cambridge, MA: MIT Press, 1991.

———. *Voyage into Substance: Art, Science, Nature, and the Illustrated Travel Account, 1760–1840*. London: MIT Press, 1984.

Stafford, Barbara Maria, and Frances Terpak. *Devices of Wonder: From the World in a Box to Images on a Screen*. Los Angeles: Getty Research Institute, 2001.

Stater, Victor. "Mordaunt, Henry, second earl of Peterborough (bap. 1623, d. 1697)." In Matthew and Harrison, *Oxford Dictionary of National Biography*, 39:21–22.

Steadman, Philip. *Vermeer's Camera: Uncovering the Truth behind the Masterpieces*. New York: Oxford University Press, 2002.

Stevenson, Christine. "Robert Hooke's Bethlem." *Journal of the Society of Architectural Historians* 55, no. 3 (September 1996): 254–75.

———. "Robert Hooke, Monuments and Memory." *Art History* 28, no. 1 (February 2005): 43–73.

Stevenson, Jane, and Peter Davidson, eds. *The Closet of the Eminently Learned Sir Kenelme Digbie Kt. Opened*. 1669. Totnes, Devon: Prospect Books, 1997.

Stewart, J. Douglas. "Kneller, Sir Godfrey, baronet (1646–1723)." In Matthew and Harrison, *Oxford Dictionary of National Biography*, 31:877–85.

Stewart, Larry. *The Rise of Public Science: Rhetoric, Technology, and Natural Philosophy in Newtonian Britain, 1660–1750*. New York: Cambridge University Press, 1992.

Stewart, Susan. "Death and Life, in That Order, in the Works of Charles Wilson Peale." In *Visual Display: Culture beyond Appearances*, edited by Lynne Cooke and Peter Wollen, 30–53. Seattle: Bay Press, 1995.

Stoesser, Allison. "Robert Hooke's Montagu House." In Cooper and Hunter, *Robert Hooke: Tercentennial Studies*, 165–80.

Stoichita, Victor I. *The Self-Aware Image: An Insight into Early Modern Meta-Painting*. Translated by A.-M. Glasheen. New York: Cambridge University Press, 1997.

Stoye, John. *The Siege of Vienna: The Last Great Trial between Cross and Crescent.* New York: Pegasus Books, 2000.

Summerson, John. *Architecture in Britain 1530–1830.* New Haven, CT: Yale University Press, 1953.

———. "The Tyranny of the Intellect." *Journal of the Royal Institute of British Architects* (February 20, 1937): 373–90.

Sutton, John. *Philosophy and Memory Traces: Descartes to Connectionism.* New York: Cambridge University Press, 1998.

Swan, Claudia. "Ad Vivum, Naer het Leven, From the Life: Defining a Mode of Representation." *Word & Image* 11 (1995): 353–72.

———. *Art, Science and Witchcraft in Early Modern Holland: Jacques de Gheyn II (1565–1629).* Cambridge: Cambridge University Press, 2005.

———. "Eyes Wide Shut: Early Modern Imagination, Demonology, and the Visual Arts." *Zeitsprünge: Forschungen zur Frühen Neuzeit* 7 (2003): 156–81.

Talley, Mansfield Kirby. *Portrait Painting in England: Studies in the Technical Literature before 1700.* London: Paul Mellon Centre for British Art, 1981.

Taylor, E. G. R. "'The English Atlas' of Moses Pitt, 1680–83." *Geographical Journal* 95, no. 4 (April 1940): 292–99.

———. "Robert Hooke and the Cartographical Projects of the Late Seventeenth Century (1666–1696)." *Geographical Journal* 90, no. 6 (December 1937): 529–40.

Tenison, Thomas. *Baconiana; or, Certain Genuine Remains of Sr. Francis Bacon.* London: J. D. for Richard Chiswell, 1679.

Thomas, Jenni. "A 'Philosophical Storehouse': The Life and Afterlife of the Royal Society Repository." PhD diss., Queen Mary and Westfields, 2009.

Thompson, Krista. "The Sound of Light: Reflections on Art History in the Visual Culture of Hip-Hop." *Art Bulletin* 91, no. 4 (December 2009): 481–505.

Thrower, Norman Joseph William, ed. *Standing on the Shoulders of Giants: A Longer View of Newton and Halley.* Los Angeles: University of California Press, 1990.

Tijou, Jean. *A New Booke of Drawings Invented and Desined by John Tijou.* London: J. Tijou, 1693.

Topsell, Edward. *The historie of foure-footed beastes: Describing the true and liuely figure of euery beast.* London: William Iaggard, 1607.

[Towneley, Richard]. "An Extract of a Letter, Written by Mr. Richard Towneley to Dr. Croon, Touching the Invention of Dividing a Foot into Many Thousand Parts, for Mathematical Purposes." *Philosophical Transactions* 25, no. 2 (May 6, 1667): 457–58.

Trachtenberg, Marvin. *Dominion of the Eye: Urbanism, Art and Power in Early Modern Florence.* New York: Cambridge University Press, 1997.

Turner, A. J. "A Forgotten Naturalist of the Seventeenth Century: William Cole of Bristol and His Collections." *Archives of Natural History* 11, no. 1 (1982): 27–41.

Turner, James Grantham. "Pepys and the Private Parts of Monarchy." In *Culture and Society in the Stuart Restoration: Literature, Drama, History*, ed. G. MacLean, 95–110. Cambridge: Cambridge University Press, 1995.

Tyson, Edward. *Phocæna; or, The Anatomy of a Porpess, dissected at Gresham College.* London: Benj. Tooke, 1680.

———. "Some Anatomical Observations of Hair Found in several parts of the Body." In [Hooke], *Philosophical Collections*, 2:11–17.

Uffenbach, Zacharias Conrad von. *London in 1710: From the Travels of Conrad von Uffenbach*. Edited and translated by W. H. Quarrell and M. Mare. London: Faber and Faber, 1936.

———. *Oxford in 1710 from the Travels of Zacharias Conrad von Uffenbach*. Edited and translated by W. H. Quarrell and W. J. C. Quarrell. Oxford: Basil Blackwell, 1928.

Unwin, Robert. "A Provincial Man of Science at Work: Martin Lister, F.R.S. and His Illustrators, 1670–1683." *Notes and Records of the Royal Society of London* 49, no. 2 (1995): 209–30.

Varey, Simon. *Space and the Eighteenth-Century English Novel*. New York: Cambridge University Press, 1990.

Vasari, Giorgio. *The Lives of the Artists*. Translated by J. C. Bondanella and P. Bondanella. New York: Oxford University Press, 1991.

Vertue, George. *Notebooks. Volume of the Walpole Society* 18 (1929–30), 20 (1931–32), 22 (1933–34), 24 (1935–36), 26 (1937–38), 30 (1948–50).

Viatte, Françoise. "Weaving a Rope of Sand." *Yale French Studies* 89 (1996): 85–102.

Vincent, Clare. "A Beam Compass by Christoph Trechsler the Elder and the Origin of the Micrometer Screw." *Metropolitan Museum Journal* 24 (1989): 209–22.

Vitruvius. *Ten Books on Architecture*. Translated by Ingrid Rowland. New York: Cambridge University Press, 1999.

Volpe, Tod. *Framed: America's Art Dealer to the Stars Tells All*. Toronto: ECW Press, 2003.

Waals, Jan van der. "The Print Collection of Samuel Pepys." *Print Quarterly* 1, no. 4 (1984): 236–57.

Walker, Matthew. "*Architectus Ingenio*: Robert Hooke, the Early Royal Society, and the Practices of Architecture." PhD diss., York University, 2009.

———. "The Limits of Collaboration: Robert Hooke, Christopher Wren and the Designing of the Monument to the Great Fire of London." *Notes and Records of the Royal Society* 65, no. 2 (2011): 121–43.

Wall, Jeff. "Photography and Liquid Intelligence." In *Jeff Wall*, edited by Thierry de Duve et al., 90–93. London: Phaidon, 1996.

Waller, Richard. "The Life of Dr. Robert Hooke." In *The Posthumous Works of Robert Hooke*, edited by Richard Waller, i–xxviii. London: Sam. Smith and Benj. Walford, 1705.

Walpole, Horace. *Anecdotes of Painting in England*. 4 vols. Strawberry Hill: Walpole, 1763.

Walton, Kendall. *Mimesis as Make-Believe: On the Foundations of the Representational Arts*. Cambridge, MA: Harvard University Press, 1990.

Ward, Ned. *The London-Spy Compleat, in Eighteen-Parts*. London: J. How, 1703.

———. *The London-Spy Compleat, in Eighteen Parts*. 4th ed. London: J. How, 1709.

Warner, Deborah J. *The Sky Explored: Celestial Cartography 1500–1800*. New York: Alan R. Liss, 1979.

Waterfield, Giles. *Mr. Cartwright's Pictures: A Seventeenth Century Collection*. London: Dulwich Picture Gallery, 1987.

Waterhouse, Ellis. *Painting in Britain, 1530–1790*. New Haven, CT: Yale University Press, 1953.

Weber, Max. "Bureaucracy." In *From Max Weber: Essays in Sociology*, translated, edited, and introduced by H. H. Gerth and C. W. Mills. London: Routledge and Kegan Paul, 1948.

Webster, Charles. "Alchemical and Paracelsian Medicine." In *Health, Medicine and Mortality in the Sixteenth Century*, edited by Charles Webster, 301–34. New York: Cambridge University Press, 1979.

———. *The Great Instauration: Science, Medicine, and Reform, 1626–1660*. New York: Holmes and Meier, 1976.

———. "Henry Power's Experimental Philosophy." *Ambix* 14, no. 3 (1967): 150–78.

Weinberg, Rosemary. *The History of the Worshipful Company of Feltmakers 1604–2004*. Chichester: Phillimore, 2004.

Wenke, Robert J. "Egypt: Origins of Complex Societies." *Annual Review of Anthropology* 18 (1989): 129–55.

Wenzel, Michael. "The Windsor Beauties by Sir Peter Lely and the Collection of Paintings at St. James's Palace, 1674." *Journal of the History of Collections* 14, no. 2 (2002): 205–13.

Westfall, Richard S. "Hooke and the Law of Universal Gravitation: A Reappraisal of a Reappraisal." *British Journal for the History of Science* 3, no. 3 (June 1967): 245–61.

———. *Never at Rest: A Biography of Isaac Newton*. New York: Cambridge University Press, 1980.

Wey Gómez, Nicolás. *The Tropics of Empire: Why Columbus Sailed South to the Indies*. Cambridge, MA: MIT Press, 2008.

Whinney, Margaret, and Oliver Millar. *English Art 1625–1714*. Oxford: Clarendon Press, 1957.

Wilding, Nick. "Graphic Technologies." In Cooper and Hunter, *Robert Hooke: Tercentennial Studies*, 123–34.

Wilkins, John. *An Essay towards a Real Character and a Philosophical Language*. London: S. Gellibrand and J. Martyn, 1668.

———. *Mercury; or, The Secret and Swift Messenger*. London: I. Norton, 1641.

Williams, Robert. *Art, Theory, and Culture in Sixteenth-Century Italy: From Techne to Metatechne*. New York: Cambridge University Press, 1997.

Willis, John. *Mnemonica; or, The Art of Memory*. London: L. Sowersby, 1661.

Willis, Thomas. *Pharmaceutica Rationalis; or, An Exercitation of the Operations of Medicines in Humane Bodies: Shewing the Signs, Causes and Cures of most Distempers incident Thereunto*. London: J. B. for T. Dring, 1684.

———. *The Remaining Medical Works of that Famous and Renowned Physician Dr. Thomas Willis*. London: T. Dring, 1681.

———. *Two Discourses concerning the Soul of Brutes*. 1683. Translated by Samuel Pordage and introduced by S. Diamond, Gainesville, FL: Scholars' Facsimiles and Reprints, 1971.

Wilson, Catherine. "Visual Surface and Visual Symbol: The Microscope and the Occult in Early Modern Science." *Journal of the History of Ideas* 49, no. 1 (1988): 85–108.

Wilson, George. "On the Early History of the Air-Pump in England." *Edinburgh New Philosophical Journal* 46 (1848–49): 330–54.

Wilson, John H. *The Court Wits of the Restoration: An Introduction*. New York: Octagon Books, 1967.

Wilton-Ely, John. "The Rise of the Professional Architect in England." In *The Architect: Chapters in the History of the Profession*, edited by Spiro Kostof, 180-208. New York: Oxford University Press, 1977.

———. "Wren, Hawksmoor and the Architectural Model." In *English Architecture Public and Private: Essays for Kerry Downes*, edited by John Bold and E. Chaney, 147–59. London: Hambledon, 1993.

Wölfflin, Heinrich. *Kunstgeschichtliche Grundbegriffe: Das Problem der Stilentwicklung in der Neuren Kunst*. Munich: Hugo Bruckmann, 1920.

———. *Principles of Art History: The Problem of the Development of Style in Later Art*. Translated by M. D. Hottinger. New York: Dover, 1950.

Wood, P. W. "Methodology and Apologetics: Thomas Sprat's *History of the Royal Society*." *British Journal for the History of Science* 13, no. 43 (1980): 1–26.

Woollett, Anne T., et al. *Rubens and Brueghel: A Working Friendship*. Los Angeles: Getty Research Institute, 2006.

[Wren, Christopher]. "The Description of an Instrument Invented Divers Years Ago by Dr. Wren, for Drawing the Out-lines of Any Object in Perspective." *Philosophical Transactions of the Royal Society of London* 4 (1669): 898–99.

———. *Wren's "Tracts" on Architecture and Other Writings*. Edited by Lydia Soo. New York: Cambridge University Press, 1998.

Wren, Christopher, Jr. *Parentalia; or, Memoirs of the Family of the Wrens*. London: T. Osborn and R. Dodsley, 1750.

Wright, John P. "Locke, Willis, and the Seventeenth-Century Epicurean Soul." In *Atoms, Pneuma, and Tranquility: Epicurean and Stoic Themes in European Thought*, edited by M. J. Osler, 239–58. New York: Cambridge University Press, 1991.

Yates, Frances. *The Art of Memory*. Chicago: University of Chicago Press, 1966.

Yeo, Richard. "Before Memex: Robert Hooke, John Locke, and Vannevar Bush on External Memory." *Science in Context* 20, no. 1 (2007): 21–47.

Zilsel, Edgar. "The Origins of William Gilbert's Scientific Method." *Journal of the History of Ideas* 2, no. 1 (January 1941): 1–32.

Zimmer, Carl. *Soul Made Flesh: The Discovery of the Brain—and How It Changed the World*. New York: Free Press, 2004.

Zittel, Claus. "Conflicting Pictures: Illustrating Descartes' *Traité de l'homme*." In *Silent Messengers: The Circulation of Material Objects of Knowledge in the Early Modern Low Countries*, ed. Sven Dupré and Christoph Lüthy, 217–60. Berlin: LIT, 2011.

Zlator, Zdenko. *Between the Double Eagle and the Crescent: The Republic of Dubrovnik and the Origins of the Eastern Question*. New York: Columbia University Press, 1992.

Zwicker, Steven. *Lines of Authority: Politics and English Literary Culture, 1649–1689*. Ithaca, NY: Cornell University Press, 1993.

INDEX